# Studies in Systems, Decision and Control

## Volume 56

**Series editor**

Janusz Kacprzyk, Polish Academy of Sciences, Warsaw, Poland
e-mail: kacprzyk@ibspan.waw.pl

*About this Series*

The series "Studies in Systems, Decision and Control" (SSDC) covers both new developments and advances, as well as the state of the art, in the various areas of broadly perceived systems, decision making and control- quickly, up to date and with a high quality. The intent is to cover the theory, applications, and perspectives on the state of the art and future developments relevant to systems, decision making, control, complex processes and related areas, as embedded in the fields of engineering, computer science, physics, economics, social and life sciences, as well as the paradigms and methodologies behind them. The series contains monographs, textbooks, lecture notes and edited volumes in systems, decision making and control spanning the areas of Cyber-Physical Systems, Autonomous Systems, Sensor Networks, Control Systems, Energy Systems, Automotive Systems, Biological Systems, Vehicular Networking and Connected Vehicles, Aerospace Systems, Automation, Manufacturing, Smart Grids, Nonlinear Systems, Power Systems, Robotics, Social Systems, Economic Systems and other. Of particular value to both the contributors and the readership are the short publication timeframe and the world-wide distribution and exposure which enable both a wide and rapid dissemination of research output.

More information about this series at http://www.springer.com/series/13304

Krzysztof Walkowiak

# Modeling and Optimization of Cloud-Ready and Content-Oriented Networks

 Springer

Krzysztof Walkowiak
Faculty of Electronics
Department of Systems and Computer
  Networks
Wrocław University of Technology
Wrocław
Poland

ISSN 2198-4182 ISSN 2198-4190 (electronic)
Studies in Systems, Decision and Control
ISBN 978-3-319-80776-8 ISBN 978-3-319-30309-3 (eBook)
DOI 10.1007/978-3-319-30309-3

Book Reviewers: Andrzej Kasprzak and Mirosław Klinkowski

Printed on acid-free paper

This Springer imprint is published by Springer Nature
The registered company is Springer International Publishing AG Switzerland

*To my wife Monika*

# Preface

Modeling and optimization of computer and communication networks has always been a significant topic to both researchers and practitioners. The interest in developing efficient optimization models and methods is motivated by the deployment of new network services and the growing numbers of users. Armed with a good set of algorithms, it is possible to design the network across various protocols, technologies, topologies, and traffic patterns in an efficient way to provide solutions that meet the expected cost and performance limits.

In the past few decades, a great deal of research has been accomplished on modeling and optimization of computer and communication networks. However, in the last few years, networks have undergone an important change driven by the emergence and rapid expansion of new services. The concepts of cloud computing and content-oriented networking are some of the most significant and disruptive technologies that have recently gained wide interest. Cloud computing together with content-oriented networking have revolutionized the way various network services are delivered to end users, bringing new research challenges. More specifically, the key issue is to model and optimize networks with new traffic patterns to allow cost-effective implementation of recent networking services with Quality of Service guarantees and high scalability. The need to develop appropriate and powerful optimization methods for these new services is enhanced by intense competition among telecoms and network service providers observed in the global market.

The fact that cloud computing and content-oriented services have proliferated has triggered a change in the network traffic patterns. In particular, in place of traditional unicast flows, anycast and multicast transmissions have been gaining wide acceptance. This trend mostly follows from the nature of cloud computing and content-oriented services; that is, highly specialized data centers distributed throughout the network are used as repository for computing and storage. In addition, the most popular and voluminous content offered in contemporary networks is video that can be efficiently delivered to end customers via multicast streaming. As a consequence, the main aim of this book is to cover various aspects of modeling and optimization that are relevant in the context of cloud computing and content-oriented services, including issues related to anycast and multicast flows. The emphasis is on formulating the considered problems as mathematical optimization models in the form of mixed-integer programming (MIP) or integer programming (IP), and on different optimization methods including branch-and-cut algorithms, Lagrangian relaxation, and various heuristic and metaheuristic algorithms. Moreover, several numerical experiments are described in order to show the performance of the optimization methods proposed and to present the know-how on experimental evaluation of networking problems.

Contemporary networks offer much flexibility in terms of new service deployment. In fact, different network layers can be utilized to implement services based on many factors including technical and economic aspects, starting from the physical layer typically implementing optical technologies, through the network layer employing connection-oriented protocols like MPLS, and ending with the application layer solutions realized by overlay networks. To this end, this book focuses on modeling and optimization of cloud-ready and content-oriented networks in the context of different layers and accounts for specific constraints following from protocols and technologies used in a particular layer. A wide range of additional constraints important in contemporary networks is addressed in this book, including various types of network flows, survivability issues, multi-layer networking, and resource location.

The contents of this book are organized as follows. Chapter 1 briefly presents information on cloud computing and content-oriented services, and introduces basic notions and concepts of network modeling and optimization. Chapter 2 covers various optimization problems that arise in the context of connection-oriented networks. Chapter 3 focuses on modeling and optimization of Elastic Optical Networks. Chapter 4 is devoted to overlay networks. The book concludes with Chap. 5, summarizing the book and presenting recent research trends in the field of network optimization. The material of each chapter is mostly self-contained.

The book presents the author's research results gained during the past eight years. Some of these results have been published in prestigious international journals and conference proceedings. As well as including additional new results, the existing results are finalized, extended, and presented in a comprehensive and cohesive way.

I would like to thank Prof. Janusz Kacprzyk for including my book in the prestigious Springer series "Studies in Systems, Decision and Control," and Springer's staff members for their kind cooperation and support during the publishing process.

The book would not have been written without the continuing support and help from a number of colleagues and students from the Department of Systems and Computer Networks, Faculty of Electronics, Wrocław University of Technology. I am especially grateful to Prof. Michał Woźniak, my former and current Ph.D. students: Michał Aibin, Damian Bulira, Grzegorz Chmaj, Jakub Gładysz, Róża Goścień, Wojciech Kmiecik, Michał Kucharzak, Adam Smutnicki, Maciej Szostak, and my M.Sc. students: Wojciech Charewicz, Maciej Donajski, Tomasz Kacprzak,

Michał Kosowski, Marek Miziołek, Krzysztof Pajak, Juliusz Skowron, Bartosz Rabiega, Michał Tarnawski and Michał Wiśniewski for their insightful comments, fruitful discussions, and help in computer experiments.

Also, I extend my warmest thanks to Prof. Andrzej Kasprzak, who first showed me the beauty of networks and always supported my academic career.

Moreover, I am indebted to my former and current collaborators: Prof. Mirosław Klinkowski, Dr. Michał Przewoźniczek, Dr. Jacek Rak, Prof. Arun Sen, and Prof. Massimo Tornatore for valuable and clarifying discussions on network optimization topics during the past few years.

My family perhaps suffered the most during my work on this book. I would like to express my gratitude to my wife Monika and my kids Zuzia and Maciek for their support, love, and understanding during the preparation of this book.

Wrocław                                                                        Krzysztof Walkowiak
November 2015

# Acknowledgments

The work was supported in part by the National Science Centre (NCN) under Grant DEC-2012/07/B/ST7/01215, the statutory funds of the Department of Systems and Computer Networks, Wroclaw University of Technology, and the European Commission and the European Commission under the 7th Framework Programme, Coordination and Support Action, Grant Agreement Number 316097, ENGINE— European research centre of Network intelliGence for INnovation Enhancement (http://engine.pwr.edu.pl/).

Wrocław                                                    Krzysztof Walkowiak
November 2015

# Contents

# Acronyms

| | |
|---|---|
| AFA | Adaptive Frequency Assignment |
| ATM | Asynchronous Transfer Mode |
| B-ISDN | Broadband Integrated Services Digital Network |
| BV-T | Bandwidth-Variable Transponder |
| CAGR | Compounded Annual Growth Rate |
| CON | Connection-Oriented Network |
| CG | Content Group |
| DC | Data Center |
| DHP | Dual Homing Protection |
| DNP | Dedicated Node Protection |
| DPP | Dedicated Path Protection |
| DTP | Dedicated Tree Protection |
| EA | Evolutionary Algorithm |
| EON | Elastic Optical Network |
| FD | Flow Deviation |
| FEC | Forwarding Equivalence Class |
| FF | First Fit |
| GMPLS | Generalized MultiProtocol Label Switching |
| GRASP | Greedy Randomized Adaptive Search Procedure |
| IETF | Internet Engineering Task Force |
| IP | Integer Programming |
| ILP | Integer Linear Programming |
| ITU | International Telecommunication Union |
| ITU-T | International Telecommunication Union—Telecommunication Standardization Sector |
| LPF | Longest Path First |
| LR | Lagrangian Relaxation |
| LS | Local Search |

| LSP | Label Switch Path |
|---|---|
| MC-OXC | Multicasting Capable Optical Cross Connect |
| MIR | Mixed-Integer Rounding |
| MIP | Mixed-Integer Programming |
| MPLS | MultiProtocol Label Switching |
| MSF | Most Subcarriers First |
| OXC | Optical Cross Connects |
| P2P | Peer-to-Peer |
| PBB | Provider Backbone Bridging |
| PBB-TE | Provider Backbone Bridging—Traffic Engineering |
| RMSA | Routing, Modulation, and Spectrum Allocation |
| RRSA | Routing, Regenerator, and Spectrum Allocation |
| RSA | Routing and Spectrum Allocation |
| RWA | Routing and Wavelength Assignment |
| SA | Simulated Annealing |
| SBPP | Shared Backup Path Protection |
| SLA | Straddling Link Algorithm |
| TS | Tabu Search |
| VLAN | Virtual Local Area Network |
| VPN | Virtual Private Network |
| WDM | Wavelength Division Multiplexing |
| WSON | Wavelength Switched Optical Networks |

# Chapter 1
# Introduction

This chapter introduces basic ideas related to cloud-ready and content-oriented networks. First of all, the concepts of cloud computing and content-oriented networking are described and examined in the context of the evolution of computer and communication networks. Next, we present how to model network flows including different approaches related to unicast, anycast and multicast flows. Moreover, a brief description of various optimization methods that can be applied to solving optimization problems related to cloud-ready and content-oriented networks is reported. Finally, the chapter is concluded with a short discussion about conventions used in the book concerning naming and numbering.

## 1.1 Cloud-Ready and Content-Oriented Networks

The presentation starts with an introduction to cloud computing and content-oriented networking. As well as providing a general description of the key aspects of these two concepts, we want to show how the shift in information technology towards cloud computing and content-oriented solutions impacts computer and communication networks and in consequence, what new research challenges appear in the field of network modeling and optimization.

### 1.1.1 Cloud Computing

Recently, the *cloud computing* paradigm has evolved from an emerging approach to a recognized solution for delivering IT services. The concept of cloud computing merging flexible networking capabilities and scalable distributed computing is an excellent response to many contemporary challenges both in business and research domains. Cloud computing—as a general idea—is not new. However, the growth and maturity

© Springer International Publishing Switzerland 2016
K. Walkowiak, *Modeling and Optimization of Cloud-Ready and Content-Oriented Networks*, Studies in Systems, Decision and Control 56, DOI 10.1007/978-3-319-30309-3_1

of cloud computing methods observed in recent years have demonstrated that cloud computing overwhelmed previous approaches related to distributed processing such as utility computing and grid computing. In fact, cloud computing is the result of a long evolution of different technology-centric developments and business-centric trends mostly focused on the concept of outsourcing. In addition, the advent of cloud computing has convinced many business and technology leaders, that a new concept of *combining and sharing* IT resources can provide a big advantage over the classic approach of *building and maintaining* them. Consequently, the cloud computing concept is now a flexible, cost-effective and settled solution for a wide range of business and customer services [1].

Numerous definitions of cloud computing have been proposed, the following definition proposed by the National Institute of Standards and Technology [2] has received the highest industry-wide acceptance:

> *Cloud computing is a model for enabling ubiquitous, convenient, on-demand network access to a shared pool of configurable computing resources (e.g., networks, servers, storage, applications, and services) that can be rapidly provisioned and released with minimal management effort or service provider interaction.*

The current popularity of the cloud computing paradigm arises mostly from several IT trends observed in the last two decades including: the dot-com boom which started a seismic explosion of interest in outsourcing IT services; popularity, maturity and scalability of the contemporary Internet; deployment of data centers by companies such as Google, Amazon and Microsoft; and virtualization technologies. In addition, we must point out another significant trend, namely, the growing need to process huge volumes of data known as the *Big Data* concept. Big Data is a general concept that is related to the tremendous growth, availability and use of various types of information coming from numerous different sources, e.g., climate information, medical records, stock ticker data, financial transactions, purchase transaction records, sensor data, social media, and digital photos and videos [1, 3–5].

In order to illustrate the influence of cloud computing services on the development of computer and communication networks, several statistics and forecasts will be reported. We use the data provided by Cisco in the *Cisco Global Cloud Index* report [6] and the *Cisco Visual Networking Index* report [7]. In these reports, Cisco categorizes the network traffic and presents various statistics including volume of the traffic per year in exabytes (EB) (where $1EB = 10^{18}$ B) and the predicted value of the CAGR (Compounded Annual Growth Rate) parameter.

First of all, the growing acceptance of cloud computing services and the former popularity of grid computing services have initiated major changes in the Internet traffic. Mostly, these trends follow from the fact that an increasing number of various workloads are served by a data centers accessed through the Internet. Therefore, the majority of the traffic can be categorized as a data center traffic. According to [6], the network traffic can be divided into the following categories:

- **Data center to user**—defined as traffic that flows from the data center to end users through the Internet; for instance, streaming video to a mobile device or PC.

- **Data center to data center**—defined as traffic that flows between data centers; for instance, moving data between clouds, or copying content to multiple data centers as a part of a content distribution network.
- **Within data center**—defined as traffic that remains within the data center; for instance, moving data from a development environment to a production environment within a data center, or writing data to a storage array.
- **Non-data center traffic**—defined as all traffic not related to data centers; for instance, P2P (Peer-to-Peer) traffic, direct moving data between two individual machines.

Figure 1.1 presents a forecast of various types of traffic for years 2013–2018. The largest part of the overall traffic refers to the traffic inside the data centers that will reach 6389 EB in year 2018. Data center to data center traffic is predicted to increase in the most rapidly with a CAGR of 29 %, while the traffic not related to data centers will remain relatively stable in the near future (only 4 % of CAGR) [6].

Figure 1.2 shows a prediction of data center traffic divided into two categories: cloud data center traffic and traditional data center traffic. More specifically, the former one is defined as the traffic associated with cloud consumer and business applications, while the latter is defined as traffic associated with non-cloud consumer and business applications. Global cloud traffic will increase nearly 3.9-fold from 2013 to 2018 with CAGR of 32 %. In turn, the traditional data center traffic will grow at a much slower rate. In addition, it is worth mentioning that by 2018, 78 % of all workloads processed in the data centers will be executed in the cloud. Moreover, the workload density defined as the number of workloads per physical server for cloud data centers will increase from 5.2 in 2013 to 7.5 by 2018. The corresponding numbers concerning traditional data centers are 2.2 in 2013 and 2.5 in 2018, which clearly shows that cloud data centers are expected to better facilitate the concepts of virtualization. These trends also follow from the fact that cloud computing solutions are promoting migration of workloads across servers, both inside data centers and among data centers located in different geographic areas. Besides that, end-user cloud computing applications are supported by several workloads distributed across servers, which creates multiple streams of traffic within and between data centers [6].

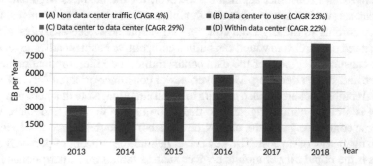

**Fig. 1.1**  Various types of traffic—prediction for years 2013–2018

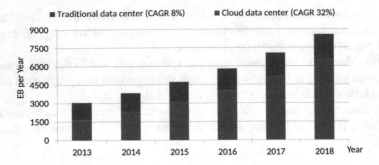

**Fig. 1.2**  Cloud data center traffic versus traditional data center traffic in years 2013–2018

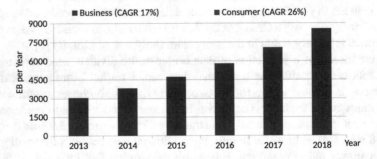

**Fig. 1.3**  Business data center traffic versus consumer data center traffic in years 2013–2018

Finally, Fig. 1.3 reports a forecast with regard to two segments of data center traffic, i.e., business and consumer. Business data centers are usually focused on organizational needs and handle traffic for business needs that may have higher security requirements. Consumer data centers typically serve a wider audience and handle traffic for the mass consumer base. The key observation is that consumer traffic will grow at a faster rate than business traffic (26 vs. 17 %) and in 2018 consumer traffic will account for approx. 67 % of the overall data center traffic. It should be pointed out that real-time and time-sensitive applications are important for both the business and consumer segments. However, for the business segment, another important service is related to fast and flexible access to large data archives, that are used—according to the Big Data concept—to enable advanced analytics of largely unstructured data archives which can build a competitive business advantage. In turn, for the consumer segment of the data center traffic, the killer application seems to be video and audio streaming. Moreover, newer consumer-oriented services such as personal content lockers are also gaining in much attention. Note that personal content lockers is a cloud computing service that allows users to store and share various electronic content (e.g., music, photos, and videos) through an easy-to-use interface at a relatively low or no cost. The key driver behind the concept of personal content lockers is the popularity of mobile devices such as tablets and smartphones [6].

A cloud delivery model denotes a specific combination of IT resources used to provide services in cloud computing systems. There are three major cloud delivery models [1, 2, 8, 9]:

- **Infrastructure-as-a-Service (IaaS)**. The IaaS delivery model assumes that the provider outsources to customers a self-contained IT environment with infrastructure including hardware, storage, servers, virtual machines, networking components and other bare IT resources. The provider is the owner of the equipment and is responsible for housing, operation, and maintenance. The key difference from traditional hosting or outsourcing services is that the IaaS model provides resources that are usually virtualized and packaged into bundles. The main goal of using the IaaS model is to offer customers a high level of control over its configuration and utilization.
- **Platform-as-a-Service (PaaS)**. In this model, customers are provided with a predefined and ready to use IT environment comprised of virtualized servers and associated services (e.g., operating system, programming language execution environment, database, web server) for developing, testing and running applications. The key benefit of the PaaS model is that customers are not involved in the administration duties such as setting up and maintaining of the bare infrastructure resources. However, a potential drawback of the PaaS concept is a lower level of control over the underlying IT resources compared to the IaaS model.
- **Software-as-a-Service (SaaS)**. The SaaS delivery model represents a case where applications are hosted by the provider and made available to customers over a network, typically the Internet. Customers do not have to manage the cloud infrastructure and platform used to run the software. Consequently, the maintenance of IT resources is significantly simplified. However, at the same time, customers have a very limited administrative control over the SaaS implementation.

The SaaS approach seems to be a dominant model, mainly due to its fast deployment of various IT technologies that support web services and service-oriented architecture (SOA). Additionally, the SaaS model provides the highest level of simplicity and elasticity from the business point of view. More specialized cloud delivery models include concepts such as Storage-as-a-Service, Database-as-a-Service, Security-as-a-Service, Communication-as-a-Service, Integration-as-a-Service, Testing-as-a-Service, and Process-as-a-Service [1].

Most of the IT services offered in the cloud computing model are provisioned by specialized data centers frequently located far away from customers; the communication network makes it possibile to exchange information between customers and IT resources providing the requested services. Therefore, the communication network is a crucial element of the cloud computing idea. The key requirements for cloud-ready networks are flexibility defined as the ability to guarantee the required capacity on demand, multilayer oriented network management, and cross-strata capabilities offering a joint optimization of the resources of the cloud-based application and the underlying network providing connectivity [10]. Furthermore, it should be emphasized that existing networks mostly account for unicast (one-to-one) transmission, whilst the variety of applications offered in the cloud computing model leads to new traffic patterns including anycast (one-to-one-of-many) and multicast (one-to-many) flows. Finally, aggregation of cloud computing processing in a relatively small number of places (data centers) makes the network traffic on links adjacent to data centers becomes voluminous, and the network must be designed properly, in order to cope with this problem [3].

As mentioned above, one of the key technology drivers for cloud computing is virtualization. The concept of virtualization can be defined as a process of converting a physical IT resource into a virtual resource. More precisely, virtualization makes it possible to abstract the details of physical resources (e.g., server, storage, network, power) and provides virtualized IT resources. An illustrative example of the virtualization concept is a *virtual machine* defined as a computing environment with an operating system and installed applications. User software being run on a virtual machine is separated from the underlying hardware resources. Therefore, the concept of virtual machines provides hardware independence. As a result, virtualization makes cloning and manipulating of virtual resources significantly easier than duplicating physical resources. For instance, virtual machines can be copied easily and moved between hosting machines. The migration of virtual machines can be made within the data center or outside the data center. In the second case, the migration has a twofold influence on the network traffic. Firstly, virtual machines and all associated data must be transmitted (migrated) between data centers. Secondly, all network traffic related to the migrated systems is transferred to/from a new data center. In consequence, the migration mechanism offered in the context of virtualization makes it possible to use anycast transmission [1, 3, 4, 8, 9, 11]. For a detailed description of cloud computing concepts as well as other information on networking aspects of cloud computing services refer to [1, 3, 4, 8–15].

## 1.1.2   Content-Oriented Networking

The concept of *content-oriented networking* has been gaining much attention in recent years. The main reason is the fact that a growing volume of the Internet traffic is generated by diverse content-oriented applications and services including various types of electronic content such as video streaming, music streaming, web objects (text, graphics, music), software distribution, on-demand media streaming, and file sharing. In consequence, efficient content delivery over the network has become a significant requirement for enhancing the web performance for both fixed and mobile users. Figure 1.4 presents the Cisco predictions of global consumer Internet traffic by sub segments between 2014 and 2019 [7]. One can observe that the video traffic will grow with a CAGR of 33 % and will account for 80 % of all consumer Internet traffic in 2019. Moreover, Cisco forecasts that video on demand traffic will increase 100 % from 2014 to 2019. These figures do not embrace video exchanged through P2P file sharing. Overall video traffic including TV, video on demand, Internet, and P2P will be in the range of 80 to 90 % percent of global consumer traffic by 2019.

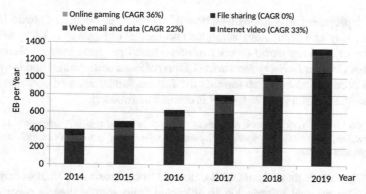

**Fig. 1.4** Global consumer Internet traffic by sub segments—prediction for years 2014–2019

In addition, Netflix—the most popular video streaming service in the US—currently generates approx. 37 % of all consumer traffic in North America.

Since existing Internet architecture designed in the 1960s and 1970s is based on the host-to-host communication model, most of the currently used solutions are inadequate when a growing number of users focus on content rather than specific hosts. Content-oriented networking is a relatively new communication approach which focuses on *what* is transmitted rather than *where* the information is transmitted (i.e., which network elements are exchanging the information). In consequence, networks are largely perceived mostly from the perspective of the content exchanged in them rather than network elements such as hosts, devices and links. Moreover, information is decoupled from its sources by means of a clear location-identity split, i.e., information is named, addressed, and matched independently of its location. Accordingly, information may be located anywhere in the network. Key mechanisms required to realize the idea of the content-oriented networking approach are in-network caching, multiparty communication through replication, and interaction models decoupling senders and receivers. The key goal of developing content-oriented networking solutions is to provide a network infrastructure that is better suited to content distribution and more resilient to disruptions and failures. Architectures based on the content-oriented networking have many advantages over classic IP networking, e.g., scalable and cost-effective content distribution, persistent and unique naming, enhanced integrity and security, mobility, multihoming, reliability improvement, and better scalability [16–18]. Note that in the literature many various terms are used instead of content-oriented networking to denote related concepts, i.e., information-centric networking, content-centric networking, and content-aware networking [16, 18].

It is worth mentioning that the concept of content-oriented networking is yet to be settled in detail. Discussion are ongoing on many technical aspects including the general service model, architecture, and protocols that will be standardized, although there is consensus on some of the features that content-oriented networking will provide. A more thorough review and analysis of content-oriented networking can be found in [16–27].

The deployment and implementation of the content-oriented networking idea requires many new technology solutions and radical changes in existing networks. Since the networking world favors evolution-based progress over revolutions, the concept of *Content Delivery Networks* (CDNs) offers a compromise and is a popular solution used online as a step towards the full implementation of content-oriented networking. According to [28], CDN is defined as follows:

> *A content delivery network represents a group of geographically dispersed servers deployed to facilitate the distribution of information generated by web publishers in a timely and efficient manner.*

In CDNs, the requested content is delivered to end users on behalf of origin web servers. The original information is offloaded from source sites to other content servers located in different locations in the network, usually in data centers. Usually, CDNs attempt to find the closest site offering the requested content. The set of content stored in particular CDN servers is selected carefully to approach the hit ratio of 100 %, which means that almost all requests to replicated servers are fulfilled [28–33].

Figure 1.5 reports the Cisco predictions of network traffic for years 2014–2019 divided into two categories: CDN traffic and non-CDN traffic [7]. By 2019, CDNs will carry over half of Internet traffic and 62 % of all Internet traffic will cross CDNs by 2019, up from 39 % in 2014. Furthermore, in this period of time CDN traffic will grow much faster (CAGR of 38 %) compared to non-CDN traffic (CAGR of 15 %). It should be noted that 72 % of all Internet video traffic will be delivered by CDNs by 2019, up from 57 % in 2014. Additionaly, Akamai—one of the most popular global CDN providers—claims to deliver between 15 % and 30 % of all web traffic reaching more than 25 Tb/s.

CDNs were proposed to overcome the limitations of Web caching solutions which were the first widely implemented network storage solution. The main drawbacks of Web caching are limited scalability, problems with content consistency and lack of transparency for end users. In addition, these drawbacks have been brought to light by the growing popularity of services such as dynamic content, video on demand and live streaming. The key benefit of CDNs is that content servers spread over

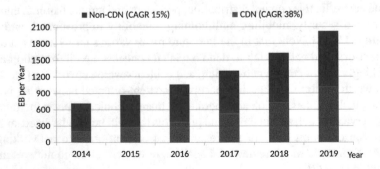

**Fig. 1.5**  CDN traffic versus non-CDN traffic—prediction for years 2014–2019

the Internet cooperate for content delivery to ensure the highest possible level of diverse QoS parameters embracing delay, throughput and availability. In turn, the main advantage of CDNs from the end user perspective is that content servers are completely transparent to users, and they proactively replicate the content at various network locations with a large number of direct interconnections to many operator networks and ISPs [19, 28, 29, 31, 33].

From the business point of view, the CDN operational model embraces three main parties, i.e., the content provider, the CDN provider and the network operator. Since there are no standard mechanisms for content publication and delivery to end users, the whole process is usually based on manual configuration according to the specific content and CDN technology being used. Another interesting issue in the cooperation between the main three parties involved in CDNs is the business model. In the most popular approach known as *content-centric*, the content provider pays the CDN (who usually owns no transport networks) to provide the content to end users by replicating it in various data centers located close to end users. The CDN ensures that their data centers have good connections to backbone networks of telecoms and ISPs to offer access to the content with low delay and high availability. Another possible scenario of the CDN business model is the *access-centric* approach, when the telecoms and ISPs pay the CDN to serve popular content from caches close to their subscribers. However, the access-centric model of CDN evolves into a scenario where the telecoms and ISPs deploy their own CDN services. Finally, a recent trend is that the largest web content providers such as Facebook, Google, Alibaba, Microsoft and Netflix build their own CDNs to offer their customers high-quality access to content [19, 28, 34].

There are two main types of CDNs: *push-based* and *pull-based* [28, 30]. In the former case, the content provider is responsible for uploading the content directly to the CDN. The main benefit of this approach is flexibility, since the content provider decides which files are uploaded to the CDN. In the latter case, the content provider just rewrites the content URLs to the CDN, and the CDN then takes responsibility for uploading the content to the network and caching it until it expires. However, it should be stressed that regardless of the delivery model (push vs. pull), the traffic pattern of the CDN is the same. In essence, each required piece of information is sent once to each network node hosting a CDN server (data center). Next, requesting users fetch the content from one of the CDN servers. The first phase usually does not contribute much to the overall network traffic, while the second phase—depending on content popularity—may generate very large volumes of traffic. For further information on CDNs, the reader is referred to [19, 23, 26, 28–33].

The key assumption behind both content-oriented networking and CDNs is that the same content requested by end users is replicated in various locations (mostly data centers) geographically dispersed in the network. This fact naturally offers the opportunity for using the anycast transmissions as the best way to ensure cost-effective and highly resilient delivery of the content. Moreover, when content is to be delivered to multiple users at the same time (e.g., live video/music streaming, software distribution), the most advantageous transmission approach is multicast, since it minimizes the consumption of network bandwidth and provides high scalability.

### *1.1.3   Network Architecture Model*

In this book, we focus on three network layers that can be used to implement and deliver cloud computing and content-oriented services. In particular, Fig. 1.6 shows a generic network architecture addressed in the book. At the top of this architecture is the application layer, which provides the cloud computing and content services using the concept of *overlay networks*. The middle layer includes basic network transport technologies and protocols, e.g., MPLS (MultiProtocol Label Switching), ATM (Asynchronous Transfer Mode) and Connection-Oriented Ethernet. The common attribute of technologies and protocols considered in the intermediate layer is *connection-oriented* networking, i.e., information belonging to the same traffic demand is transmitted along a predefined connection from the source to the destination along one routing path. Finally, the lower layer refers to an optical backbone network based on transparent or translucent solutions, where traffic remains in the optical domain as it transits the network nodes. It is assumed that the Elastic Optical Network (EON) is implemented in the optical layer. The concept of EON is a natural successor to the most popular optical backbone technology at the moment: Wavelength Switched Optical Networks (WSONs) based on the WDM technology.

Following the presented network architecture, the book is divided into three main chapters. In Chap. 2, we start with a discussion on connection-oriented networks, since the optimization models and algorithms developed in the context of this type of networks are the most straightforward. Chapter 3 covers topics related to the optimization of EONs. Finally, Chap. 4 focuses on the optimization of overlay networks.

The proposed architecture accounts for the most popular scenarios commonly used to realize cloud computing and content-oriented services. In essence, the default layer applied by the majority of users to provide cloud computing and content-

**Fig. 1.6**  Three-layered
network architecture model

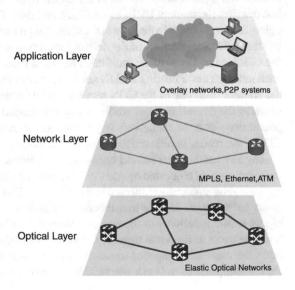

oriented services is the network layer. The reason is that the majority of business and consumer users access the Internet using technologies and protocols included in the network layer only, since most networking services offered by telecoms and ISPs (Internet Service Providers) are implemented in this layer. However, in some cases provisioning of cloud computing or content-oriented services must account for additional requirements and constraints.

As some cloud computing and content-oriented services may require very high volumes of network traffic related to data centers, network transfers with bit-above beyond 100 Gb/s are expected. The most effective way to provide such high- band- width services is to use the optical layer directly, since higher layers are either not able to ensure such transfers or cannot provide them in a cost-effective manner. Data centers that provide almost all cloud computing and content-oriented services are connected to the Internet by optical connections, and optimization of network resources in the optical layer directly should provide substantial benefits including a reduction of administration overheads that occur when many network layers are involved in provisioning network services.

On the other hand, some cloud computing and content-oriented services require extra features and functions that are not available in the network or optical layers, e.g., extra security requirements, new routing protocols and specific QoS parame- ters. Moreover, when implementing cloud computing and content-oriented services in the network or optical layer, the user is highly limited to the existing offering of telecoms and ISPs, since it is difficult to convince institutional network operators to implement new solutions in their networks or the cost of these new solutions is not affordable for the users. In consequence, network services accessible in the net- work and optical layers may lack the expected flexibility and scalability. In addition, enhancing interworking among different layers in order to improve network func- tionality may involve the adoption of complex and inflexible interfaces between the layers, and thus this solution is not efficient enough to provide the required QoS and to meet the limited budget. An ideal answer to these problems is the concept of overlay networks implemented in the application layer. Overlay networks overcome various limitations of the lower layers in a relatively simple and cost-effective way. Moreover, as overlay services are developed using only the default, point-to-point connectivity provided by underlying layers, it is not necessary to cooperate with parties operating the underlying networks when a new overlay service is started or an existing overlay service is modernized.

Each of the three analyzed layers is characterized with various technological and operational attributes and constraints. More details regarding each layer are presented in the following chapters.

## 1.2  Flow Modeling

In terms of topology, computer and communication networks can be modeled as graphs with possible additional constraints (e.g., link capacity constraint). However, the main goal of networks is to send information. Therefore, network modeling must account for the flow of data (information) between network nodes. Consequently, a pure graph-oriented approach is not sufficient for detailed modeling and optimization of computer and communication networks. Therefore this section introduces and analyzes a range of concepts of flow modeling. The majority of the presented basic models are based on a general concept of *multicommodity flows* that is broadly used to model various kinds of network flows. Three types of network flows, namely unicast, anycast and multicast, will be addressed. Unicast flow is the basic approach to sending data in computer and communication networks widely applied in many services and applications. In contrast, anycast and multicast flows are relatively new approaches that have been gaining momentum recently due to the proliferation of cloud computing and content-oriented services.

The theory of multicommodity flows was originally developed in the mid-20th century in the context of transport networks. A *commodity* is simply defined by a source node, destination node and volume (bit-rate) [35, 36]. The main idea behind multicommodity flows is the assumption that the bit-rate of each commodity (demand) expressed in b/s (bits per second) is constant. More specifically, all demands to be provisioned in the network are defined with a constant intensity of information arrival. In the context of a transport (backbone) network carrying aggregated traffic consisting of numerous single sessions, the assumption that demands have a constant rate is quite reasonable. Another example of network flows with a fixed bit-rate are streaming services such as video streaming, IPTV, radio, etc. However, it should be noted that single transmission between individual users is usually characterized with the bit-rate changing over the time. Modeling and optimization of such traffic is a major challenge [37, 38]. For a comprehensive survey on the topic of multicommodity flows, the reader is referred to [35–45].

## 1.2.1 Unicast Flows

This section focuses on unicast flows defined as *one-to-one* transmissions, i.e., every commodity has exactly one source node and one destination node [37, 44, 46, 47].

**One Commodity Flow**

First, a basic concept of one commodity flows is introduced. The network is modeled as a graph $G = (V, E)$, where $V$ is a set of nodes (vertices) and $E$ is a set of edges (directed links). Let $\delta^+(v) = \{(v, w) : (v, w) \in E\}$ be a set of links that originate at node $v$. Similarly, let $\delta^-(v) = \{(w, v) : (w, v) \in E\}$ denote a set of all links that terminate in node $v$. A single commodity (demand) is defined by a source node $s$, destination node $t$ and volume $h$. The flow of one commodity is defined as a function $x : E \to R^1$:

$$\sum_{e\in\delta^+(v)} x(e) - \sum_{e\in\delta^-(v)} x(e) = \begin{cases} +h & \text{if } v = s \\ -h & \text{if } v = t, \quad v \in V \\ 0 & \text{otherwise} \end{cases} \tag{1.2.1a}$$

$$x(e) \geq 0, \quad e \in E. \tag{1.2.1b}$$

Function $x(e)$ denotes the flow of the commodity sent on link $e$. Notice that the left-hand side of (1.2.1a) is the difference between the flow leaving and entering a particular node $v$. If $v$ is the source node ($v = s$), this value must be $h$ (volume of the commodity), since the flow of value $h$ must leave node $s$ considering all links leaving and entering node $s$. In case of the destination node ($v = t$), the same value must equal $-h$, since the flow of volume $h$ must enter the considered node $v$ again taking into account all links leaving and entering node $v$. Finally, if the node $v$ is neither the source nor the destination node of the commodity ($v \neq s, t$), the flow balance at node $v$ (left-hand side of (1.2.1a)) must be 0 and such nodes are called *transit* nodes. Note that the constraint (1.2.1a) is known as a flow conservation law [37, 41]. Constraint (1.2.1b) ensures that the flows cannot be negative.

It should be noted due to technological constraints, links of communication networks have a limited. Let $c_e$ denote the capacity of link $e$ expressed in the same quantity (e.g., b/s, Kb/s, Mb/s, Gb/s, etc.) as a commodity volume. Typically, the definition of one commodity flow must incorporate a following link capacity constraint (i.e., flow on a particular link cannot exceed the link capacity):

$$x(e) \leq c_e, \quad e \in E. \tag{1.2.1.c}$$

Note that for the sake of simplicity the symbol $x_e$ will be used from now on instead of $x(e)$.

**Node-Link Formulation**

The presentation proceeds with a *node-link* formulation of multicommodity flows. Multicommodity means that multiple commodities are transmitted in the network simultaneously. A single commodity (also referred to as demand) is defined as a set of information having the same source node and destination node. Let $h_{ij}$ denote the demand volume of traffic from node $i$ do node $j$. It is assumed that all commodities (demands) are included in set $D$. Let $s_d$ and $t_d$ denote the source and destination of demand $d \in D$, respectively. Let $h_d$ be the volume (bit-rate) of demand $d$, i.e., $h_d = h_{ij}$ for $i = s_d$ and $j = t_d$. There are two ways of formulating multicommodity flows: node-link notation and link-path notation. The multicommodity flow formulated using the node-link notation is defined by the following model.

---

**Unicast/Bifurcated/Node-Link**
**sets**

| | |
|---|---|
| $V$ | nodes |
| $E$ | links |

$\delta^+(v)$    links leaving node $v$
$\delta^-(v)$    links entering node $v$
$D$        demands

**constants**

$h_d$    volume (requested bit-rate) of demand $d$
$s_d$    source node of demand $d$
$t_d$    destination node of demand $d$
$c_e$    capacity of link $e$

**variables**

$x_{ed}$    flow realizing demand $d$ allocated to link $e$; 0, otherwise (continuous, non-negative)

**constraints**

$$\sum_{e\in\delta^+(v)} x_{ed} - \sum_{e\in\delta^-(v)} x_{ed} = \begin{cases} +h_d & \text{if } v = s_d \\ -h_d & \text{if } v = t_d, \quad v \in V, d \in D \\ 0 & \text{otherwise} \end{cases} \tag{1.2.2a}$$

$$\sum_{d\in D} x_{ed} \le c_e, \quad e \in E. \tag{1.2.2b}$$

Note that the flow conservation law of the above model given by (1.2.2a) is very similar to (1.2.1a). The only difference between these two equations is an additional lower index $d$ related to demands. More specifically, $x_{ed}$ denotes the flow of commodity $d$ in link $e$. In (1.2.2a), for every demand $d$ the balance of flow in each node $v$ is checked (left-hand side of (1.2.2a)). As above, in the case of the source node of a particular demand $d$ ($v = s_d$), the value must be equal to the demand volume $h_d$. In the case of the destination node of demand $d$ ($v = t_d$), the right-hand side of (1.2.2a) must be $-h_d$. Finally, for all transit nodes ($v \ne s_d, t_d$) the flow balance is 0. Moreover, it is clear that $\sum_{d\in D} x_{ed}$ denotes the overall flow in link $e$ calculated as a of all demands included in set $D$. Using this definition of a link flow, the capacity constraint is formulated as (1.2.2b).

There are two types of multicommodity flows [37, 38]:

- Bifurcated flows. The commodity (demand) can be split and sent using many different paths (routes) in the network.
- Non-bifurcated (unsplittable, single-path) flows. The whole commodity (demand) is sent along a single path (route).

Notice that model (1.2.2) is formulated for bifurcated multicommodity flows, since the flow variables $x_{ed}$ are defined as continuous and non-negative. As a consequence, the flow of each demand can be realized by different paths. More precisely, according to constraint (1.2.2a) for all links leaving the source node of demand $d$

(included in set $\delta^+(s_d)$), the sum of $x_{ed}$ must be equal to $h_d$. However, since variables $x_{ed}$ are continuous and non-negative, the flow of volume $h_d$ can be allocated to different links from set $\delta^+(s_d)$ in any configuration and the only requirement is that the whole flow leaving node $s_d$ must be exactly $h_d$.

The next model assumes non-bifurcated flows, i.e., the whole demand must be realized on a single path. The key difference from the previous model (1.2.2) is that variables $x_{ed}$ are defined as binary and denote simply whether demand $d$ uses link $e$ or not.

---

**Unicast/Non-Bifurcated/Node-Link**

**sets**

$V$        nodes
$E$        links
$\delta^+(v)$    links leaving node $v$
$\delta^-(v)$    links entering node $v$
$D$       demands

**constants**

$h_d$    volume (requested bit-rate) of demand $d$
$s_d$    source node of demand $d$
$t_d$    destination node of demand $d$
$c_e$    capacity of link $e$

**variables**

$x_{ed}$    $= 1$, if demand $d$ uses link $e$; 0, otherwise (binary)

**constraints**

$$\sum_{e\in\delta^+(v)} x_{ed} - \sum_{e\in\delta^-(v)} x_{ed} = \begin{cases} +1 & \text{if } v = s_d \\ -1 & \text{if } v = t_d, \quad v \in V, d \in D \\ 0 & \text{otherwise} \end{cases} \tag{1.2.3a}$$

$$\sum_{d\in D} x_{ed}h_d \leq c_e, \quad e \in E. \tag{1.2.3b}$$

As well as the fact that variables $x_{ed}$ are defined as binary, another difference between the non-bifurcated model (1.2.3) and the bifurcated model (1.2.2) can be observed in constraints. In particular, the right-hand side of (1.2.3a) is equal to 1, $-1$ or 0, depending on the type of node being considered. Moreover, the left-hand side of the link capacity constraint (1.2.3b) is modified, since the demand volume $h_d$ is included.

**Link-Path Formulation**

Multicommodity flows can also be defined using a *link-path* formulation. First, a notation of a *path* must be defined. Let $v_1, v_2, \ldots, v_a$, $(a > 1)$ be a sequence of various nodes where $(v_i, v_{i+1})$ is an oriented link for each $i = 1, \ldots, (a - 1)$. A sequence of nodes and links $v_1, (v_1, v_2), v_2, \ldots, v_{a-1}, (v_{a-1}, v_a), v_a$ is called a *path*. For each commodity (demand) $d \in D$, there is a set of candidate paths connecting nodes $s_d$ and $t_d$ (end nodes of the demand). Let $P(d)$ be a set of candidate paths for demand $d$. Note that the set of candidate paths can contain either all possible paths that can be constructed in the graph or only selected subset of paths. To limit the number of candidate paths various approaches can be used including the k-shortest path algorithm and the hop-limit approach [48]. To define each path, the constant $\delta_{edp}$ is used. To be more specific, $\delta_{edp}$ is set to 1 if path $p$ of demand $d$ includes link $e$ and 0 otherwise. The flow variable is $x_{dp}$ ($0 \leq x_{dp} \leq h_d$) and denotes the flow of demand $d$ allocated to path $p$. The link-path formulation of bifurcated multicommodity flows is as follows.

---

**Unicast/Bifurcated/Link-Path**

**sets**

$E$      links

$D$      demands

$P(d)$    candidate paths for flows realizing demand $d$

**constants**

$\delta_{edp}$    $= 1$, if link $e$ belongs to path $p$ realizing demand $d$; 0, otherwise

$h_d$     volume (requested bit-rate) of demand $d$

$c_e$     capacity of link $e$

**variables**

$x_{dp}$    flow realizing demand $d$ allocated to path $p$; 0, otherwise (continuous, non-negative)

**constraints**

$$\sum_{p \in P(d)} x_{dp} = h_d, \quad d \in D \tag{1.2.4a}$$

$$\sum_{d \in D} \sum_{p \in P(d)} \delta_{edp} x_{dp} \leq c_e, \quad e \in E. \tag{1.2.4b}$$

Equation (1.2.4a) is in the model to ensure that the whole volume of demand $d$ is realized in the network using candidate paths included in set $P(d)$. Since bifurcated flows are assumed, the demand can be provisioned using more than one path. The second condition (1.2.4b) is a link capacity constraint. Note that the left hand side of (1.2.4b) calculated the overall flow on link $e$.

The corresponding link-path formulation for non-bifurcated flows is obtained by changing the flow variable $x_{dp}$ to a binary one.

---

**Unicast/Non-Bifurcated/Link-Path**

**sets**

$E$      links
$D$      demands
$P(d)$   candidate paths for demand $d$

**constants**

$\delta_{edp}$   = 1, if link $e$ belongs to path $p$ realizing demand $d$; 0, otherwise
$h_d$            volume (requested bit-rate) of demand $d$
$c_e$            capacity of link $e$

**variables**

$x_{dp}$   = 1, if path $p$ is used to realize demand $d$; 0, otherwise (binary)

**constraints**

$$\sum_{p \in P(d)} x_{dp} = 1, \quad d \in D \tag{1.2.5a}$$

$$\sum_{d \in D} \sum_{p \in P(d)} \delta_{edp} x_{dp} h_d \leq c_e, \quad e \in E. \tag{1.2.5b}$$

In the above model (1.2.5)—as in the case of node-link formulations—both constraints have to be adjusted slightly in comparison with the bifurcated formulation (1.2.4). Firstly, the right-hand side of (1.2.5a) is now 1, what ensures that exactly one candidate path is selected to realize a particular demand. Secondly, the demand volume $h_d$ is added to the left-hand side of (1.2.5b).

**Comparison of Formulations**

We now present a short comparison and discussion of the presented modeling approaches, namely, node-link and link-path. The numbers of variables and constraints for both formulations are shown in Table 1.1.

The key difference between the formulations lies in the fact that using the link-path formulation, the model can be tuned to find a tradeoff between accuracy and complexity. More specifically, the node-link formulation by default includes all possible routing paths for each considered demand (commodity), which follows directly

**Table 1.1** Comparison of unicast flow formulations

| Model | Number of variables | Number of constraints |
|---|---|---|
| Node-link formulation | $|D||E|$ | $|D||V| + |E|$ |
| Link-path formulation | $|D||P|$ | $|D| + |E|$ |

from the flow conservation constraints. On the other hand, the link-path formulation can use only a limited set of routing paths. In consequence, the solution of the optimization problem calculated using the link-path modeling with a limited set of candidate paths is not guaranteed to be optimal in a global sense, which is guaranteed if the node-link modeling is applied. However, the link-path modeling requires $|D||P|$ decision variables, while the node-link modeling needs $|D||E|$ decision variables, where $|D|$ is the number of demands, $|E|$ is the number of links, and $|P|$ is the number of candidate paths for one demand. Additionally, in terms of the number of constraints, there is a clear difference in terms of the complexity. In fact, the link-path formulation uses $|D|$ constraints to ensure the routing, while the node-link formulation in the flow conservation law constraints requires $|D||V|$ constraints. In addition, both formulations need $|E|$ constraints for the link capacity control. Note that for a larger network, we can set the number of candidate paths $|P|$ to a much lower value compared to the number of links $|E|$ or number of nodes $|V|$, which can significantly reduce the model size. In consequence, the link-path model can be solved much faster than the node-link model by using either exact methods such as branch-and-bound or branch-and-cut algorithms or heuristic algorithms. Moreover, in some cases the node-link model—according to a very large size—cannot be solved at all due to a lack of computer memory required to create and next to solve this model.

### 1.2.2 Anycast Flows

This section introduces two formulations of anycast flows: node-link and link-path. Since most network technologies and protocols currently utilized as services using the anycast paradigm are connection-oriented, only non-bifurcated flows will be considered in the models. However, the following models can be easily adapted to bifurcated flows in the same way as the unicast formulations presented in Sect. 1.2.1.

Anycast is an *one-to-one-of-many* transmission technique of delivering information to or from one of many hosts providing the same information. Anycasting was proposed in [49] in the context of the IP protocol to address a situation where a service requested by a host, application or user is supported by several servers, and the requesting party does not care, which server is used. Since then, anycasting has been applied in the context of various network protocols and technologies as a typical approach in situations when the same content/service is replicated in many different locations in the network [3, 50–73]. For instance, a CDN system using many data centers (DCs) providing the same content (e.g., Akamai, YouTube) can be considered to be a good example of applying anycasting to content delivery. In turn, cloud computing services provisioned by various data centers illustrate the case when anycasting can be applied to providing a range of services.

It should be stressed that anycasting—as a whole—is a complex approach and its implementation in communication networks triggers the need to cope with many problems such as replica location, replica ranking, replica consistency, redirection of requests, accounting, security, and routing [28, 29, 33].

Firstly, some basic assumptions on the network model in the context of anycast transmission are introduced. More specifically, anycasting is used to provide to clients (end users) some content or services. To enable anycasting, the content/service is available in many locations in the network and it is provisioned by data centers of replica servers. For the sake of simplicity, the local connection between the data center and the backbone network node is not taken into account, i.e., a network node is equivalent with the connected data center. This is because usually local access links used to connect data centers are of a very large bandwidth and therefore capacity of these links is not included as a constraint in the model. Similarly, aggregation of client requests is assumed and the model includes only the network node to which the client is connected, not individual clients [70].

An anycast request is defined by one network node known as *client* node. The second node required to provision the transmission is selected from candidate nodes hosting data centers that provide the required content/service. In general, the anycast request is bidirectional since two transmissions in opposite directions are required to establish an anycast request. One transmission is from the client node to the data center node (upstream) and the second is from the data center node to the client (downstream). For instance, in the context of CDN systems, the upstream transmission is applied to send a client request to obtain some content, while the downstream transmission is used to send the requested content to the client [70].

According to [70], anycast demands (requests) in connection-oriented networks can be modeled in two ways: *reduced* and *standard*. In the former case, it is assumed that the anycast transmission is only in one direction, i.e., from or to the data center (replica server). This assumption follows directly from an important feature of many anycast systems, i.e., asymmetry of flow. Let's recall that one of the key applications of anycasting is caching and replicating of the content in the network. In most cases, the access to this content is asymmetric since the users usually fetch much more data from the content server than they send in the opposite direction. Moreover, the traffic asymmetry is observed in everyday use of the Internet—most of ISPs offer asymmetric access lines (e.g., ADSL). In the reduced model, the anycast request is established as a single demand in the network (usually downstream), while transmission in the opposite direction (usually upstream) is ignored due to the fact that the volume of the upstream transmission is much lower than the downstream. Therefore, in the reduced model, an anycast demand is defined by the following triple: client node, set of admissible data center nodes and downstream bit-rate. In contrast, the unicast demand is defined by the following triple: origin node, destination node and bit-rate.

In the standard model, the anycast request is realized by two demands: upstream (from the client to the data center) and downstream (in the opposite direction). Both downstream and upstream demands of the same anycast request described as *associated*. Let $\tau(d)$ be the index of the demand associated with demand $d$. Both associated demands $d$ and $\tau(d)$ have to be connected to the same data center node. This is because the upstream demand sends data related to client requests of the content/service provided by the data center. In order to reduce the latency and complexity of the system, the requested data center is responsible for answering the received requests by using the downstream demand. However, we assume that both associated anycast demands do not have to use the same routing path.

Provisioning of anycast demands can be divided into two phases. The first step is the data center (server) selection process—the client must choose one node from admissible data centers that provide the requested content/service. When the data center node is selected, both end nodes of the anycast demand are determined, which means that the anycast demands can be processed in a similar way to classical unicast demands. Therefore, the second step is the selection of routing paths for both associated demands.

As in the case of unicast flows, anycast flows can be formulated in two ways: node-link and link path. In the following, we define both formulations in the context of the standard model of anycasting. The reduced model can be simply obtained by eliminating all demands related to a particular direction (upstream or downstream) from the formulation. The models presented below assume that all data centers located in the network can serve each request, i.e., they can provide the requested content/service for each anycast client. However, the models can easily be modified to address a scenario where there are several types of data centers provisioning and each type is assigned to a particular subset of content/service requests.

Formulations of anycast flows have been presented in numerous papers, e.g., [53–57, 61, 65, 74–76]. However, all these papers consider a significant simplification assuming that anycast flows are unidirectional. In turn, papers [58, 59, 63, 64, 67, 69–73, 77, 78] account for bidirectional anycast flows.

**Node-Link Formulation**

In general, in the node-link formulation anycast flows are modeled in the same way as in the context of unicast flows, i.e., the binary variable $x_{ed}$ denotes whether or not demand $d$ uses link $e$. However, due to the fact that for anycasting only one of the end nodes is given, while the second end node must be selected from candidate servers, a new variable is required. To this end, the variable $z_{vd}$ denotes whether the data center located at node $v$ is selected for demand $d$.

---

**Anycast/Non-Bifurcated/Node-Link**
**sets**

$V$       nodes
$R$       data center nodes

| $E$ | links |
|---|---|
| $\delta^+(v)$ | links leaving node $v$ |
| $\delta^-(v)$ | links entering node $v$ |
| $D$ | demands |
| $D^{DS}$ | anycast downstream demands |
| $D^{US}$ | anycast upstream demands |

**constants**

| $h_d$ | volume (requested bit-rate) of demand $d$ |
|---|---|
| $s_d$ | source node of demand $d$ (client node for upstream demand) |
| $t_d$ | destination node of demand $d$ (client node for downstream demand) |
| $c_e$ | capacity of link $e$ |
| $\tau(d)$ | index of a demand associated with demand $d$. If $d$ is a downstream demand, then $\tau(d)$ must be an upstream demand and vice versa |

**variables**

$x_{ed}$ = 1, if demand $d$ uses link $e$; 0, otherwise (binary)

$z_{vd}$ = 1, if data center located at node $v$ is selected for demand $d$; 0, otherwise (binary)

**constraints**

$$\sum_{e \in \delta^+(v)} x_{ed} - \sum_{e \in \delta^-(v)} x_{ed} = \begin{cases} +z_{vd} & \text{if } v \in R \\ -1 & \text{if } v = t_d \\ 0 & \text{otherwise} \end{cases} \quad v \in V, \; d \in D^{DS} \qquad (1.2.6a)$$

$$\sum_{e \in \delta^+(v)} x_{ed} - \sum_{e \in \delta^-(v)} x_{ed} = \begin{cases} +1 & \text{if } v = s_d \\ -z_{vd} & \text{if } v \in R \\ 0 & \text{otherwise} \end{cases} \quad v \in V, \; d \in D^{US} \qquad (1.2.6b)$$

$$\sum_{d \in D} x_{ed} h_d \le c_e, \quad e \in E \qquad (1.2.6c)$$

$$z_{vd} = z_{v\tau(d)}, \quad d \in D^{DS}, v \in R \qquad (1.2.6d)$$

$$\sum_{v \in R} z_{vd} = 1, \quad d \in D. \qquad (1.2.6e)$$

The first two constraints (1.2.6a) and (1.2.6b) are flow conservation conditions formulated for downstream and upstream anycast demands, respectively. Let's recall that two such constraints are required, since each type of anycast demand (downstream and upstream) is defined by a different fixed end node, i.e., for downstream demands the destination node is defined by the client node, while for upstream demands the origin node is fixed at the client node. Accordingly, in the case of constraint (1.2.6a), for all nodes $v \in R$ the value of the flow balance (left-hand side of 1.2.6a) must be equal to $z_{vd}$. As only one node can be selected as the server node for demand $d$, only in the case of this one node $v \in R$ the left-hand side is 1 (equivalent to the demand source node) and for all other server nodes not selected in variable $z_{vd}$, the left-hand side is simply 0, which denotes that these nodes are transit nodes for demand $d$. For the destination node $v = t_d$ the left-hand side of (1.2.6a) is $-1$, since this is the destination node of demand $d$. Finally, for all remaining cases denoting transit nodes, the flow balance must be 0. The second constraint (1.2.6b) formulated for upstream demands is generally analogous to the previous one. The difference can be observed on the left-hand side where—according to upstream transmission—for the source node the value is fixed to 1, while in the case of the destination node the flow balance is 1 only for the selected server node $v$, given by $z_{vd} = 1$. The condition (1.2.6c) includes the link capacity constraint. Equality (1.2.6d)—known as *anycast constraint*—ensures that two associated demands $d$ and $\tau(d)$ defined by the same client must connect the same pair of nodes: the client node and the selected server node. Thus, it is guaranteed that both associated demands are assigned to the same server node. The last constraint (1.2.6e) is in the model to ensure that each demand is assigned to exactly one server node.

**Link-Path Formulation**

Link-path formulation of anycast flows uses pre-calculated routing paths as in the case of unicast flows. However, in the context of anycast flows the set of candidate paths includes routing paths connecting the client node to all feasible server nodes. Consequently, selection of one of the candidate paths determines the selection of the data center (server node). More specifically, if $d$ is an upstream demand, set $P(d)$ includes candidate paths that originate at one of the server nodes and terminate at the client node. If $d$ is a downstream demand, candidate paths in set $P(d)$ connect the client node and one of the server nodes. Note that in the context of unicast flows, candidate paths always connect the same pair of nodes.

------------

**Anycast/Non-Bifurcated/Link-Path**

**sets**

| | |
|---|---|
| $E$ | links |
| $D$ | demands |
| $D^{DS}$ | anycast downstream demands |
| $P(d)$ | candidate paths for demand $d$. If $d$ is an upstream demand, path $p$ connects client node and server node. If $d$ is a downstream demand, path $p$ connects server node and client node |

**constants**

| | |
|---|---|
| $\delta_{edp}$ | = 1, if link $e$ belongs to path $p$ realizing demand $d$; 0, otherwise |
| $h_d$ | volume (requested bit-rate) of demand $d$ |
| $c_e$ | capacity of link $e$ |
| $s(p)$ | source node of path $p$ |
| $t(p)$ | destination node of path $p$ |

$\tau(d)$   index of a demand associated with demand $d$. If $d$ is a downstream demand, then $\tau(d)$ must be an upstream demand and vice versa

**variables**

$x_{dp} = 1$, if path $p$ is used to realize demand $d$; 0, otherwise (binary)

**constraints**

$$\sum_{p \in P(d)} x_{dp} = 1, \quad d \in D \tag{1.2.7a}$$

$$\sum_{d \in D} \sum_{p \in P(d)} \delta_{edp} x_{dp} h_d \leq c_e, \quad e \in E \tag{1.2.7b}$$

$$\sum_{p \in P} x_{dp} s(p) = \sum_{p \in P(\tau(d))} x_{\tau(d)p} t(p), \quad d \in D^{DS}. \tag{1.2.7c}$$

The above formulation is very similar to the unicast link-path formulation (1.2.5). The first two constraints (1.2.7a) and (1.2.7b) are identical as above. The former one defines the non-bifurcated multicommodity flows and ensures that exactly one path is selected for each demand, while the latter one is capacity constraint. We must stress that set $P(d)$ with candidate paths for demand $d$ is constructed in a different way compared to the unicast formulation and constraint (1.2.7a) ensures the selection of a routing path and a server node. The last equality (1.2.7c) is the *anycast constraint* analogous to (1.2.6d). Note that the left-hand side of (1.2.7c) is equal to the index of the source node selected for demand $d$. In the same way, the right-hand side of (1.2.7c) is equal to the index of the destination node chosen for demand $\tau(d)$.

**Comparison of Formulations**

Table 1.2 reports the size of both anycast formulations in terms of the number of variables and constraints. It is notable that—compared to unicast formulations—the size of the node-link model increases to a greater extent compared to the link-path formulation. This is mainly a consequence of the fact that the node-link formulation requires additional variable $z_{vd}$ to control the data center selection. In addition, similarly to unicast formulations, the anycast link-path formulation provides more flexibility, since this model can be tuned according to find the best tradeoff between accuracy and complexity by using a set of selected routing paths.

**Table 1.2** Comparison of anycast flow formulations

| Model | Number of variables | Number of constraints |
|---|---|---|
| Node-link formulation | $|D||E| + |D||V|$ | $|D||V| + |E| + 0.5|D||R| + |D|$ |
| Link-path formulation | $|D||P|$ | $1.5|D| + |E|$ |

## *1.2.3   Multicast Flows*

This section focuses on modeling multicast flows. In computer and communication networks, the most commonly used basic techniques for routing of data are unicast (one-to-one) and broadcast (one-to-all). However, these methods are not efficient in scenarios where the same data is to be delivered to a relatively large group of users, geographically separated and with similar interest levels in the delivered content. The multicast transmission—defined as a *one-to-many* transmission from one source node (usually known as the root) to a group of receiving nodes (terminals)—is intended to be an efficient approach to realizing group transmission in communication networks. Group transmission can be provisioned by the unicast approach. More specifically, content provided by the root node can be delivered to all receivers using a set of separate unicast demands for each node pair including the root and receiver. However, this approach can lead to a situation where the same data is transmitted on a particular link many times, which increases network traffic and in consequence increases network costs in terms of both CAPEX and OPEX expenditures. In contrast, multicasting consumes network resources in a more economical way, since a special tree topology is constructed to ensure that the same volume of information is transmitted on a particular link, even if multiple receivers use this link to connect to the root. This is achieved by a special function implemented in network nodes that copies incoming data to many outgoing interfaces. Recently, we have seen a growing popularity of multicasting due to the development of many new services focused on streaming services such as IPTV, Video on Demand (VoD), radio streaming, CDNs, distance learning, software distribution, monitoring, and result distribution in computing systems [29, 33, 38, 79–82].

It should be noted that multicast modeling can refer to two classical network problems [38]:

- Minimum Steiner tree problem. Given a subset of nodes (vertices) $W$ included in the original graph $G = (V, E)$ that consist of $V$ nodes and $E$ links, interconnect nodes in $W$ by a subgraph $T$ of the shortest length (defined as a sum of the lengths of all links included in the subgraph). This subgraph $T$ is a tree (no loops) that can also contain nodes not included in set $W$. These new nodes introduced to the subgraph to decrease the total length of the tree are known as Steiner points or Steiner vertices.
- Minimum Spanning Tree (MST) problem. The minimum spanning tree is defined as a subgraph $T$ of the original graph $G = (V, E)$ with the shortest length (defined as a sum of the lengths of all links included in the subgraph). Moreover, $T$ is a tree (no loops) that includes all nodes in $V$.

The key difference between these problems is that in the Steiner tree problem extra intermediate nodes (Steiner vertices) may be added to the created subgraph to reduce the length of the spanning tree.

Multicast flows can be modeled in several ways. In this section, four formulations of multicast flows are presented: canonical formulation, flow formulation, level formulation and candidate tree formulation. Each formulation models a single multicast session; however, the models can be easily modified to incorporate a case with multiple multicast sessions in a similar way as in the context of unicast or anycast flows. The considered graph is modeled as $G = (V, E)$. The multicast session is defined by root node $s \in V$ and a set of receivers included in set $R \subseteq V$.

## Canonical Formulation

The first approach proposed for modeling multicast flows in communication networks is the *canonical* formulation [83, 84]. The multicast transmission is modeled using the Steiner tree problem. In particular, for each network link (edge) $e \in E$, there is a variable $x_e$ indicating whether link $e$ is in the Steiner tree ($x_e = 1$) or not ($x_e = 0$). The formulation uses cuts of the original network graph $G = (V, E)$. Let $\eta(W)$ define a graph cut induced by $W \subseteq V$, i.e., $\eta(W)$ includes a set of links with the source node in set $W$ and the destination node in its complement set $(V \setminus W)$. In the context of multicast transmission, a subset of all possible cuts $\eta(W)$ must fulfil the following condition: set $W$ contains root node $s$ and set $(V \setminus W)$ embraces at least one receiving node from set $R$. Each such a cut must contain at least one link included in the tree defined by variables $x_e$ to provide a routing path from root node to each receiver.

---

**Multicast/Canonical**

**sets**

| | |
|---|---|
| $E$ | links |
| $V$ | nodes |
| $R$ | receivers |
| $\eta(W)$ | set of links defined by a graph cut induced by $W \subseteq V$, i.e., $\eta(W)$ includes links with the source node in set $W$ and the destination node in its complement set $(V \setminus W)$ |

**constants**

| | |
|---|---|
| $s$ | source (root) node of multicast session |
| $h$ | volume (requested bit-rate) of multicast session |
| $c_e$ | capacity of link $e$ |

**variables**

| | |
|---|---|
| $x_e$ | $= 1$, if multicast tree uses link $e$; 0, otherwise (binary) |

**constraints**

$$x(\eta(W)) \geq 1, \quad W \subset V : s \in W, (V \setminus W) \cap R \neq \emptyset \qquad (1.2.8a)$$

$$x(\eta(W)) = \sum_{e \in \eta(W)} x_e, \qquad (1.2.8b)$$

$$x_e h \leq c_e, \quad e \in E. \qquad (1.2.8c)$$

The key elements of the model are included in constraints (1.2.8a) and (1.2.8b). To be more specific, these constraints ensure that for every cut $\eta(W)$ that embraces all links connecting two sets of nodes, namely nodes in $W$ and nodes in the complement set $(V \setminus W)$ at least one link is included in the constructed tree ($x_e = 1$). Moreover, the set of considered cuts $\eta(W)$ must satisfy the following conditions: set $W$ includes the root node $s$ and set $(V \setminus W)$ includes at least one receiving node. Note that constraint (1.2.8a) ensures that each receiver included in set $R$ is connected to the root node by a routing path. This is because if constraint (1.2.8a) is not satisfied (i.e., $x(\eta(W)) = 0$) for at least one of the considered cuts, there is no connection between the root node and at least one of the receiving nodes. In consequence, the principal condition of multicasting that each receiver must be connected by a routing path to the root node is not satisfied. Inequality (1.2.8c) is the link capacity constraint that ensures that the link flow (given by $x_e h$) cannot exceed the link capacity. It should be underlined that the number of possible cuts in a graph that must be considered in the above formulation grows exponentially with the network size given by the number of nodes. This is undoubtedly the main drawback of the canonical formulation, which significantly limits the scalability of this model and reduces the applicability of this formulation in network optimization problems.

**Flow Formulation**

The next formulation—known as *flow* formulation—is based on the multicommodity node-link formulation used in the context of unicast flows. The unicast node-link formulation can be adapted to model multicast transmission as follows. The general idea behind flow formulation is to define a unicast path connecting the root node $s$ and each receiving node $r \in R$. For this purpose, binary variable $x_{er}$ is introduced to denote whether multicast flow from the root node to receiver $r$ uses link $e$. This modeling can lead to a situation, in which a particular link is used by more than one unicast path connecting the source node and the receivers, which means that the same data is sent several times on this link. Consequently, an additional binary variable $x_e$ denotes whether link $e$ is included in the multicast tree, which ensures that the multicast flow goes through a link at most once [84–93].

---

**Multicast/Flow**

**sets**

| | |
|---|---|
| $V$ | nodes |
| $E$ | links |
| $R$ | receivers |
| $\delta^+(v)$ | links leaving node $v$ |
| $\delta^-(v)$ | links entering node $v$ |

**constants**

$s$  source (root) node of multicast session

   $h$  volume (requested bit-rate) of multicast session
   $c_e$  capacity of link $e$

**variables**

$x_{er}$  $=1$, if multicast flow to receiver $r$ uses link $e$; 0, otherwise (binary)
$x_e$   $= 1$, if multicast tree uses link $e$; 0, otherwise (binary)

**constraints**

$$\sum_{e \in \delta^+(v)} x_{er} - \sum_{e \in \delta^-(v)} x_{er} = \begin{cases} +1 & \text{if } v = s \\ -1 & \text{if } v = r, \\ 0 & \text{otherwise} \end{cases} \quad v \in V, r \in R \qquad (1.2.9a)$$

$$x_{er} \leq x_e, \quad e \in E, r \in R \qquad (1.2.9b)$$

$$x_e h \leq c_e, \quad e \in E. \qquad (1.2.9c)$$

Constraint (1.2.9a) defines—using the node-link formulation—unicast paths connecting root node $s$ and each receiver $r \in R$. Recall from model (1.2.3) that the left-hand side of (1.2.9a) is the total number of outgoing links minus the total number of incoming links of the non-bifurcated unicast path defined for each network node $v$ and each receiver $r$. Therefore, if the considered node $v$ is the root node $s$, then the right-hand side of (1.2.9a) must be 1. If node $v$ is a receiving node, it must be $-1$. All other nodes are transit nodes and the flow balance must be 0. Inequality (1.2.9b) ensures that each link is used in the multicast tree at most one time. Notice that variable $x_e$ is set to 1, only if the particular link $e$ is included in a unicast path to at least one receiver $r \in R$. The final condition (1.2.9c) denotes the link capacity constraint.

**Level Formulation**

The next model—referred to as the *level* formulation—is based on the fact that multicast trees can be divided into subsequent levels [38, 94, 95]. More specifically, the root node of the tree is assumed to be located on level 1. All direct children of the root (nodes that have a direct link from the root) are located on level 2, etc. More generally, it is assumed that if a parent node of $v$ is on level $l$, then node $v$ is located on level $l + 1$. Set $L = \{1, 2, \ldots, |L|\}$ includes all possible levels defined for a particular optimization problem. Comparing the level formulation against the flow formulation, network links are represented differently, i.e., a pair of nodes $(v, w)$ defines a network link from node $v$ to node $w$. To model the multicast tree, a binary variable $x_{vwl}$ denotes whether link $(v, w)$ is used in the multicast tree and $v$ is located on level $l$ of the tree. As in the flow formulation, an additional variable $x_{vw}$ indicates, if link $(v, w)$ is included in the multicast tree. Note that the level formulation is also referred to as layered graphs [94].

---

**Multicast/Level**

**sets**

| | |
|---|---|
| $V$ | nodes |
| $E$ | links |
| $R$ | receivers |
| $L$ | levels, root node is located on level 1, children of the root node are located on level 2, etc. |
| $\delta^+(v)$ | links leaving node $v$ |
| $\delta^-(v)$ | links entering node $v$ |

**constants**

| | |
|---|---|
| $s$ | source (root) node of multicast session |
| $h$ | volume (requested bit-rate) of multicast session |
| $c_{vw}$ | capacity of link $(v, w)$ |

**variables**

| | |
|---|---|
| $x_{vwl}$ | $= 1$, if the link $(v, w)$ is used in multicast tree and node $v$ is located on level $l$ of the tree; 0, otherwise (binary) |
| $x_{vw}$ | $= 1$, if multicast tree uses link $(v, w)$; 0, otherwise (binary) |

**constraints**

$$\sum_{v:(v,s)\in E} \sum_{l\in L} x_{vsl} = 0, \tag{1.2.10a}$$

$$\sum_{v:(v,r)\in E} \sum_{l\in L} x_{vrl} = 1, \quad r \in R \tag{1.2.10b}$$

$$\sum_{w:(v,w)\in E} x_{vw1} = 0, \quad v \in W \setminus \{s\} \tag{1.2.10c}$$

$$x_{vwl} \le \sum_{u:(u,v)\in E} x_{uv(l-1)}, \quad (v, w) \in E, l \in L \setminus \{1\} \tag{1.2.10d}$$

$$\sum_{l\in L} x_{vwl} \le x_{vw}, \quad (v, w) \in E \tag{1.2.10e}$$

$$x_{vw} \le \sum_{l\in L} x_{vwl}, \quad (v, w) \in E \tag{1.2.10f}$$

$$x_{vw}h \le c_{vw}, \quad (v, w) \in E. \tag{1.2.10g}$$

The first constraint (1.2.10a) imposes that root node $s$ cannot download multicast flows, i.e., the total flow on all links $(v, s)$ entering node $s$ must be zero taking into account all possible levels. Equation (1.2.10b) ensures that every receiving node $r \in R$ must be connected to the multicast tree. More precisely, the number of links entering each node $r \in R$ must be exactly 1 considering all possible levels and all possible parent nodes $v$ such as $(v, r) \in E$. The next condition (1.2.10c) is in the model to express the requirement that only root node $s$ can be a parent node on level 1. In other words, nodes $v$ except node $s$ cannot be the parent of the first level link. Constraint (1.2.10d) ensures that node $v$ cannot upload multicast flows to any other node $w$ on level $l$ such as $(v, w) \in E$ if node $v$ is not located on level $l - 1$ of the multicast tree. More specifically, the right-hand side of (1.2.10d) denotes the number of multicast links entering node $v$ on level $l - 1$ considering all possible parent nodes $u$ such as $(u, v) \in E$. If there is no such a link, the right-hand side of (1.2.10d) is 0, which in consequence imposes that $x_{vwl}$ is set to 0 for all links $(v, w) \in E$. It should be pointed out that inequality (1.2.10d) plays a very important role in the level model, since it ensures—together with constraints (1.2.10a) and (1.2.10b)—that a path from the root node to each receiver must be constructed to form a multicast transmission. Constraints (1.2.10e) and (1.2.10f) are used to bind variables $x_{vwl}$ and $x_{vw}$. In particular, condition (1.2.10e) ensures that if for any level $l$ there is a link $(v, w)$ included in the multicast tree ($\sum_{l \in L} x_{vwl} = 1$), then $x_{vw}$ must be 1. In turn, condition (1.2.10f) guarantees that if there is no link between nodes $v$ and $w$ on any level $l$ ($\sum_{l \in L} x_{vwl} = 0$), consequently $x_{vw}$ is 0. Condition (1.2.10g) denotes the link capacity constraint.

The level formulation enables a simple modeling of hop-constrained multicasting assuming that there is an upper limit on the number of hops between the root node and any receiving node [86]. The main reason for this additional hop-limit constraint is to improve QoS (Quality of Service) parameters of the multicasting including network reliability and transmission delay.

**Candidate Tree Formulation**

The last multicast formulation known as *candidate tree*, takes inspiration from the link-path modeling used in the context of unicast or anycast flows. More precisely, there is a set of candidate trees—defined as subgraphs of the considered network graph—that includes the root node and all receiving nodes of the considered multicast session. In other words, in each candidate tree for every node pair root-receiver there is exactly one path connecting these two nodes. Let $P$ denote a set that includes candidate trees calculated for the considered multicast session. Binary decision variable $x_p$ denotes which candidate tree is selected to provision the multicast session [38, 96].

---

**Multicast/Candidate Tree**
**sets**

$E$     links
$P$     candidate trees for multicast session

**constants**

$\delta_{ep}$    $= 1$, if link $e$ belongs to tree $p$; 0, otherwise

$h$     volume (requested bit-rate) of multicast session

$c_e$    capacity of link $e$

**variables**

$x_p$    $= 1$, if tree $p$ is used to realize multicast session; 0, otherwise (binary)

**constraints**

$$\sum_{p \in P} x_p = 1, \tag{1.2.11a}$$

$$\sum_{p \in P} \delta_{ep} x_p h \le c_e, \quad e \in E. \tag{1.2.11b}$$

Compared to the multicast formulations discussed above, the candidate tree model is quite simple and contains only two constraints. The first (1.2.11a) states that exactly one candidate tree must be selected to provision the multicast session, while the second (1.2.11b) is the link capacity constraint. The candidate tree formulation has the same weaknesses as the analogous link-path formulation. More precisely, if the set of candidate trees includes a limited set of all possible trees, the solution is not guaranteed to be optimal in a global sense, which is guaranteed when the canonical, flow or level formulations are applied.

**Comparison of Formulations**

As in the case of unicast and anycast flow formulations, we also present a comparison of multicast models. In particular, Table 1.3 shows the number of variables and constraints in three basic models applied to modeling of multicast flows in computer and communication network optimization problems: flow formulation, level formulation and candidate tree formulation. The canonical formulation is excluded from the comparison due to the exponential number of constraints required to list all cuts of a graph.

The key advantage of the level formulation when compared to the flow formulation, is the lower complexity of the model expressed by the number of variables. Let's recall that the level formulation uses $(|E||L| + |E|)$ variables while the flow formulation includes $(|E||R| + |E|)$ variables, where $|E|$ is the number of links, $|R|$ is the number of receivers, and $|L|$ is the number of levels. Note that in most cases

**Table 1.3** Comparison of multicast flow formulations

| Model | Number of variables | Number of constraints |
|---|---|---|
| Flow formulation | $|E||R| + |E|$ | $|V||R| + |E||R| + |E|$ |
| Level formulation | $|E||L| + |E|$ | $|R| + |V| + |E||L| + 4|E| + 1$ |
| Candidate tree formulation | $|P|$ | $|E| + 1$ |

the number of possible levels $|L|$ is set to a value lower than the number of receivers $|R|$ in the session. However, the lowest complexity is offered by the candidate tree formulation, which is especially important if larger networks are analyzed. Moreover, the candidate tree formulation—similar to the link-path formulations of unicast and anycast flows—makes it possible to tune the size of the model by using various candidate trees.

## 1.3   Optimization Methods

Since optimization methods applied to network optimization have been comprehensively addressed in numerous textbooks, monographs and other publications, in this section we present a brief overview of optimization methods that are commonly applied to computer and communication network optimization.

A common approach to optimization of computer and communication networks is *mathematical programming*. Most of mathematical programs formulated in the context of networking problems are either *mixed-integer programming* (MIP) formulations or *integer programming* (IP) formulations, i.e., either some or all variables used in the formulation are integers (binary). Frequently, the term *integer linear programming* (ILP) is used to refer to integer programming formulations. In fact, the only general exact optimization method of solving MIP and IP problems is the branch-and-bound algorithm accompanied by an efficient enhancement known as the branch-and-cut algorithm. For more details on mathematical programming, there are numerous textbooks providing exhaustive surveys of mathematical programming and combinatorial optimization, e.g., [37, 97–100].

Having formulated a network optimization problem as a mathematical program, the problem can be solved relatively easily using dedicated optimization software that provides very an efficient implementation of state-of-the-art mathematical programming algorithms including Simplex and branch-and-cut. Therefore, researchers do not have to allocate a great deal of time to implementing these methods, and the only effort required is preparing the optimization model in a form suitable for the optimization solver. Perhaps the most popular packages applied in this context are the IBM ILOG CPLEX Optimizer [101] and the Gurobi Optimizer [102]. Another benefit of mathematical programming methods is that they yield optimal results. However, one of the outcomes of the development of computer and communication networks, including the introduction of new technologies and the general increase of network complexity, is that network design has become more complicated in recent years. Therefore, the major drawback of mathematical programming methods in the context network optimization problem is a lack of scalability; this means that for greater numbers of problem, the exact methods cannot provide optimal or even feasible results. In consequence, other optimization methods are required for tackling complicated problems and large instances, namely heuristic and metaheuristic algorithms.

Heuristic and metaheuristic methods represent a family of approximate optimization techniques that can solve various problems in a reasonable time; however, the solution does not have the optimality guarantee that is only ensured in exact algorithms. The word *heuristic* comes from the ancient Greek word *heuriskein*, which means 'the art of discovering new strategies (rules) to solve the problems'. In turn, the prefix *meta* also is a Greek word and denotes an upper level methodology. The key difference between heuristic and metaheuristic approaches is the complexity of the method. In essence, heuristic algorithms are relatively simple methods based on trial and error or other constructive approaches which produce acceptable solutions to complex problems in a reasonably practical time. For instance, a heuristic method can use a greedy approach, in which decision variables are selected in a single run of the algorithm and a single decision variable is assigned at each step. In contrast, metaheuristic methods use more complex strategies including randomization, local search, memory structures that keep information extracted during the search, and population-based search. Examples of metaheuristic methods popular in the context of network optimization problems are simulated annealing, tabu search, evolutionary and genetic algorithms, GRASP, ant colony optimization, and particle swarm optimization. It should be emphasized that there are no agreed definitions of heuristics and metaheuristics in literature. Some authors apply the terms heuristics and metaheuristics interchangeably. However, since most researchers tend to name all stochastic algorithms with randomization and local search as metaheuristic, this book follows this convention. Once again, for more thorough information on various types of algorithms including heuristic and metaheuristic methods, we refer to [37, 103–109].

Note that many interesting topics related to the application of various optimization methods in the context of computer and communication networks are addressed in [105, 106, 110, 111]. Moreover, aspects related to network optimization in the context of survivability and reliability requirements can be found in [37, 112–114].

## 1.4   Naming and Numbering Conventions

This section introduces and discusses conventions used in the book with regard to naming and numbering. In general, we tried to follow the standards used in [37], which is one of the most popular books on the optimization of computer and communication networks. In particular, the equation numbering convention contains three elements: chapter number, section number and equation number. For instance, Eq. (2.4.1) refers to the first equation in Chap. 2, Sect. 2.4.

We name each optimization problem formulation when it is first stated. The goal of the naming scheme is to denote the most important details about the specific problem. In a nutshell, the naming convention is **N/F/P/T/F/I**, where:

- **N** denotes the considered network context in terms of technology or protocol, i.e., **CON**—Connection-Oriented Network, **EON**—Elastic Optical Network, **OVR**—Overlay Network, **MLN**—Multi Layer Network.
- **F** represents the type of the network flow: **A**—anycast, **M**—multicast, **U**—unicast. Note that different types of network flows can occur in one problem, e.g., **AU** indicates joint optimization of anycast and unicast flows.
- **P** denotes the type of optimization problem, namely, **FA**—flow allocation, **ND**—network design, **NDL**—network design and location, **LD**—location design, **FAL**—flow allocation and location, **RSA**—Routing and Spectrum Allocation, **RMSA**—Routing, Modulation and Spectrum Allocation.
- **T** (not obligatory) provides some additional information on the problem type, i.e., **DPP**—Dedicated Path Protection, **SBPP**—Shared Backup Path Protection, **PCycle**—p-Cycles, **DTP**—Dedicated Tree Protection, **DHP**—Dual Homing Protection, **DNP**—Dedicated Node Protection.
- **F** denotes the objective function of the optimization, e.g., **Cost**, **Spectrum**, **Average Spectrum**, **Time**, **Throughput**.
- **I** (optional) provides some additional information on the problem, e.g., type of the flow modeling approach.

For instance, the model **EON/AU/RSA/Cost/Link-path/Channel-based** refers to a RSA problem in EONs with joint anycast and unicast flows, objective function of cost, using link-path modeling of flows and channel-based modeling of spectrum. Similar naming convention is applied in the context of algorithms presented.

Note that due the high diversity of optimization problems presented in this book, the notation may be misleading in some cases. More specifically, the same symbol can have different meanings in separate optimization models. Therefore, in most cases, an optimization model is self-contained and includes a description of all notation and symbols used to formulate the model. When a successive model includes many elements common with previous models presented in the same section, it is presented in a compressed way.

# References

1. Erl, T., Puttini, R., Mahmood, Z.: Cloud Computing: Concepts. Technology & Architecture, 1st edn. Prentice Hall Press, Upper Saddle River (2013)
2. Mell, P., Grance, T.: The nist definition of cloud computing. Technical report, Gaithersburg, MD, United States, pp. 800–145 (2011)
3. Klinkowski, M., Walkowiak, K.: On the advantages of elastic optical networks for provisioning of cloud computing traffic. IEEE Netw. **27**(6), 44–51 (2013)
4. Venters, W., Whitley, E.: A critical review of cloud computing: researching desires and realities. J. Inf. Technol. **27**(3), 179–197 (2012)
5. Walkowiak, K.: M.; Wozniak, M. Klinkowski, and W. Kmiecik. Optical networks for cost-efficient and scalable provisioning of big data traffic. Int. J. Parallel Emerg. Distrib. Syst. **30**(1), 15–28 (2015)

6. Cisco. Cisco global cloud index: Forecast and methodology, 2013–2018, 2014. White paper. http://www.cisco.com/c/en/us/solutions/service-provider/global-cloud-index-gci/index.html
7. Cisco. Cisco visual networking index: Forecast and methodology, 2014–2019, 2015. White paper. http://www.cisco.com/c/en/us/solutions/service-provider/visual-networking-index-vni/index.html
8. Develder, C., De Leenheer, M., Dhoedt, B., Pickavet, M., Colle, D., De Turck, F., Demeester, P.: Optical networks for grid and cloud computing applications. Proc. IEEE **100**(5), 1149–1167 (2012)
9. Qi Zhang, Lu, Cheng, R.Boutaba: Cloud computing: state-of-the-art and research challenges. J. Internet Serv. Appl. **1**(1), 7–18 (2010)
10. Contreras, L.M., Lopez, V., De Dios, O.G., Tovar, A., Munoz, F., Azanon, A., Fernandez-Palacios, J.P., Folgueira, J.: Toward cloud-ready transport networks. IEEE Commun. Mag. **50**(9), 48–55 (2012)
11. Jain, R., Paul, S.: Network virtualization and software defined networking for cloud computing: a survey. IEEE Commun. Mag. **51**(11), 24–31 (2013)
12. Duan, Q., Yan, Y., Vasilakos, A.V.: A survey on service-oriented network virtualization toward convergence of networking and cloud computing. IEEE Trans. Netw. Serv. Manag. **9**(4), 373–392 (2012)
13. Kachris, C., Tomkos, I.: A survey on optical interconnects for data centers. IEEE Commun. Surv. Tutor. **14**(4), 1021–1036 (Fourth 2012)
14. Marston, S., Li, Z., Bandyopadhyay, S., Zhang, J., Ghalsasi, A.: Cloud computing âĂŤ the business perspective. Decis. Support Syst. **51**(1), 176–189 (2011)
15. Rimal, B., Choi, E., Lumb, I.: A taxonomy and survey of cloud computing systems. In: Proceedings of the Fifth International Joint Conference on INC, IMS and IDC (NCM 2009), pp. 44–51 (2009)
16. Ahlgren, B., Dannewitz, C., Imbrenda, C., Kutscher, D., Ohlman, B.: A survey of information-centric networking. IEEE Commun. Mag. **50**(7), 26–36 (2012)
17. Bari, M.F., Chowdhury, S., Ahmed, R., Boutaba, R., Mathieu, B.: A survey of naming and routing in information-centric networks. IEEE Commun. Mag. **50**(12), 44–53 (2012)
18. Xylomenos, G., Ververidis, C.N., Siris, V.A., Fotiou, N., Tsilopoulos, C., Vasilakos, X., Katsaros, K.V., Polyzos, G.C.: A survey of information-centric networking research. IEEE Commun. Surv. Tutor. **16**(2), 1024–1049 (Second 2014)
19. Carofiglio, G., Morabito, G., Muscariello, L., Solis, I., Varvello, M.: From content delivery today to information centric networking. Comput. Netw. **57**(16), 3116–3127 (2013). Information Centric Networking
20. Fang, C., Yu, F.R., Huang, T., Liu, J., Liu, Y.: A survey of green information-centric networking: research issues and challenges. IEEE Commun. Surv. Tutor. **17**(3), 1455–1472 (thirdquarter 2015)
21. Jiang, X., Bi, J., Nan, G., Li, Z.: A survey on information-centric networking: rationales, designs and debates. China Commun. **12**(7), 1–12 (2015)
22. Liu, W., Shun-Zheng, Y., Gao, Y., Wei-Tao, W.: Caching efficiency of information-centric networking. IET Netw. **2**(2), 53–62 (2013)
23. Mangili, M., Martignon, F., Capone, A.: A comparative study of content-centric and content-distribution networks: performance and bounds. In: Proceeding of the IEEE Global Communications Conference (GLOBECOM 2013), pp. 1403–1409 (2013)
24. Trossen, D., Parisis, G.: Designing and realizing an information-centric internet. IEEE Commun. Mag. **50**(7), 60–67 (2012)
25. Tyson, G., Bodanese, E., Bigham, J., Mauthe, A.: Beyond content delivery: can icns help emergency scenarios? IEEE Netw. **28**(3), 44–49 (2014)
26. Zhang, G., Li, Y., Lin, T.: Caching in information centric networking: a survey. Computer Netw. **57**(16), 3128–3141 (2013). Information Centric Networking
27. Zhang, M., Luo, H., Zhang, H.: A survey of caching mechanisms in information-centric networking. IEEE Commun. Surv. Tutor. **17**(3), 1473–1499 (thirdquarter 2015)

28. Held, G.: A Practical Guide to Content Delivery Networks. Auerbach Publications, Boston (2005)
29. Hofmann, M., Beaumont, L.: Content Networking: Architecture, Protocols, and Practice (The Morgan Kaufmann Series in Networking). Morgan Kaufmann Publishers Inc., San Francisco (2005)
30. Pallis, G., Vakali, A.: Insight and perspectives for content delivery networks. Commun ACM 49(1), 101–106 (2006)
31. Peng, G.: CDN: content distribution network. CoRR, cs.NI/0411069 (2004)
32. Plagemann, T., Goebel, V., Mauthe, A., Mathy, L., Turletti, T., Urvoy-Keller, G.: From content distribution networks to content networks âĂŤ issues and challenges. Comput. Commun. 29(5), 551–562 (2006). Networks of Excellence
33. Tarkoma, S.: Overlay Networks: Toward Information Networking, 1st edn. Auerbach Publications, Boston (2010)
34. Wetzel, R.: Cdn business models-the drama continues. Bus. Commun. Rev. 32(4), 51 (2002)
35. Ford, L., Fulkerson, D.: A suggested computation for maximal multi-commodity network flows. Manag. Sci. 5(1), 97–101 (1958)
36. Robacker, J.T.: Notes on linear programming: part xxxvii concerning multicommodity networks. Technical report, Rand Corporation, Research Memorandum RM-1799 (1956)
37. Pioro, M., Medhi, D.: Routing, Flow, and Capacity Design in Communication and Computer Networks. Morgan Kaufmann, Burlington (2004)
38. Walkowiak, K.: Modeling and Optimization of Computer Networks. Wroclaw University of Technology, Poland (2011)
39. Assad, A.: Multicommodity network flows-a survey. Networks 8(1), 37–91 (1978)
40. Cantor, D.G., Gerla, M.: Optimal routing in a packet-switched computer network. IEEE Trans. Comput. C–23(10), 1062–1069 (1974)
41. Ford, L., Fulkerson, D.: Flows in Networks. Princeton University Press, New Jersey (1962)
42. Frank, H., Chou, W.: Routing in computer networks. Networks 1(2), 99–112 (1971)
43. Iri, M.: On an extension of the maximum-flow minimum-cut theorem to multicommodity flows. J. Op. Res. Soc. Jpn. 13(3), 129–135 (1971)
44. Kennington, J.: A survey of linear cost multicommodity network flows. Op. Res. 26(2), 209–236 (1978)
45. Tomlin, J.: Minimum-cost multicommodity network flows. Op. Res. 14(1), 45–51 (1966)
46. Gendron, B., Crainic, T., Frangioni, A.: Multicommodity capacitated network design. In: Sanso, Brunilde, Soriano, Patrick (eds.) Telecommunications Network Planning. Centre for Research on Transportation, pp. 1–19. Springer, US (1999)
47. Minoux, M.: Discrete cost multicommodity network optimization problems and exact solution methods. Ann. Op. Res. 106(1–4), 19–46 (2001)
48. Herzberg, M., Bye, S.J., Utano, A.: The hop-limit approach for spare-capacity assignment in survivable networks. IEEE/ACM Trans. Netw. 3(6), 775–784 (1995)
49. Partridge, C., Mendez, T., Milliken W.: Rfc 1546 host anycasting service (1993)
50. Agarwal, G., Shah, R., Walrand, J.: Content distribution architecture using network layer anycast. In: Proceedings of The Second IEEE Workshop on Internet Applications (WIAPP 2001), pp. 124–132 (2001)
51. Basturk, E., Engel, R., Haas, R., Peris, V., Saha, D.: Using network layer anycast for load distribution in the internet. In Tech. Rep, IBM TJ Watson Research Center (1997)
52. Bhattacharjee, S., Ammar, M., Zegura, E., Shah, V., Fei, Z.: Application-layer anycasting. In: Proceedings of the IEEE Sixteenth Annual Joint Conference of the IEEE Computer and Communications Societies (INFOCOM 1997), vol. 3, pp. 1388–1396. IEEE (1997)
53. Bui, M., Jaumard, B., Develder, C.: Anycast end-to-end resilience for cloud services over virtual optical networks. In: Proceedings of the 15th International Conference onTransparent Optical Networks (ICTON 2013), pp. 1–7 (2013)
54. Buysse, J., De Leenheer, M., Develder, C., Dhoedt, B.: Exploiting relocation to reduce network dimensions of resilient optical grids. In: Proceedings of the 7th International Workshop on Design of Reliable Communication Networks (DRCN 2009), pp. 100–106 (2009)

55. Der-Rong, D.: Anycast routing and wavelength assignment problem on wdm network. IEICE Trans. Commun. **88**(10), 3941–3951 (2005)
56. Develder, C., Buysse, J., Dhoedt, B., Jaumard, B.: Joint dimensioning of server and network infrastructure for resilient optical grids/clouds. IEEE/ACM Trans. Netw. **22**(5), 1591–1606 (2014)
57. Develder, C., Buysse, J., Shaikh, A., Jaumard, B., De Leenheer, M., Dhoedt, B.: Survivable optical grid dimensioning: anycast routing with server and network failure protection. In: Proceedings of the IEEE International Conference on Communications (ICC 2011), pp. 1–5 (2011)
58. Gladysz, J., Walkowiak, K.: Modeling of survivable network design problems with simultaneous unicast and anycast flows. In: Proceedings of the 2nd International Logistics and Industrial Informatics (LINDI 2009), pp. 1–6 (2009)
59. Goscien, R., Walkowiak, K., Klinkowski, M.: Joint anycast and unicast routing and spectrum allocation with dedicated path protection in elastic optical networks. In: Proceedings of the 10th International Conference on the Design of Reliable Communication Networks (DRCN 2014), pp. 1–8 (2014)
60. Guyton, J.D., Michael, F.: Schwartz. Locating nearby copies of replicated internet servers. In: Proceedings of the Conference on Applications, Technologies, Architectures, and Protocols for Computer Communication (SIGCOMM 1995), SIGCOMM '95, ACM, pp. 288–298, New York (1995)
61. Habib, M.F., Tornatore, M., De Leenheer, M., Dikbiyik, F., Mukherjee, B.: Design of disaster-resilient optical datacenter networks. J. Lightwave Technol. **30**(16), 2563–2573 (2012)
62. Hyytia, E.: Heuristic algorithms for the generalized routing and wavelength assignment problem. In: Proceedings of the 17th Nordic Teletraffic Seminar(NTS-17), pp. 373–386 (2004)
63. Rak, J., Walkowiak, K.: Survivability of anycast and unicast flows under attacks on networks. In: Proceedings of the 201 International Congress on Ultra Modern Telecommunications and Control Systems and Workshops (ICUMT 2010), pp. 497–503 (2010)
64. Rak, J., Walkowiak, K.: Reliable anycast and unicast routing: protection against attacks. Telecommun. Syst. **52**(2), 889–906 (2013)
65. Shaikh, A., Buysse, J., Jaumard, B., Develder, C.: Anycast routing for survivable optical grids: scalable solution methods and the impact of relocation. IEEE/OSA J. Opt. Commun. Netw. **3**(9), 767–779 (2011)
66. Smutnicki, A., Walkowiak, K.: Optimization of p-cycles for survivable anycasting streaming. In: Proceedings of the 7th International Workshop on Design of Reliable Communication Networks (DRCN 2009), pp. 227–234 (2009)
67. Walkowiak, K.: Heuristic algorithm for anycast flow assignment in connection-oriented networks. In: Sunderam, V., van Albada, G., Sloot, P.M.A., Dongarra, J., (ed.) Computational Science (ICCS 2005), vol. 3516 of Lecture Notes in Computer Science, pp. 1092–1095. Springer, Berlin (2005)
68. Walkowiak, K.: Lagrangean heuristic for anycast flow assignment in connection-oriented networks. Computational Science (ICCS 2006). vol. 3991 of Lecture Notes in Computer Science, pp. 626–633. Springer, Berlin, (2006)
69. Walkowiak, K.: Survivable routing of unicast and anycast flows in mpls networks. In: Proceedings of the 3rd EuroNGI Conference on Next Generation Internet Networks (NGI 2007), pp. 72–79 (2007)
70. Walkowiak, K.: Anycasting in connection-oriented computer networks: models, algorithms and results. Int. J. Appl. Math. Comput. Sci. **20**(1), 207–220 (2010)
71. Walkowiak, K., Klinkowski, M.: Joint anycast and unicast routing for elastic optical networks: modeling and optimization. In: Proceedings of the 2013 IEEE International Conference on Communications (ICC 2013), pp. 3909–3914 (2013)
72. Walkowiak, K., Rak, J.: Joint optimization of anycast and unicast flows in survivable optical networks. In: Proceedings of the 14th International Telecommunications Network Strategy and Planning Symposium (NETWORKS 2010), pp. 1–6 (2010)

73. Walkowiak, K., Rak, J.: Simultaneous optimization of unicast and anycast flows and replica location in survivable optical networks. Telecommun. Syst. **52**(2), 1043–1055 (2013)

74. Buysse, J., De Leenheer, M., Dhoedt, B., Develder, C.: On the impact of relocation on network dimensions in resilient optical grids. In: Proceedings of the 14th Conference on Optical Network Design and Modeling (ONDM 2010), pp. 1–6 (2010)

75. Develder, C., Buysse, J., De Leenheer, M., Jaumard, B., Dhoedt, B.: Resilient network dimensioning for optical grid/clouds using relocation. In: Proceedings of the IEEE International Conference on Communications (ICC 2012), pp. 6262–6267 (2012)

76. Jaumard, B., Buysse, J., Shaikh, A., De Leenheer, M., Develder, C.: Column generation for dimensioning resilient optical grid networks with relocation. In: Proceddings of the IEEE Global Telecommunications Conference (GLOBECOM 2010), pp. 1–6 (2010)

77. Smutnicki, A., Walkowiak, K.: A heuristic approach to working and spare capacity optimization for survivable anycast streaming protected by p-cycles. Telecommun. Syst. **56**(1), 141–156 (2014)

78. Walkowiak, K., Rak, J.: Shared backup path protection for anycast and unicast flows using the node-link notation. In: Proceedings of the 2011 IEEE International Conference on Communications (ICC 2011), pp. 1–6 (2011)

79. Buford, J., Yu, H., Lua, E.: P2P Networking and Applications. Morgan Kaufmann Publishers Inc., San Francisco (2008)

80. Joseph, V., Mulugu, S.: Deploying Next Generation Multicast-enabled Applications: Label Switched Multicast for MPLS VPNs, VPLS, and Wholesale Ethernet, 1st edn. Morgan Kaufmann Publishers Inc., San Francisco (2011)

81. Minoli, D.: IP Multicast with Applications to IPTV and Mobile DVB-H. Wiley-IEEE Press, New Jersey (2008)

82. Steinmetz, R., Wehrle, K.: Peer-to-Peer Systems and Applications. Lecture Notes in Computer Science. vol. 3485, Springer, Berlin (2005)

83. Koch, T., Martin, A.: Solving steiner tree problems in graphs to optimality. Networks **32**, 207–232 (1998)

84. Wong, R.: A dual ascent approach for steiner tree problems on a directed graph. Math. Program. **28**(3), 271–287 (1984)

85. Charbonneau, N., Vokkarane, V.: Static routing and wavelength assignment for multicast advance reservation in all-optical wavelength-routed wdm networks. IEEE/ACM Trans. Netw. **20**(1), 1–14 (2012)

86. Dahl, G., Gouveia, L., Requejo, C.: On formulations and methods for the hop-constrained minimum spanning tree problem. In: Resende, M., Pardalos, P. (eds.) Handbook of Optimization in Telecommunications, pp. 493–515. Springer, New York (2006)

87. Li, Z., Li, B., Jiang, D., Lau, L.C.: On achieving optimal throughput with network coding. In: Proceedings of the IEEE 24th Annual Joint Conference of the IEEE Computer and Communications Societies (INFOCOM 2005), vol. 3, pp. 2184–2194 (2005)

88. Noronha, C., Tobagi, F.: Optimum routing of multicast streams. In: Proceedings of the 13th IEEE Networking for Global Communications (INFOCOM 1994), vol. 2, pp. 865–873 (1994)

89. Oliveira, C., Pardalos, P.: A survey of combinatorial optimization problems in multicast routing. Comput. Op. Res. **32**(8), 1953–1981 (2005)

90. Oliveira, C., Pardalos, P., Resende, M.: Optimization problems in multicast tree construction. In: Resende, M., Pardalos, P. (eds.) Handbook of Optimization in Telecommunications, pp. 701–731. Springer, New York (2006)

91. Wu, C., Li, B.: Optimal peer selection for minimum-delay peer-to-peer streaming with rateless codes. In: Proceedings of the ACM Workshop on Advances in Peer-to-peer Multimedia Streaming (P2PMMS 2005), P2PMMS'05, pp. 69–78, New York (2005) ACM

92. Wu, C., Li, B.: Optimal rate allocation in overlay content distribution. In: Akyildiz, I., Sivakumar, R, Ekici, E., Oliveira, J., McNair, J. (eds.), NETWORKING 2007. Ad Hoc and Sensor Networks, Wireless Networks, Next Generation Internet, Lecture Notes in Computer Science, vol. 4479, pp. 678–690, Springer, Berlin, Heidelberg (2007)

93. Yan, S., Ali, M., Deogun, J.: Route optimization of multicast sessions in sparse light-splitting optical networks. In: Proceedings of the IEEE Global Telecommunications Conference (GLOBECOM 2001), vol. 4, pp. 2134–2138 (2001)
94. Gouveia, L., Simonetti, L., Uchoa, E.: Modeling hop-constrained and diameter-constrained minimum spanning tree problems as steiner tree problems over layered graphs. Math. Program. **128**(1–2), 123–148 (2011)
95. Walkowiak, K.: Network design problem for p2p multicasting. In: Proceedings of the International Network Optimization Conferenc (INOC 2009) (2009)
96. Lin, F.Y.S., Wang, J.L.: A minimax utilization routing algorithm in netwoks with single-path routing. In: Proceedings of the IEEE Global Telecommunications Conference, (GLOBECOM 1993), vol. 2, pp. 1067–1071 (1993)
97. Luenberger, D., Ye, Y.: Linear and nonlinear programming, vol. 116, Springer Science & Business Media (2008)
98. Minoux, M.: Mathematical Programming: Theory and Algorithms. Wiley, New Jersey (1986)
99. Nemhauser, G., Wolsey, L.: Integer and combinatorial optimization (1999)
100. Paul Williams, H.: Model Solving in Mathematical Programming. Wiley, New Jersey (1993)
101. IBM. ILOG CPLEX optimizer. http://www.ibm.com
102. Inc., Gurobi Optimization. Gurobi optimizer reference manual
103. Bertsekas, D.: Network Optimization: Continuous and Discrete Models. Athena Scientific, Massachusetts (1998)
104. Bhandari, R.: Survivable Networks: Algorithms for Diverse Routing. Kluwer Academic Publishers, Norwell (1998)
105. Corne, D., Oates, M., Smith, G., (eds). Telecommunications Optimization: Heuristic and Adaptive Techniques (2000)
106. Donoso, Y., Fabregat, R.: Multi-Objective Optimization in Computer Networks Using Meta-heuristics. Auerbach Publications, Boston (2007)
107. Kleinberg, J., Tardos, E.: Algorithm Design. Addison-Wesley Longman Publishing Co. Inc, Boston (2005)
108. Talbi, E.: Metaheuristics—From Design to Implementation. Wiley, New Jersey (2009)
109. Yang, X.: Nature-Inspired Optimization Algorithms. Elsevier, Amsterdam (2014)
110. Cormode, G., Thottan, M.: Algorithms for Next Generation Networks, 1st edn. Springer Publishing Company, Berlin (2010)
111. Koster, A., Muoz, X.: Graphs and Algorithms in Communication Networks: Studies in Broadband, Optical. Wireless and Ad Hoc Networks, 1st edn. Springer Publishing Company, Berlin (2009)
112. Grover, Wayne D.: Mesh-based Survivable Transport Networks: Options and Strategies for Optical, MPLS. SONET and ATM Networking. Prentice Hall PTR, NJ (2003)
113. Kalmanek, Ch., Misra, S., Yang, R.: Guide to Reliable Internet Services and Applications, 1st edn. Springer Publishing Company, Berlin (2010)
114. Vasseur, J., Pickavet, M., Demeester, P.: Network Recovery: Protection and Restoration of Optical, SONET-SDH, IP, and MPLS. Morgan Kaufmann Publishers Inc., San Francisco (2004)

# Chapter 2
# Connection-Oriented Networks

This chapter focuses on the optimization of connection-oriented networks (CONs). Following a brief introduction to the basics of technologies and protocols used in CONs, we formulate several optimization problems that arise in the context of connection-oriented networking applied to cloud computing and content-oriented services. To address specific attributes of cloud computing and content-oriented services, the optimization problems presented—besides including classical unicast flows—also embrace anycast and multicast network flows. All presented optimization problems are formulated as ILP models. Moreover, for selected optimization problems, we propose and analyze solution algorithms and report results of numerical experiments.

## 2.1 Introduction

In a connection-oriented network, information (packets, frames, bits) is forwarded along a predefined connection from the source to the destination. More precisely, all data belonging to a particular demand is transmitted along one routing path determined and established before the transmission begins. The telephone network is notable as probably the first CON developed for communication. Following the deployment of telephone networks, the concept of connection has been extended and employed in many network technologies, including MPLS (MultiProtocol Label Switching), ATM (Asynchronous Transfer Mode), WDM (Wavelength Division Multiplexing), and Connection-Oriented Ethernet [1]. We outline these technologies below.

The MPLS architecture was proposed by the Internet Engineering Task Force (IETF) in the standard RFC3013 [2]. MPLS is designed to carry IP packets, but with more efficient traffic engineering and QoS guarantees than classical IP networks. MPLS networks consist of two types of devices: LER (Label Edge Router) located

© Springer International Publishing Switzerland 2016
K. Walkowiak, *Modeling and Optimization of Cloud-Ready and Content-Oriented Networks*, Studies in Systems, Decision and Control 56, DOI 10.1007/978-3-319-30309-3_2

at the entry and exit points of the MPLS network, and LSR (Label Switch Router) located inside the MPLS network. In the MPLS network, packets are transmitted along a LSP (Label Switch Path) between LERs and LSRs. Each packet header includes a piece of information known as the *label* which indicates a particular LSP. When an MPLS packet enters the LSR, the label is used to find the outgoing port of the LSR where the packet should be forwarded according to the packet's FEC (Forwarding Equivalence Class). It should be noted that in MPLS the labels are local, i.e., each LSR can change the label of a particular LSP for each subsequent link along the path, which is explicitly included in the switching tables of each LSR. This procedure ensures that MPLS is a connection-oriented technology, i.e., all packets assigned to the same FEC follow exactly the same routing path. However, it should be noted that MPLS can also be used in a connectionless mode, if LDP (Label Distribution Protocol) is used for signaling. To classify a packet to a particular FEC class, an IP address or other element of the IP packet header can be used. FEC classes are characterized with various QoS parameters, therefore it is possible to provision different types of traffic in a different ways. Thus, packets included in different FEC classes but with the same source and destination nodes can be assigned to different routing paths. This is the key attribute of MPLS that enables efficient traffic engineering provisioning. For more information on MPLS refer to [1–5].

The ATM architecture was standardized by the ITU-T (International Telecommunication Union—Telecommunication Standardization Sector) in 1987 as the preferred solution for implementing the B-ISDN (Broadband Integrated Services Digital Network) concept. In the late 1980s, ATM was one of the most promising proposals for high-speed networking in backbone networks supported by demand-switched, semipermanent, and permanent broadband connections for both point-to-point and point-to-multipoint applications. The general idea of ATM is very similar to MPLS, which in fact can be regarded as a successor of ATM. More specifically, ATM—like MPLS—is based on the concept of label-switching. In ATM, fixed-size packets known as *cells* arrive in one port of the ATM switch and the label included in the cell header is used to indicate the outgoing port of the switch, while a new label number is assigned to the cell. However, in spite of the fact that ATM offered many sophisticated, mature and robust solutions for high speed networking, its popularity has been declining. According to experts, the main reasons for this trend are ATM's relatively high complexity, costly equipment and poor compatibility with legacy IP networks. For further information on ATM, see [1, 3, 4, 6]

WDM is an optical technology which enables the multiplexing of multiple optical signals on a single optical fiber by using different wavelengths (colors) of laser light to carry different signals. The concept of WDM was first published in the late 1970s. Since late 1990s, WDM has been the most popular technology implemented in backbone transport networks. A WDM network is formed by OXCs (Optical Cross Connects) interconnected by fibers. Note that WDM is a connection-oriented technique, since the entire signal is transmitted along a single connection (routing path) known as a *lightpath*. The frequency grid of WDM is divided with 50 GHz channel spacing. Optical devices generally cannot convert the wavelength, therefore the whole WDM connection must use the same color on every link along the path. This additional requirement, known as the *wavelength continuity* constraint, complicates the demand provisioning in WDM networks compared to aforementioned technologies such as MPLS or ATM. More precisely, an optimization problem known as the Routing and Wavelength Assignment (RWA) appears in WDM networks, where both the routing path and wavelength must be selected for each demand. In contrast, optimization of flows in MPLS and ATM networks involves selecting the routing paths only, since there are no additional constraints on the link capacity resources. However, it is possible to use *opaque* OXC in WDM networks. In particular, the OXC is equipped with converters that transform the optical signal transmitted over a wavelength to another wavelength. This is achieved by first demultiplexing the optical signals and then converting them into electronic signals. Next, electronic signals are switched using electronic switching and then converted back into optical signals. Finally, these signals are multiplexed again into the output optical fiber. With this capability, optimization of connections in WDM networks is exactly the same as in MPLS and ATM networks. Note that the optimization problems presented in this chapter can be applied in the context of WDM networks with full wavelength conversion capabilities. For problems assuming the wavelength continuity constraint, when the wavelength conversion is not available, the reader is referred to Chap. 3. For a good survey on the topic of the WDM networks see [1, 5–9]

The supremacy of Ethernet in enterprise networking and in local area networks is a direct consequence of well understood operational practices, low-cost solutions and plug-and-play features. As a result, applications of Ethernet appear in larger environments, including backbone transport networks. However, the reduced use of Ethernet networks has brought the need to enhance Ethernet to improve network scalability and answer the challenge of using Ethernet in larger networks. The key additions have been PBB (Provider Backbone Bridging) proposed in 2008 [10] and PBB-TE (Provider Backbone Bridging—Traffic Engineering) defined in 2009 [11]. Both standards provide traffic-engineered, resource-managed transport using special tunnels for transmitting information (frames) belonging to the same VLAN (Virtual Local Area Network). Moreover, using the PBB-TE standard means that customer VLAN traffic is encapsulated in a PBB header, which includes Mac-in-Mac encapsulation. Next, the traffic is forwarded through the network using static database entries. As a result, the Ethernet transmission is connection-oriented, since all frames included in the same VLAN are transported by one Ethernet switched path that follows the same routing path [3, 12, 13].

The remainder of this chapter is devoted to various aspects of modeling and optimization of connection-oriented networks in the context of cloud computing and content-oriented services. The key assumption following from the connection-oriented nature of the considered networks is that non-bifurcated multicommodity flows are applied to model network traffic.

It should be stressed that all optimization problems presented in this section are $\mathcal{NP}$-complete. This follows directly from the fact that the problems use non-bifurcated flows, and in the case of network design problems modular link capacity is applied. Therefore, since the authors of [4] show that the unicast flow allocation problem with non-bifurcated flows and the network design problem with unicast flows and modular link capacity are $\mathcal{N}\mathcal{P}$-complete, the anycast and multicast versions of these problems are also $\mathcal{N}\mathcal{P}$-complete.

## 2.2  Allocation of Anycast Flow

This section starts with a relatively simple problem related to the allocation of anycast flows. Flow allocation problems assume that only network flows need be optimized, since an existing network is studied and the capacity of the network links is fixed. This assumption stems from by the fact that the considered network is in an operational phase and augmenting its physical resources, such as link capacity, is not achievable in a short time perspective. However, at the same time, there is a need to improve network performance, which can only be achieved by an improved optimization of network flows [4].

### 2.2.1  Formulation

The general notation is the same as in Sect. 1.2. To recall briefly, the network is modeled as a directed graph $G = (V, E)$, where $V$ is a set of nodes (vertices) and $E$ is a set of edges (directed links) with limited capacity $c_e$. A number of data centers (DCs) are located in the network. For simplicity, the local connection between the data center and the backbone network node is not included in the model, i.e., it is assumed that the network node is equivalent with the data center connected to that node. This is because local access links used to connect data centers are usually of a high bandwidth and thus capacity of this link is not incorporated in the model.

A set of anycast demands denoted by $D$ is given. A standard model of anycasting is employed [14]; a single anycast request is realized by two associated demands: upstream (from the client to the data center) and downstream (in the opposite direction). Let $\tau(d)$ be the index of a demand associated with demand $d$. The demand volume is described by the constant $h_d$. Link-path modeling is assumed, and therefore for each demand $d \in D$, a set of candidate paths $P(d)$ is given. If $d$ is an upstream demand, set $P(d)$ contains candidate paths that originate at one of the DC nodes and terminate at the client node. In turn, if $d$ is a downstream demand, candidate paths in set $P(d)$ connect the client node and one of the DC nodes. Paths are defined using constant $\delta_{edp}$, i.e., $\delta_{edp}$ equals 1 if path $p$ of demand $d$ includes link $e$ and 0 otherwise. There is one decision variable $x_{dp}$ that denotes, which path $p \in P(d)$ is selected to realized demand $d$. The objective of the flow allocation problem is to find a routing path for each demand that minimizes a selected performance metric and at the same time provides a feasible solution in terms of the link capacity constraint. Moreover, the anycast constraint must be satisfied to ensure that both associated anycast demands are assigned to the same DC node.

---

**CON/A/FA/Cost**

**sets**

$E$       links
$D$      anycast demands
$D^{DS}$   anycast downstream demands
$P(d)$    candidate paths for flows realizing demand $d$. If $d$ is an anycast upstream
          demand, candidate path connects client node and DC node. If $d$ is a down-
          stream demand, candidate path connects the DC node and the client node

**constants**

$\delta_{edp}$     =1, if link $e$ belongs to path $p$ realizing demand $d$; 0, otherwise
$h_d$       volume (bit-rate) of demand $d$
$c_e$       capacity of link $e$
$\zeta_e$       unit routing cost on link $e$
$\tau(d)$     index of a demand associated with demand $d$. If $d$ is a downstream demand,
          then $\tau(d)$ must be an upstream demand and vice versa
$s(p)$    origin node of path $p$
$t(p)$     destination node of path $p$

**variables**

$x_{dp}$     =1, if path $p$ is used to realize demand $d$; 0, otherwise (binary)
$f_e$       flow on link $e$ (continuous non-negative)

**objective**

$$\text{minimize} \quad F = \sum_{e \in E} \zeta_e f_e \tag{2.2.1a}$$

**constraints**

$$f_e = \sum_{d \in D} \sum_{p \in P(d)} \delta_{edp} x_{dp} h_d, \quad e \in E \tag{2.2.1b}$$

$$\sum_{p \in P(d)} x_{dp} = 1, \quad d \in D \tag{2.2.1c}$$

$$f_e \leq c_e, \quad e \in E \tag{2.2.1d}$$

$$\sum_{p \in P(d)} x_{dp} s(p) = \sum_{p \in P(\tau(d))} x_{\tau(d)p} t(p), \quad d \in D^{DS}. \tag{2.2.1e}$$

Objective (2.2.1a) is to minimize the overall routing cost. However, other performance metrics can be considered as the optimization goal, e.g., delay or network
congestion. Moreover, the allocation problem can be formulated without the objective function, i.e., the challenge is to find a feasible routing configuration that satisfies

the link capacity constraint (see [4] for more details). Equality (2.2.1b) defines the value of variable $f_e$. Constraint (2.2.1c) ensures that exactly one routing path is selected to realize demand $d$. Condition (2.2.1d) is a link capacity constraint to meet the requirement that the flow of each link given as a sum of all demands that use this link cannot exceed the link capacity. Finally, equality (2.2.1e) defines the anycast constraint guaranteeing that two associated demands are assigned to the same DC node.

## 2.2.2  Algorithms

This section is devoted to heuristic and metaheuristic algorithms applicable to the optimization problem of anycast flow allocation formulated as (2.2.1a)–(2.2.1e). Note that it can be modified easily to enable joint optimization of anycast flows and other types of flows such as unicast flows using the link-path modeling and multicast flows applying the candidate tree modeling.

### Greedy Algorithm

The first proposed algorithm named GR/A/FA (Greed Algorithm for Anycast Flows), is based on a very simple greedy approach. More precisely, the general idea is to allocate anycast demands one by one according to a selected ordering of demands. However, due to the anycast constraint (2.2.1e), both associated anycast demands must be processed jointly. Algorithm 2.1 shows the scheme of the GR/A/FA algorithm.

---

**Algorithm 2.1** GR/A/FA (Greedy Algorithm for Anycast Flow Allocation)

---

**Require:** set of edges $E$, set of anycast demands $D$, sets $P(d)$ including candidate paths for each
    demand $d \in D$
**Ensure:** path selection (routing) for each demand $d \in D$ included in set $X$, value of objective
    function
1: **procedure** $GR/A/FA(D, P(d))$
2: **for** $d \in D$ **do**
3:    **if** $Is\_Not\_Allocated(\tau(d))$ **then**
4:       $p := Find\_Best\_Path(d)$
5:    **else**
6:       $p := Find\_Best\_Path\_DC(d, DC\_Node(\tau(d)))$
7:    **end if**
8:    $X := X \cup \{x_{dp}\}$
9:    $D := D \setminus \{d\}$
10: **end for**
11: **end procedure**

---

Let set $X$ (known as the solution) include all variables $x_{dp}$ that are equal to 1. Solution $X$ determines the unique set of selected routing paths for each demand $d \in D$. The algorithm performs a single loop that analyzes all demands (lines 2–10). Demands are processed according to a selected ordering. For instance, the demands can be sorted in decreasing order of the bit-rate denoted as $h_d$. The rationale of this approach is that larger demands should be processed first, since they need more capacity resources. Next, when the network becomes more congested, it is easier to allocate smaller demands than larger ones. Another possible metric for ordering is a multiplication of the bit-rate and the hop count of the shortest path available for a particular demand. Note that the construct of the greedy method makes it possible to use a wide range of ordering strategies.

As mentioned above, due to the anycast constraint, both associated demands $d$ and $\tau(d)$ must be connected to the same DC node, therefore there are two possible ways to process a demand. In particular, if the associated demand $\tau(d)$ is yet to be allocated, the current demand $d$ can be assigned to any DC node and function $Find\_Best\_Path(d)$ can use all candidate paths available for demand $d$ (line 4). On the other hand, if the associated demand $\tau(d)$ is already processed by the algorithm, demand $d$ can only be assigned to the path connected to the same DC node as demand $\tau(d)$, i.e., function $DC\_Node(\tau(d))$ returns the DC node selected for demand $\tau(d)$ (line 6). Note that the functions $Find\_Best\_Path$ and $Find\_Best\_Path\_DC$ are generic and can implement various strategies to find a routing path. For instance, the shortest path in terms of kilometer distance can be selected in a residual graph that only includes links with a residual capacity greater than the requested bit-rate of the considered demand. The maximum complexity of algorithm GR/A/FA is $O(|D|\,|P|)$, where $|D|$ denotes the number of demands and $|P|$ is the number of candidate paths for each demand.

Note that the main advantages of the greedy algorithm are its adaptability to the means of ordering and routing strategies and its relatively low complexity, which should result in a short execution time. On the other hand, the algorithm's main drawback is that in a congested network, it may have trouble finding a feasible solution due to a shortage of capacity resources. Therefore, below we introduce more advanced methods of solving the problem of anycast flow allocation.

## Flow Deviation

The Flow Deviation (FD) algorithm was proposed for optimization of bifurcated and non-bifurcated flows in [15], and has been widely applied to various optimization problems [4, 14, 16–27].

The FD method was developed following the expansion of store-and-forward communications networks together with the ARPANET [19]. The FD algorithm is based on the concept of a *shortest route* flow applied to successive flow deviations that lead to local minima. More precisely, for each demand in the network, the shortest route is determined according to a selected objective function and based on the current routing (flow allocation) in the network. Next, we attempt to deviate the flow of the demand to the shortest route. In the context of bifurcated flows, the flow deviation can be applied to a part of the demand flow, since multipath routing is allowed. However, for non-bifurcate flows with single path routing, the flow deviation must affect the whole demand [15].

Since this chapter focuses on connection-oriented networks, the non-bifurcated version of the FD method is applied. An FD algorithm for optimization of anycast flows known as FD/A/FA (Flow Deviation for Anycast Flow Allocation) is formulated and discussed. Note that the first time the FD method was applied in the context of anycast flows was in [26], while in [14, 27] the same author proposed an FD method for joint optimization of anycast and unicast flows. The key innovation of the FD/A/FA algorithm, in comparison to the unicast version proposed in [15], is that the new method processes associated anycast demands together, which satisfies the anycast constraint (2.2.1e). Algorithm 2.2 reports the pseudocode of the FD/A/FA heuristic.

---

**Algorithm 2.2** FD/A/FA (Flow Deviation for Allocation of Anycast Flows)

---

**Require:** set of edges $E$, set of anycast demands $D$, sets $P(d)$ including candidate paths for each demand $d \in D$

**Ensure:** path selection (routing) for each demand $d \in D$ included in set $X_i$, value of objective function

1: **procedure** $FD/A/FA(D, P(d))$
2:   $X_1 := Find\_Intial\_Solution(D, P(d))$, $i := 1$
3:   $test := 0$
4:   **repeat**
5:     $SR(X_i) := Find\_Shortest\_Paths(D, P(d), X_i)$
6:     $H := X_i$
7:     **for** $d \in D$ **do**
8:       $K := (H - \{x_{dp}\}) \cup \{x_{dq}\}$, where $x_{dp} \in H$ and $x_{dq} \in SR(X_i)$
9:       $K := (K - \{x_{\tau(d)p'}\}) \cup \{x_{\tau(d)q'}\}$, where $x_{\tau(d)p'} \in H$ and $x_{\tau(d)q'} \in SR(X_i)$
10:      $F(H) = Find\_Objective(H)$
11:      $F(K) = Find\_Objective(K)$
12:      **if** $F(K) < F(H)$ **then** $H := K$
13:      $D := D \setminus \{d, \tau(d)\}$
14:    **end for**
15:    **if** $X_i \neq H$ **then**
16:      $i := i + 1$
17:      $X_i := H$
18:    **else**
19:      $test := 1$
20:    **end if**
21:  **until** $test < 1$
22: **end procedure**

---

Let $X_i$ denote a solution of the problem obtained in iteration $i$. In turn, sets $H$ and $K$ are temporary solutions used in the algorithm. A special flag $test$ denotes the stopping condition of the algorithm. The algorithm starts with a feasible initial solution $X_1$ (line 2). To find a such a solution, an algorithm based on Phase 1 of the original FD method [15] can be applied with the modifications required to process anycast flows; for more details, see [14]. Next, the flag *test* is initialized. The main loop of the algorithm (lines 4–21) is repeated until a new solution $H$ generated in the current iteration provides a modification of the solution $X_i$ created in the previous iteration. Function *Find_Shortest_Paths*$(D, P(d), X_i)$ calculates set $SR(X_i)$ including the shortest routes for all demands (line 5). This function ensures that the anycast constraint (2.2.1e) is always satisfied. More specifically, for each pair of associated demands $d$ and $\tau(d)$ a demand with a large value of volume $h_d$ is selected first. Without losing generality, let us assume that $h_d \geq h_{\tau(d)}$. In the case of demand $d$, the *Find_Shortest_Paths* function finds the shortest route under a selected metric taking into account all available routing paths included in $P(d)$. To ensure that both associated demands use the same DC node, in the case of demand $\tau(d)$ the function *Find_Shortest_Paths* considers only the paths from $P(\tau(d))$ that are connected to the DC node selected for demand $d$. Note that usually a partial derivative of the objective function is used as a link metric applied in the selection of the shortest path in function *Find_Shortest_Paths*.

Next, the current selection $X_i$ is saved as $H$ (line 6). The loop defined in lines 7–14 processes all demands included in set $D$ according to a selected ordering as in the case of Algorithm 2.1. However, to ensure the anycast constraint, associated demands $d$ and $\tau(d)$ are processed together. A new selection $K$ is obtained by using the flow deviation operation, i.e., demands $d$ and $\tau(d)$ are switched to shortest paths included in set $SR(X_i)$ (lines 8 and 9). Solutions $H$ and $K$ are compared in terms of the objective function value; the new solution K is saved as the current solution (lines 10–12) only if it decreases the objective function value. When all demands are processed, we check whether solution $H$ obtained in the current iteration brings about a change relative to the previous solution $X_i$ (lines 15–20). If the solutions differ, the algorithm is continued, otherwise the algorithm stops.

Note, that the algorithm converges in a finite number of steps, since there is a finite number of non-bifurcated flows. Repetitions of the same solution are impossible due to the stopping condition. The maximum number of the FD/A/FA algorithm iterations can be estimated as the number of all possible routing path combinations. In particular, let $k$ denote the number of candidate paths defined between a pair of nodes and let $r$ denote the number of DC nodes in the network. In consequence, $kr$ defines the number of candidate paths for each anycast demand. However, since associated demands $d$ and $\tau(d)$ must be processed jointly due to the anycast constraint (2.2.1e), for each $kr$ routing paths available for $d$, only $k$ routing paths are available for $\tau(d)$, since both $d$ and $\tau(d)$ must use the same DC node. Thus, the possible number of path combinations for a pair of associated anycast demands is defined as $rk^2$. In consequence, the number of all possible path combinations for a pair of associated anycast demands is $(rk^2)^{\frac{|D|}{2}}$.

## Lagrangian Relaxation

The next algorithm proposed for optimization of anycast flows in connection-oriented networks is the Lagrangian relaxation (LR) method combined with a subgradient optimization approach. The main aim of this approach is to iteratively solve the optimization problem using a heuristic algorithm FD/A/FA that uses results given by solving dual problems as the initial solutions. The LR technique with the subgradient optimization has been successfully used for solving various network optimization problems, e.g., [4, 6, 14, 28–36].

To the best of our knowledge, the Lagrangian relaxation technique for anycast flows in connection-oriented flows was introduced for the first time in [35]. In turn, the authors of [14] report how to apply Lagrangian relaxation for joint optimization of anycast and unicast flows.

The key aim of the LR decomposition algorithm is to consider the dual problem of the original optimization problem by relaxing constraints in order to obtain a simpler subproblem. This procedure makes it possible to move iteratively towards the optimal solution of the original problem. Consequently, selecting a suitable constraint to be relaxed is key.

In order to formulate a dual problem to model (2.2.1), the same approach as in [14, 29, 35] is applicable. More specifically, the problem (2.2.1) is first transformed into an equivalent formulation, which is better suited for the LR procedure. The key observation is that the objective function (2.2.1a) does not decrease with $f_e$, therefore the equality (2.2.1b) one be replaced with the following inequality:

$$f_e \leq \sum_{d \in D} \sum_{p \in P(d)} \delta_{edp} x_{dp} h_d, \quad e \in E. \tag{2.2.1f}$$

For ease of reference, the modified optimization problem defined as (2.2.1a), (2.2.1c)–(2.2.1f) will be referred to as CON/A/FA/Cost/2. It should be noted that problems CON/A/FA/Cost (2.2.1a)–(2.2.1e) and CON/A/FA/Cost/LR (2.2.1a), (2.2.1c)–(2.2.1f) are equivalent, i.e., the optimal solution obtained to one of these problems guarantees the optimal solution to the other problem.

A popular approach of Lagrangian relaxation in the context of network optimization is to relax the capacity constraint [4, 6, 32, 33]. The method shown below is based on the concept proposed in [29] and constraint (2.2.1f) is relaxed using vector $\boldsymbol{\lambda} = (\lambda_1, \lambda_2, \lambda_{|E|})$ of positive Lagrangian multipliers $\lambda_e$ for each link $e \in E$. Accordingly, the following Lagrangian relaxation of problem CON/A/FA/Cost/2 is formulated as follows.

---

**CON/A/FA/Cost/LR**

**objective**

$$\text{minimize} \quad \varphi(\boldsymbol{\lambda}) = \sum_{e \in E} \zeta_e f_e + \sum_{e \in E} \lambda_e \left( \sum_{d \in D} \sum_{p \in P(d)} \delta_{edp} x_{dp} h_d - f_e \right) \tag{2.2.2a}$$

**constraints**

$$\sum_{p\in P(d)} x_{dp} = 1, \quad d \in D \tag{2.2.2b}$$

$$f_e \le c_e, \quad e \in E \tag{2.2.2c}$$

$$\sum_{p\in P(d)} x_{dp}s(p) = \sum_{p\in P(\tau(d))} x_{\tau(d)p}t(p), \quad d \in D^{DS} \tag{2.2.2d}$$

$$f_e \ge 0, \quad e \in E. \tag{2.2.2e}$$

The objective function (2.2.2a) follows directly from the relaxation of condition (2.2.1f). Constraints (2.2.2b)–(2.2.2d) are copied from the original problem. A new constraint (2.2.2e) is added to the problem. Note that condition (2.2.2e) in model CON/A/FA/Cost/2 is guaranteed by constraint (2.2.1f). However, since (2.2.1f) is relaxed and incorporated to the dual function (2.2.2a), this new constraint (2.2.2e) is required to guarantee variables $f_e$ to be positive.

The objective function (2.2.2a) of the Lagrangian problem can be rewritten as

$$\text{minimize} \quad \varphi(\lambda) = \left( \sum_{e\in E} (\zeta_e f_e - \lambda_e f_e) \right) + \left( \sum_{e\in E} \sum_{d\in D} \sum_{p\in P(d)} \lambda_e \delta_{edp} x_{dp} h_d \right). \tag{2.2.3}$$

Since there are no coupling constraints between variables $f_e$ and $d_dp$, the problem (2.2.2) can be divided into two independent subproblems CON/A/FA/Cost/LR/1 and CON/A/FA/Cost/LR/2, defined below.

---

**CON/A/FA/Cost/LR/1**

**objective**

$$\text{minimize} \quad \varphi_1(\lambda) = \sum_{e\in E} (\zeta_e f_e - \lambda_e f_e) \tag{2.2.4a}$$

**constraints**

$$f_e \le c_e, \quad e \in E \tag{2.2.4b}$$

$$f_e \ge 0, \quad e \in E. \tag{2.2.4c}$$

It should be noted that problem CON/A/FA/Cost/LR/1 (2.2.4) can be separated into $|E|$ subproblems, each of which is solved by the following formula:

$$f_e = \begin{cases} c_e & \text{if } \zeta_e - \lambda_e < 0 \\ 0 & \text{if } \zeta_e - \lambda_e \ge 0 \end{cases} \quad e \in E. \tag{2.2.5}$$

**CON/A/FA/Cost/LR/2**

**objective**

$$\text{minimize} \quad \varphi_2(\lambda) = \sum_{e \in E} \sum_{d \in D} \sum_{p \in P(d)} \lambda_e \delta_{edp} x_{dp} h_d \qquad (2.2.6a)$$

**constraints**

$$\sum_{p \in P(d)} x_{dp} = 1, \quad d \in D \qquad (2.2.6b)$$

$$\sum_{p \in P(d)} x_{dp} s(p) = \sum_{p \in P(\tau(d))} x_{\tau(d)p} t(p), \quad d \in D^{DS}. \qquad (2.2.6c)$$

In turn, problem CON/A/FA/Cost/LR/2 defined in (2.2.6) can be separated into $|D^{DS}|$ subproblems, i.e., one problem for a pair of associated demands $d$ and $\tau(d)$. To solve such a problem, we need to find the shortest pair of paths under metric $\lambda_e h_d$ for demands $d$ and $\tau(d)$ considering each DC node individually. Next, the pair of paths with the lowest value of path lengths is selected as the final solution. This ensures the optimal solution of problem CON/A/FA/Cost/LR/2.

The above method solves the dual problem; the next step is the subgradient search [4, 37]. Let $x_{dp}(\lambda)$ be the optimal solution of the Lagrangian relaxation for a fixed vector of multipliers $\lambda$. Let $X$ denote the set of all variables $x_{dp}$ equal to one. The corresponding subgradient of the dual function (2.2.6) at $\lambda$ is defined as:

$$\gamma_e(\lambda) = \left( \sum_{d \in D} \sum_{p \in P(d)} \delta_{edp} x_{dp}(\lambda) h_d \right) - f_e(\lambda) \quad e \in E. \qquad (2.2.7)$$

The Lagrangian multipliers in subsequent iterations of the subgradient procedure (denoted as $i$) are updated as follows:

$$\lambda_e^{i+1} = \max(0, \lambda_e^i + t_i \gamma_e^i) \quad e \in E. \qquad (2.2.8)$$

The step-size $t_i$ can be defined as proposed in [4, 29]:

$$t_i = \frac{\rho(\bar{\varphi} - \varphi(\lambda^i))}{\|\gamma^i\|^2}. \qquad (2.2.9)$$

Note that $\bar{\varphi}$ denotes the upper bound of the dual function (2.2.2a), which can be calculated using a heuristic algorithm that yields a feasible solution of the primal problem. Moreover, $\rho$ is commonly used in the range $0 \leq \rho \leq 2$ [4]. Algorithm 2.3 reports the pseudocode of the subgradient optimization procedure applied to Lagrangian relaxation of the anycast flow allocation problem defined in (2.2.1a)–(2.2.1e).

First, tuning parameters used in algorithm LR/A/FA are initialized (line 2). The values selected for the initialization are as proposed in [14, 35]. The maximum number of iterations is set to $i_{max} := 100$ as the stopping condition for dual iterations. Additionally, it is possible to select another value depending on specific problem characteristics. Parameter $\rho_{iter}$ counts the number of subsequent iterations that do not improve the dual function value, while parameter $\rho_{maxiter}$ defines the limit of such iterations. If $\rho_{iter}$ reaches $\rho_{maxiter}$, then parameter $\rho$ used to calculate the step size according to (2.2.9) decreases. The vector of Lagrangian multipliers $\lambda^1$ is initialized with all values equal to 1 (line 3), but again— depending on the specific problem— different assignment can be made. The upper bound of the dual function $\bar{\varphi}$ is obtained as the solution of the FD/A/FA method shown in Algorithm 2.2 (line 4). The main loop of the algorithm including subsequent iterations of the subgradient search is defined in lines 5–26. In summary, first the dual problem is solved according to the

---

**Algorithm 2.3** LR/A/FA (Lagrangian Relaxation for Anycast Flows)

**Require:** set of edges $E$, set of anycast demands $D$, sets $P(d)$ including candidate paths for each demand $d \in D$

**Ensure:** path selection (routing) for each demand $d \in D$ included in set $X^{best}$, value of objective function

1: **procedure** $LR/A/FA(D, P(d))$
2:  $\rho := 2$, $\rho_{min} := 0.005$, $\rho_{iter} := 0$, $\rho_{maxiter} := 3$, $i_{max} := 100$, $F^{best} = \infty$, $\varphi^{best} = -\infty$
3:  $\lambda^1 = 1$
4:  $\bar{\varphi} := FD/A/FA(D, P(d))$
5:  **for** $i := 1$ **to** $i_{max}$ **do**
6:   $\rho_{iter} := \rho_{iter} + 1$
7:   $\varphi(\lambda^i) = Solve\_Dual\_Problems(\lambda^i)$
8:   **if** $\varphi(\lambda^i) > \varphi^{best}$ **then**
9:    $\varphi^{best} := \varphi(\lambda^i)$
10:    $\rho_{iter} := 0$
11:   **end if**
12:   $X := FD/A/Init\_Sol(D, P(d), X(\lambda^i))$
13:   $F = Find\_Objective(X)$
14:   **if** $F < F^{best}$ **then**
15:    $F^{best} := F$
16:    $X := X^{best}$
17:    $\bar{\varphi} := F$
18:   **end if**
19:   **if** $\rho_{iter} > \rho_{maxiter}$ **then**
20:    $\rho := max(\rho/2, \rho_{min})$
21:    $\rho_{iter} := 0$
22:   **end if**
23:   $\gamma^i := Subgradient(\lambda^i)$      ▷refer to (2.2.7)
24:   $t_i := Step\_Size(\lambda^i, \gamma^i)$      ▷ refer to (2.2.9)
25:   $\lambda^{i+1} := Update\_Lambda(\lambda^i, \gamma^i)$      ▷ refer to (2.2.8)
26: **end for**
27: **end procedure**

---

methodology described above (line 7). If the value of the dual function $\varphi(\lambda^i)$ is greater than the previous best result, the algorithm saves this value and resets counter $\rho_{iter}$ (lines 8–11). Solution $X(\lambda^i)$ obtained by solving the dual problem is used to initialize the FD/A/FA algorithm, which then finds a primal feasible solution of the problem denoted as $X$ with the value of the objective function denoted as $F$ (lines 12–13). If the new solution outperforms the previous best one, it is saved (lines 14–18). Next, it is checked whether $\rho_{iter}$ reaches the limit defined by $\rho_{maxiter}$ and parameter $\rho$ is updated accordingly (lines 19–22). Finally, the vector of Lagrangian multipliers $\lambda^i$ is updated using the subgradient optimization approach and formulas defined in (2.2.7)–(2.2.9). Note that the complexity and the execution time of the LR/A/FA method mainly depends on the number of iterations given by parameter $i_{max}$, since the most time complex part of each iteration is the execution of the FD/A/FA algorithm.

The optimization problem (2.2.1a)–(2.2.1e) uses the link-path notation to model multicommodity flows. Some papers that apply Lagrangian relaxation to routing problems use the node-link formulation [4, 28, 32–34]. However, in both cases the dual is constructed analogously, i.e., we obtain an optimization problem that can be decoupled into two independent subproblems, where one is the minimum cost routing problem with link metrics given by Lagrangian multipliers.

### Evolutionary Algorithm

The evolutionary algorithm (EA) is a stochastic heuristic method widely applied in the context of various optimization problems related to computer and communication networks. For an extensive survey of EAs and their application to network problems refer to [4, 38–42].

The key issue in designing the EA is to develop a method of encoding the optimization problem into chromosomes. To recall, in the anycast flow allocation problem (2.2.1a)–(2.2.1e), the decision is to select routing paths for each demand. A solution of the problem is denoted as a set that includes all variables $x_{dp}$ equal to 1, which directly indicates the selected routing paths. Regarding connection-oriented networks, a common approach of EA encoding is to assume that each allele in the chromosome represents one demand. The value of each allele is an index of a path selected for a particular demand (denoted as $p_d$) [4, 43–49]. Therefore, the chromosome is defined as follows:

$$X = [p_1, p_2, \ldots, p_{|D|}].  \tag{2.2.10}$$

For example, chromosome $X = [3, 1, 2]$ means that demand $d = 1$ uses path $p = 3$ ($x_{13} = 1$), demand $d = 2$ is allocated to path $p = 1$ ($x_{21} = 1$) and demand $d = 3$ is realized on path $p = 2$ ($x_{32} = 1$). Using encoding (2.2.10), it is straightforward to calculate flow on each link (2.2.1b) and next to calculate the value of the objective function (2.2.1a). It should be noted that other ways of encoding the multicommodity flow optimization problems in EAs have been proposed in literature, e.g., see [50–54].

Originally, EA was designed to solve optimization problems without constraints and with the aim of maximizing the objective (fitness) function. Therefore, to address the fact that the problem includes constraints and the objective function is to minimize the routing cost, the following modifications are required. There are three ways of incorporating processing of constraints in EA. Firstly, the encoding scheme can ensure that particular constraints are directly satisfied. Secondly, a *penalty function* can be employed, i.e., the fitness function contains not only the problem objective function, but also a special term including a measure of violation of the constraints scaled by a penalty parameter. It is assumed that the measure of violation is nonzero if the constraint is violated, and zero in the region where the constraint is not broken. Thirdly, a *repair function* can be used when the current solution (chromosome) is not feasible.

Note that in this problem (2.2.1), condition (2.2.1c) is included directly in the encoding scheme (2.2.10). More specifically, since a single allele is assigned to exactly one demand in the chromosome, the single path routing is imposed.

To address the link capacity constraint (2.2.1d), the penalty function approach is applied. More specifically, let $F(X)$ return the value the objective function (2.2.1c) of the solution encoded in chromosome $X$. Next, let $F^{PEN}(X)$ denote the value of the objective function with an additional penalty function for chromosome $X$ defined as follows:

$$F^{PEN}(X) = F(X) + Pn \sum_{e \in E} CapCon(X, e). \qquad (2.2.11)$$

Function $CapCon(X, e)$ represents the violation of capacity constraint (2.2.1c) according to network flows defined in $X$. If the capacity constraint for link $e$ is not violated (i.e., $f_e \leq c_e$), then $CapCon(X, e) = 0$, otherwise, $CapCon(X, e) = f_e - c_e$. In turn, $Pn$ denotes the penalty scaling parameter that tunes the penalty function.

Finally, to tackle the anycast constraint (2.2.1e), a simple repair function is developed (Algorithm 2.4). For an input demand $d$ it is checked whether the associated demand $\tau(d)$ uses the same DC node as the DC node of demand $d$ in the current solution $X$ (line 2). If this condition is not fulfilled, a new path for demand $\tau(d))$ needs to be selected, but using only candidate paths from set $P(\tau(d))$ connected to the DC node of demand $d$ (lines 3–5). To repair the whole solution, a function described as Algorithm 2.4 must be executed for all associated demand pairs. In order to improve the performance of the repair procedure for the whole solution including all demands, the demands can be analyzed in a particular order, e.g., by decreasing value of the requested volume (bitrate).

To address the fact that EA requires the objective function to be maximized, the fitness function is defined as follows

$$Fitness(X) = M(F^{MAX} - F^{PEN}(X)) \qquad (2.2.12)$$

where $F^{MAX}$ denotes the maximum possible value function $F^{PEN}(X)$ taking into account all chromosomes $X$ in a given population, while $M$ is a scaling parameter that conducts additional tuning of the algorithm during the optimization process.

---

**Algorithm 2.4** RFAD (Repair Function for Anycast Demand)

---

**Require:** set of edges $E$, associated anycast demands $d$ and $\tau(d)$, set $P(\tau(d))$ including candidate
    paths for demands $d$ and $\tau(d)$, current solution $X$
**Ensure:** feasible path selection (routing) for demand $\tau(d)$
1: **procedure** $RFAD(X, d, P(d), P(\tau(d)))$
2: **if** $DC\_Node(d) \neq DC\_Node(\tau(d))$ **then**
3:     $X := X \setminus \{x_{\tau(d)p}\}$
4:     $q := Find\_Best\_Path\_DC(\tau(d), DC\_Node(d))$
5:     $X := X \cup \{x_{\tau(d)q}\}$
6: **end if**
7: **end procedure**

---

It is worth noting that the proposed encoding scheme and fitness function formulation apply the classical crossover and mutation operators. For instance, to
make the one-point crossover of two parent solutions $X^1 = [p_1^1, p_2^1, \ldots, p_{|D|}^1]$
and $X^2 = [p_1^2, p_2^2, \ldots, p_{|D|}^2]$, it is necessary to select at random an integer $d$
between 1 and $|D|$. In consequence, the following two children are obtained $X^3 =
[p_1^1, p_2^1, \ldots, p_d^1, p_{d+1}^2, \ldots, p_{|D|}^2]$ and $X^4 = [p_1^2, p_2^2, \ldots, p_d^2, p_{d+1}^1, \ldots, p_{|D|}^1]$. Next, the
repair function defined in Algorithm 2.4 must be applied to both new chromosomes
to ensure the anycast constraint. The mutation operation assumes that for a randomly
selected demand $d$, the routing path is changed by a random selection of an integer
between 1 and $|P(d)|$. Again, the repair function is required to fix the potential prob-
lem with the anycast constraint. Additionally, for the described encoding scheme and
fitness function formulation, more complex approaches of crossover and mutation
operators can be used.

Finally, we discuss the issue of the initial solution of EA. More specifically,
in many cases the size of a feasible solution space in the flow allocation problem
is very small compared to the total solution space. Consequently, it is likely that
EA can encounter significant difficulties in finding a feasible solution, even using
mechanisms such as the penalty function and repair function. In such a case, the
performance of EA can be improved by using a feasible solution provided by another
method, e.g., the FD algorithm. Thereafter, using the *elite* member concept (i.e., the
best feasible solution(s) are copied directly to the next generation) [38, 40] ensures
that EA method yields a feasible solution.

### Neighborhood Search Methods

There is a wide family of stochastic heuristics based on the concept of *neighborhood search*, including Local Search (LS), Tabu Search (TS), Simulated Annealing
(SA), and Greedy Randomized Adaptive Search Procedure (GRASP) [4, 39, 40, 42,
55–58]. The neighborhood $N(X)$ of a particular solution $X$ is a subset of the total solution space including solutions $Y \in N(X)$ that can be generated from $X$ by a simple
modification of the solution. The most common approach to generating the neighborhood solution for unicast flow allocation problems in connection-oriented networks
is to simply change a routing path for one demand. The number of neighborhood

solutions can then be estimated as $\sum_{d \in D}(|P(d)| - 1)$, since for each demand $d \in D$, the currently selected path can be changed to any of the remaining paths included in set $P(d)$ [4].

However, to address the additional anycast constraint (2.2.1e) that arises in anycast flow allocation problems, the default procedure must be modified. More precisely, if the created neighborhood solution $Y \in N(X)$ does not satisfy the anycast constraint, i.e., a new path selected for demand $d$ uses a different DC node than the DC node of demand $\tau(d)$, the repair function shown in Algorithm 2.4 must be executed for demand $d$.

In addition, the neighborhood solution $Y \in N(X)$ may be not feasible in terms of the link capacity constraint (2.2.1d). In such a case, there are two options. First, the penalty function approach can be used similarly to (2.2.11). As a result, the algorithm based on the neighborhood search is able to examine unfeasible solutions as well. Secondly, all solutions that are not feasible cannot be analyzed, i.e., they are excluded from set $N(X)$.

Using this definition of the neighborhood solution, it is easy to develop range of neighborhood search methods such as LS, TS, SA and GRASP for the anycast flow allocation problem. As in the context of EA, the algorithm performance can be improved by using an initial solution yielded by another algorithm.

## 2.3 Network Design Problems for Anycast, Multicast and Unicast Flows

This section focuses on a network design problem with joint optimization of link capacity assignment and anycast, multicast and unicast flow allocation. Note that the network design problem is also referred to as the capacity and flow allocation (CFA) problem. Network design problems are among the most common optimization problems in networks. They need to be resolved when a new network is being designed from scratch or an existing network needs to be updated. The main goal of the optimization process is to select the capacity of network links and the routing configuration in order to realize all demands. The most common objective function that occurs in network design problems is the CAPEX/OPEX cost, although other network performance metrics (e.g., delay, survivability, throughput) can be applied. The link capacity constraint which ensures that the total flow on each link cannot exceed the assigned link capacity appears to be the key element of all network design problems [4, 49].

The majority of earlier network design problems have focused on the optimization of unicast flows only. Relatively few papers address joint optimization of two types of flows, i.e., anycast and unicast or multicast and unicast. The only paper that focuses on joint optimization of anycast, multicast and unicast flows in connection-oriented networks is [59], where a static RWA problem with unicast, anycast and multicast connections is examined and some heuristics are proposed and evaluated. In the case of multicast flows, the routing is fixed since only one tree is created using a minimal spanning tree algorithm, which is then iteratively pruned to remove all unnecessary leaves.

## 2.3.1   Formulation

The routing is modeled in a similar way to that presented in Sect. 2.2. However, since anycast, multicast and unicast network flows are to be provisioned in the network, some modifications are required. In particular, for each demand $d \in D$, the set $P(d)$ includes candidate structures. When $d$ is a unicast demand, the candidate structure is simply a path connecting the source and destination nodes of the demand. When $d$ is an anycast upstream demand, the candidate structure is a path connecting the client node and one of the DC nodes. In turn, when $d$ is a downstream demand, the candidate structure is a path connecting one of the DC nodes and the client node. Finally, when $d$ is a multicast demand, the candidate structure is a tree that includes a root node and all receivers of the demand.

The link capacity is given in modules, and the capacity assigned to each link is expressed as a multiple of one of the modules. This assumption follows from the fact that in most network technologies such as Ethernet, SDH/SONET and WDM the network uses some predefined values of the potential capacity available on each link. Constant $M$ denotes the size of the link capacity module given in the same unit as the demand volume $h_d$, e.g., in bits per seconds. Moreover, constant $\xi_e$ denotes the cost of using one capacity module on link $e$. Note that constant $\xi_e$ can represent various types of costs, such as CAPEX cost, OPEX cost, power consumption, etc. Integer variable $y_e$ denotes the number of capacity modules allocated to link $e$ [4, 49].

------------

**CON/AMU/ND/Cost/Link-path**

**sets**

$E$        links
$D$        demands (anycast, multicast, unicast)
$D^{DS}$    anycast downstream demands
$P(d)$     candidate structures for flows realizing demand $d$. If $d$ is a unicast demand, candidate structure is a path connecting end nodes of the demand. If $d$ is an anycast upstream demand, candidate structure is a path connecting the client node and the DC node. If $d$ is a downstream demand, candidate structure is a path connecting the DC node and the client node. If $d$ is a multicast demand, candidate structure is a tree that includes the root node and all receivers of the demand

**constants**

$\delta_{edp}$    =1, if link $e$ belongs to structure $p$ realizing demand $d$; 0, otherwise
$h_d$      volume of demand $d$
$\xi_e$      unit cost of link $e$
$M$     size of the link capacity module
$\tau(d)$    index of a demand associated with demand $d$. If $d$ is a downstream demand,
       then $\tau(d)$ must be an upstream demand and vice versa
$s(p)$    origin node of path $p$
$t(p)$    destination node of path $p$

**variables**

$x_{dp}$   =1, if structure $p$ is used to realize demand $d$; 0, otherwise (binary)
$y_e$    capacity of link $e$ as the number of capacity modules (integer)

**objective**

$$\text{minimize} \quad F = \sum_{e \in E} \xi_e y_e \tag{2.3.1a}$$

**constraints**

$$\sum_{p \in P(d)} x_{dp} = 1, \quad d \in D \tag{2.3.1b}$$

$$\sum_{d \in D} \sum_{p \in P(d)} \delta_{edp} x_{dp} h_d \leq M y_e, \quad e \in E \tag{2.3.1c}$$

$$\sum_{p \in P(d)} x_{dp} s(p) = \sum_{p \in P(\tau(d))} x_{\tau(d)p} t(p), \quad d \in D^{DS}. \tag{2.3.1d}$$

The objective function (2.3.1a) is to minimize the total network cost defined as the overall cost of all capacity allocated to network links. Equality (2.3.1b) ensures that for each demand $d \in D$ exactly one routing structure is selected. Condition (2.3.1c) is the link capacity constraint, but since a network design problem is considered, the right-hand side of (2.3.1c) is not fixed and follows from the selection of link variable $y_e$. Finally, constraint (2.3.1d) imposes the anycast constraint that holds for anycast demands only.

Model (2.3.1) is a generic formulation which can be modified to address various additional constraints that can occur in different types of network design problems. For instance, the network cost given by formula $\sum_{e \in E} \xi_e y_e$ can be incorporated as a budget constraint that limits the maximum cost to be spent on network deployment, and another function can be used as the objective, e.g., network delay, proportion of the realized demand volumes. For more details see [4].

## 2.3.2   Algorithms

The network design problem formulated as (2.3.1) is generally comparable to the
flow allocation problem addressed in Sect. 2.2 with a key additional element of link
capacity optimization. Therefore, the general concepts of heuristic algorithms out-
lined in Sect. 2.2.2 can be adapted to the new constraints.

Firstly, we show how to use the FD method proposed in [15] for the network
design problem with anycast, multicast and unicast flows. Algorithm 2.5 presents
the FD/AMU/ND method. The key assumption behind this heuristic is that the algorithm directly decides on the routing variables $x_{dp}$, while the values of link capacity
variables $y_e$ are calculated indirectly according to a particular flow allocation. More
specifically, having selected a routing structure for each demand $d \in D$ (represented
in solution $X$ that includes all variables $x_{dp}$ equal to 1) the flow on each link $f_e$ can
be calculated as described in Sect. 2.2.1 and formulated in (2.2.1b). Next, the link
capacity variable $y_e$ for each link $e \in E$ is simply selected as the minimum value
that satisfies constraint $My_e \geq f_e$. In order to evaluate the quality of the routing
assignment given in solution $X$, the network cost is calculated as $\sum_{e \in E} \xi_e y_e$ using the
obtained values of $y_e$. The main advantage of this concept is that any flow allocation
algorithm such as FD/A/FA shown in Algorithm 2.2 can be applied in this context
almost directly.

The FD/AMU/ND algorithm starts with a calculation of an initial solution using
any flow allocation algorithm (line 2). A flag *test* used in the stopping condition of
the algorithm is initialized next (line 3). The main loop of the algorithm (lines 4–30)
is repeated until the new solution improves the network cost function. First, metric $l_e$
is calculated for each link $e \in E$ according to the routing solution given in the current
solution $X_i$ (line 5); using this metric, the shortest structure for each demand $d \in D$
is calculated. In the case of an anycast demand, the anycast constraint is satisfied in
this function. Next, all demands are examined according to a predefined order (lines
8–23). In particular, each demand attempts to be rerouted to the shortest structure
included in set $SR(X_i)$. In the case of anycast demands, both associated demands must
be processed jointly (lines 9–11), while multicast and unicast demands are processed
individually (line 13). The new solution $K$ is evaluated against the previous solution
$H$, i.e., the link capacity cost is calculated accordingly to the flow allocation (lines
15–17). If the obtained solution is the same as the solution from the previous iteration
after processing all demands, the algorithm stops.

A Lagrangian relaxation method is also used in solving the network design problem with anycast, multicast and unicast flows (2.3.1). A general framework of solving
network design problems using this method is described in [30]. As in the case of
the flow allocation problem, the key issue is the formulation of a dual problem of the
original optimization problem by relaxing constraints in order to obtain a simpler
subproblem. For more details see [30].

Regarding the EA method described in Sect. 2.2.2, the following modifications are
necessary to solve the network design problem with anycast, multicast and unicast
flows. First, since various types of flows are considered, the chromosome must be

---

**Algorithm 2.5** FD/AMU/ND (Flow Deviation for Network Design with Anycast, Multicast and Unicast Flows)

---

**Require:** set of edges $E$, set of anycast, multicast and unicast demands $D$, sets $P(d)$ including candidate structures for each demand $d \in D$, capacity module size $M$, link cost

**Ensure:** routing structure selection (routing) for each demand $d \in D$ included in set $X_i$, value of objective function

1: **procedure** $FD/AMU/ND(D, P(d))$
2:  $X_1 := Find\_Intial\_Solution(D, P(d))$, $i := 1$
3:  $test := 0$
4:  **repeat**
5:   **for** $e \in E$ **do** $l_e := \xi_e \min \{y_e : My_e \geq f_e(X_i)\}$
6:   $SR(X_i) := Find\_Shortest\_Structure(D, P(d), X_i, L)$
7:   $H := X_i$
8:   **for** $d \in D$ **do**
9:    **if** $Type(d) = ANYCAST$ **then**
10:     $K := (H - \{x_{dp}\}) \cup \{x_{dq}\}$, where $x_{dp} \in H$ and $x_{dq} \in SR(X_i)$
11:     $K := (K - \{x_{\tau(d)p'}\}) \cup \{x_{\tau(d)q'}\}$, where $x_{\tau(d)p'} \in H$ and $x_{\tau(d)q'} \in SR(X_i)$
12:    **else**
13:     $K := (H - \{x_{dp}\}) \cup \{x_{dq}\}$, where $x_{dp} \in H$ and $x_{dq} \in SR(X_i)$
14:    **end if**
15:    $F(H) = Find\_Objective(H)$
16:    $F(K) = Find\_Objective(K)$
17:    **if** $F(K) < F(H)$ **then** $H := K$
18:    **if** $Type(d) = ANYCAST$ **then**
19:     $D := D \setminus \{d, \tau(d)\}$
20:    **else**
21:     $D := D \setminus \{d\}$
22:    **end if**
23:   **end for**
24:   **if** $X_i \neq H$ **then**
25:    $i := i + 1$
26:    $X_i := H$
27:   **else**
28:    $test := 1$
29:   **end if**
30:  **until** $test < 1$
31: **end procedure**

---

divided into three parts, with each part encoding a selection of routing structures for a particular type of flow. This approach involves a straightforward implementation of the crossover operator, which selects the crossover point(s) independently for each part (demand type) of the chromosome. In consequence, the obtained child solution is feasible in terms of constraint (2.3.1b). Note that after the crossover operation, the repair function reported in Algorithm 2.4 is applied only for the chromosome part that refers to anycast demands. In turn, the mutation operator does not require any changes. The second adjustment of the EA algorithm refers to chromosome evaluation. More specifically, since the goal of the optimization is to minimize network cost in terms of capacity cost, to evaluate a particular solution $X$, link flow $f_e$ for each $e \in E$ is first calculated according to (2.2.1b). Next, the link capacity variable $y_e$

is calculated for each link $e \in E$ as the minimum value that satisfies constraint $My_e \geq f_e$ and the network cost is obtained. Note that because link capacities are determined according to the flow allocation encoded in the chromosome, the link capacity constraint (2.3.1c) is always satisfied and thus there is no need to use the penalty function mechanism. This approach means that link capacity variables $y_e$ are not directly encoded in the chromosome, but they are selected indirectly according to the formulation of the optimization problem.

Concerning the metaheuristics based on the neighborhood search, e.g., LS, TS, SA, and GRASP, the approach to tackling the network design problem is similar to the EA method. In particular, the solution encodes the routing decision variable $x_{dp}$ only. The link capacity variables $y_e$ are selected according to the flow allocated on each link, which yields a value of the objective function (network cost) that is used to evaluate a particular solution. Consequently, it is not necessary to use the penalty function associated with the link capacity constraint. Again, in the case of anycast demands, the repair function shown in Algorithm 2.4 is required to cope with the anycast constraint. Note that another approach to encoding the solution in the context of neighborhood search methods is proposed in [60] where the authors use two vectors with variables $x_{dp}$ and $y_e$ to represent the solution. To control the feasibility of the solution, a penalty function is used.

## 2.4 Location Problems for Anycast and Unicast Flows

In this section, we address optimization problems related to the location of DC nodes for a network realizing anycast and unicast flows. Two types of location problems are formulated in this section. Firstly, it is assumed that each installed DC provides the same content/service and in consequence each DC can serve every anycast demand. The goal is to decide on the location of DC nodes, link capacity and routing in order to minimize the cost related to both DC and link capacity resources. In the second problem, DCs offer various types of content divided into content groups (CGs). The optimization process involves locating particular content groups and selecting routing paths.

### 2.4.1 Data Center Location and Network Design

Location problems are presented in depth in earlier studies, mainly in the context of CDNs including web proxy placement, cache location, replica location, e.g., [49, 61–73]. However, the majority of papers on location problems with anycast flows do not tackle the routing problem, i.e., it is assumed that each client is simply assigned to the closest DC node using the shortest path. This assumption considerably reduces the complexity of the optimization problem, since link capacity is not required in the model. Therefore, the location and network design (NDL) optimization problem presented below, involving joint optimization of location, link capacity and flow allocation, results in significantly more detailed modeling and optimization of real networks.

The NDL problem for anycast and unicast flows is formulated for both link-path and node-link notations. In order to write the models, some new notation must be introduced. First, let $R$ denote a set of network nodes that can host a data center. The decision variable that determines the location of DC nodes in the network is defined as $u_v$, which equals 1 if node $v$ hosts a DC node and 0 otherwise. The cost of locating a DC at node $v \in R$ is given by constant $\eta_v$.

**Link-Path Formulation**

To recall, in the link-path notation of anycast flows, the DC node is selected for a demand directly by choosing the routing path. To ensure that the path selected to realize an anycast demand a proper (installed) DC as a DC node, constant $\pi_{vdp}$ denotes whether node $v \in R$ is a DC node for path $p$ realizing anycast demand $d$.

---

**CON/AU/NDL/Cost/Link-path**

**sets**

$E$      links
$R$      candidate nodes for data center location
$D$      demands (anycast, unicast)
$D^{AN}$      anycast demands
$D^{DS}$      anycast downstream demands
$P(d)$      candidate paths for flows realizing demand $d$. If $d$ is a unicast demand, the candidate path connects end nodes of the demand. If $d$ is an anycast upstream demand, the candidate path connects the client node and a node included in set $R$ (DC candidate node). If $d$ is a downstream demand, the candidate path connects a node from set $R$ (DC candidate node) and the client node

**constants**

$\delta_{edp}$      $=1$, if link $e$ belongs to path $p$ realizing demand $d$; 0, otherwise
$\pi_{vdp}$      $=1$, if node $v \in R$ is a DC node for path $p$ realizing anycast demand $d$; 0, otherwise
$h_d$      volume of demand $d$
$\xi_e$      unit cost of link $e$
$M$      size of the link capacity module
$\tau(d)$      index of a demand associated with demand $d$. If $d$ is a downstream demand, then $\tau(d)$ must be an upstream demand and vice versa
$s(p)$      origin node of path $p$

$t(p)$     destination node of path $p$
$r$        number of data centers to be located in network
$\eta_v$   cost of location of a data center at node $v$, if opened

**variables**

$x_{dp}$  $=1$, if path $p$ is used to realize demand $d$; 0, otherwise (binary)
$y_e$     capacity of link $e$ as the number of capacity modules (integer)
$u_v$     $=1$, if node $v$ is selected to host a data center; 0, otherwise (binary)

**objective**

$$\text{minimize} \quad F = \sum_{e \in E} \xi_e y_e + \sum_{v \in R} \eta_v u_v \tag{2.4.1a}$$

**constraints**

$$\sum_{p \in P(d)} x_{dp} = 1, \quad d \in D \tag{2.4.1b}$$

$$\sum_{d \in D} \sum_{p \in P(d)} \delta_{edp} x_{dp} h_d \leq M y_e, \quad e \in E \tag{2.4.1c}$$

$$\sum_{p \in P(d)} x_{dp} s(p) = \sum_{p \in P(\tau(d))} x_{\tau(d)p} t(p), \quad d \in D^{DS} \tag{2.4.1d}$$

$$\sum_{p \in P(d)} \pi_{vdp} x_{dp} \leq u_v, \quad d \in D^{AN}, v \in R \tag{2.4.1e}$$

$$\sum_{v \in R} u_v = r. \tag{2.4.1f}$$

The goal of optimization (2.4.1a) is to minimize the sum of link capacity cost and data center location cost. Equality (2.4.1b) ensures that exactly one routing path is selected to realize demand $d$. Condition (2.4.1c) defines the link capacity constraint to meet the requirement that the flow of each link cannot exceed the allocated link capacity. Equality (2.4.1d) denotes the anycast constraint. Condition (2.4.1e) ensures that an anycast demand can be assigned to a routing path $p$ that includes a DC node installed in the network. More precisely, if a data center is not located at node $v$ (i.e., $u_v = 0$), then the right-hand side of (2.4.1e) must be zero, which guarantees that path $p$ selected to realize demand $d$ cannot use node $v$ as the DC node. The final constraint (2.4.1f) is in the model to ensure that exactly $r$ data centers are located in the network.

Note that heuristic algorithms described in Sects. 2.2.2 and 2.3.2 can be adapted to solve the CON/AU/NDL problem defined as (2.4.1a)–(2.4.1f). To achieve this, the problem must be divided to two separate subproblems: a DC location problem (without capacity and flow allocation) and a network design problem. The first subproblem can be solved using methods developed in the context of cache/replica location problems; see [61–64, 68, 69, 71–73]. When the DCs are located in the network, the pure network design problem with anycast and unicast flows can be solved.

### Node-Link Formulation

New notation is required to write the node-link formulation. Let $x_{ed}$ denote whether the routing path selected for demand $d$ uses link $e$. For anycast demands, variable $z_{vd}$ denotes whether node $v$ is selected as a DC for demand $d$.

---

### CON/AU/NDL/Cost/Node-link

**sets (additional)**

| | |
|---|---|
| $V$ | nodes |
| $\delta^+(v)$ | links leaving node $v$ |
| $\delta^-(v)$ | links entering node $v$ |
| $D^{UN}$ | unicast demands |
| $D^{US}$ | anycast upstream demands |

**constants (additional)**

| | |
|---|---|
| $s_d$ | source node of demand $d$ |
| $t_d$ | destination node of demand $d$ |

**variables**

| | |
|---|---|
| $x_{ed}$ | $=1$, if demand $d$ uses link $e$; 0, otherwise (binary) |
| $y_e$ | capacity of link $e$ as the number of capacity modules (integer) |
| $u_v$ | $=1$, if node $v$ is selected to host a data center; 0, otherwise (binary) |
| $z_{vd}$ | $=1$, if DC node $v$ is selected as a DC for demand $d$; 0, otherwise (binary) |

**objective**

$$\text{minimize} \quad F = \sum_{e \in E} \xi_e y_e + \sum_{v \in R} \eta_v u_v \qquad (2.4.2a)$$

**constraints**

$$\sum_{e\in\delta^+(v)} x_{ed} - \sum_{e\in\delta^-(v)} x_{ed} = \begin{cases} +1 & \text{if } v = s_d \\ -1 & \text{if } v = t_d, \\ 0 & \text{otherwise} \end{cases} \quad v \in V, d \in D^{UN} \qquad (2.4.2\text{b})$$

$$\sum_{e\in\delta^+(v)} x_{ed} - \sum_{e\in\delta^-(v)} x_{ed} = \begin{cases} +z_{vd} & \text{if } v \in R \\ -1 & \text{if } v = t_d \\ 0 & \text{otherwise} \end{cases} \quad v \in V, d \in D^{DS} \qquad (2.4.2\text{c})$$

$$\sum_{e\in\delta^+(v)} x_{ed} - \sum_{e\in\delta^-(v)} x_{ed} = \begin{cases} +1 & \text{if } v = s_d \\ -z_{vd} & \text{if } v \in R \\ 0 & \text{otherwise} \end{cases} \quad v \in V, d \in D^{US} \qquad (2.4.2\text{d})$$

$$\sum_{d\in D} h_d x_{ed} \le M y_e, \quad e \in E \qquad (2.4.2\text{e})$$

$$z_{vd} = z_{v\tau(d)}, \quad d \in D^{DS}, v \in R \qquad (2.4.2\text{f})$$

$$\sum_{v\in R} z_{vd} = 1, \quad d \in D^{AN} \qquad (2.4.2\text{g})$$

$$z_{vd} \le u_v, \quad d \in D^{AN}, v \in V \qquad (2.4.2\text{h})$$

$$\sum_{v\in R} u_v = r. \qquad (2.4.2\text{i})$$

The objective function (2.4.2a) is formulated in the same way as in model (2.4.1). Condition (2.4.2b) is a node-link formulation of unicast demands. Equalities (2.4.2c)–(2.4.2d) define the flow conservation constraints for downstream and upstream anycast demands, respectively. More specifically, if node $v$ is included in set $R$ (i.e., node $v$ can be selected as a DC node), the left-hand side of (2.4.2c) must be $z_{vd}$. Note that if $v$ is not selected as the DC node of downstream demand $d$ ($z_{vd} = 0$), the left-hand side of (2.4.2c) is 0 and $v$ is a transit node. In turn, if node $v$ is selected as the DC node of demand $d$ ($z_{vd} = 1$), the left-hand side of (2.4.2c) is 1 and $v$ is the source node of the demand. If the considered node $v$ is the destination (client) node ($v = t_d$) of downstream demand $d$, the left-hand side of (2.4.2c) is $-1$. Finally, in all other cases, $v$ is a transit node and the left-hand side of (2.4.2c) is 0. Constraint (2.4.2d) defines the flow conservation for upstream demands. Inequality (2.4.2e) denotes the link capacity constraint. Equality (2.4.2f) ensures that two associated demands $d$ and $\tau(d)$ use the same DC node. Constraint (2.4.2g) is in the model to guarantee that every anycast demand uses exactly one DC node. Constraint (2.4.2h) imposes the requirement that an anycast demand can be assigned to node $v$ that hosts a DC. The last inequality (2.4.2i) limits the number of DCs to be installed in the network.

## 2.4.2  Content Location and Flow Allocation

This section presents the problem of content location and flow allocation in a network with anycast and unicast flows. The content location problem is similar to other location problems, reviewed in depth in literature, such as the data placement problem [74, 75] and object placement problem [61, 62, 76–81]. However, the majority of previous research in this area considers location problems assuming that the optimization of routing (flow allocation) is not included in the model, which significantly simplifies the formulation and the optimization process.

The optimization model addressed in this section uses a concept of a *content group*, which follows directly from the observation that online content can be divided into groups according to popularity [61, 76, 77, 82–87]. More specifically, set $B$ includes content groups and constant $\psi_b$ denotes the size of data related to content group $b$ given in GB. Due to the fact that popularity of a particular content group can depend on the geographical location of users and on the population of users located at node $v$, constant $h_{vb}$ defines the volume (bit-rate) of content group $b$ requested by a client located at node $v$. This means that network flow associated with a particular content group $b$ can vary for different network nodes. Content groups are located at DCs node that are already placed in the network. Binary variable $u_{vb}$ determines whether the DC located at node $v$ stores content group $b$. However, DCs have a limited storage capacity, therefore the whole content cannot be placed in every DC node. To make the problem more realistic, it is assumed that some background unicast traffic is transmitted in the network and $h_{vw}$ denotes the volume (bit-rate) of unicast traffic from node $v$ to node $w$.

Since anycast demands are defined for each node, route modeling is a little different than in the previous models. In particular, set $P(v, w)$ includes candidate paths for flows realizing demand from node $v$ to node $w$. The assignment of clients to a DC providing the requested content is controlled by a binary variable $z_{vwb}$ that denotes whether node $v$ downloads content group $b$ from a DC located at node $w$. Due to traffic asymmetry, the upstream traffic from clients to DCs is not included directly in the model. However, the bit-rate associated with the upstream traffic is built-in unicast flows between particular nodes (constant $h_{vw}$). To reduce the number of connections to be established in the network, traffic grooming is assumed. More specifically, the whole traffic between a particular pair of nodes (both anycast traffic serving downloads of content groups and unicast traffic) is provisioned using a single connection. Without this assumption, due to a potential large number of content groups, the number of connections serving individual requests to single content groups may be high. In consequence, binary variable $x_{vwp}$ denotes whether path $p$ is used to realize the demand from node $v$ to node $w$. Similarly, variable $f_{vwp}$ represents the volume of flow from node $v$ to node $w$ allocated to path $p$.

## CON/AU/FAL/Content Groups/Cost

**sets**

| | |
|---|---|
| $E$ | links |
| $V$ | nodes (clients) that request content |
| $R$ | nodes with a data center |
| $P(v, w)$ | candidate paths for flows realizing demand from node $v$ to node $w$ |
| $B$ | content groups |

**constants**

| | |
|---|---|
| $\delta_{evwp}$ | $=1$, if link $e$ belongs to path $p$ realizing demand between from node $v$ to node $w$; 0, otherwise |
| $h_{vw}$ | volume (bit-rate) of unicast traffic from node $v$ to node $w$ |
| $g_{vb}$ | volume (bit-rate) of content group $b$ requested by node $v$ |
| $s_v$ | storage capacity of a DC located at node $v$ (GBytes) |
| $\psi_b$ | size of data related to content group $b$ (GBytes) |
| $c_e$ | capacity of link $e$ |
| $\zeta_e$ | unit routing cost on link $e$ |
| $M$ | large number |

**variables**

| | |
|---|---|
| $x_{vwp}$ | $=1$, if path $p$ is used to realize demand from node $v$ to node $w$; 0, otherwise (binary) |
| $z_{vwb}$ | $=1$, if node $v$ downloads content group $b$ from a DC located at node $w$; 0, otherwise (binary) |
| $f_{vwp}$ | volume of flow from node $v$ to node $w$ allocated to path $p$ (continuous non-negative) |
| $u_{vb}$ | $=1$, if DC located at node $v$ stores content group $b$; 0, otherwise (binary) |

**objective**

$$\text{minimize} \quad F = \sum_{e \in E} \sum_{v,w \in V} \sum_{p \in P(v,w)} \delta_{evwp} \zeta_e f_{vwp} \qquad (2.4.3a)$$

**constraints**

$$\sum_{p \in P(v,w)} x_{vwp} = 1, \quad v, w \in V \qquad (2.4.3b)$$

$$f_{vwp} \leq M x_{vwp}, \quad v, w \in V, p \in P(v, w) \qquad (2.4.3c)$$

$$\sum_{p \in P(v,w)} f_{vwp} \geq h_{vw} + \sum_{b \in B} z_{vwb} g_{wb}, \quad v, w \in V \qquad (2.4.3d)$$

$$\sum_{v,w \in V} \sum_{p \in P(d)} \delta_{evwp} f_{vwp} \leq c_e, \quad e \in E \tag{2.4.3e}$$

$$z_{vwb} \leq u_{vb}, \quad v \in R, w \in V, b \in B. \tag{2.4.3f}$$

$$\sum_{b \in B} \psi_b u_{vb} \leq s_v, \quad v \in R \tag{2.4.3g}$$

$$\sum_{v \in R} u_{vb} \geq 1 \quad b \in B. \tag{2.4.3h}$$

The goal of optimization (2.4.3a) is to minimize the routing cost of network flows required to transmit content from DCs and the background unicast traffic. Equality (2.4.3b) imposes a single path routing. Constraint (2.4.3c) binds variables $x_{vwp}$ and $f_{vwp}$ to ensure that flow on path $p \in P(v, w)$ is zero if path $p$ is not selected to realize traffic from node $v$ to node $w$. Condition (2.4.3d) implements traffic grooming, i.e., the volume allocated to a path selected for node pair $v$ and $w$ must be enough to serve both unicast traffic ($h_{vw}$) and content traffic ($\sum_{b \in B} z_{vwb} g_{wb}$). Inequality (2.4.3e) is the link capacity constraint. Condition (2.4.3f) ensures that anycast client at node $w$ can download a content group $b$ from DC located at node $v$, only if DC $v$ stores content group $b$. The next inequality (2.4.3g) controls the storage capacity limit of each DC. Finally, constraint (2.4.3h) ensures that every content group is allocated to at least one DC node. Note that to introduce additional survivability constraints, the right-hand side of constraint (2.4.3h) could be changed to 2 or a larger number to enable content replication.

The most straightforward way of solving the above problem using heuristic algorithms is to consider two separate subproblems. The decision on content group placement needs to be determined first, which makes the problem a pure flow allocation problem and methods proposed in Sect. 2.2.2 can be applied.

## 2.5 Survivable Allocation of Anycast and Unicast Flows

This section focuses on the flow allocation problem with additional survivability constraints. Two types of network flows are addressed: anycast and unicast. We start with a short discussion on survivability of connection-oriented networks.

The key idea of providing survivability in a connection-oriented network is to establish two failure-disjoint paths for each demand, i.e., *working* path used in a normal, failure-free state of the network and *backup* path used when the working path is not available due to a network failure. Two popular protection methods based on this approach and considered in the context of connection-oriented networks are Dedicated Path Protection (DPP) and Shared Backup Path Protection (SBPP). The key difference between DPP and SBPP is the fact that SBPP makes it possible to share capacity between backup paths belonging to different demands under the condition that these resources can be used by a single demand in a given failure scenario. In

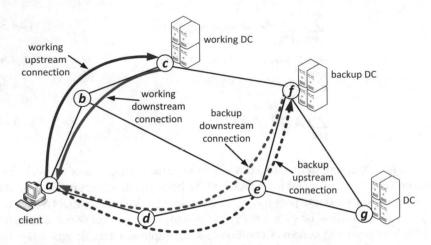

**Fig. 2.1** An example of a survivable anycast connection with different working and backup DCs

contrast, backup paths in DPP have their own dedicated capacity. This means that SBPP brings capacity savings compared to DPP.

Additionaly, there are two categories of DPP schemes: 1+1 and 1:1. In the former, traffic is permanently duplicated on both the working path and the backup path. The receiving node selects the signal with the best quality. The 1 + 1 approach is very efficient in terms of recovery time, since the reaction to a network failure is almost immediate. However, the 1+1 method is relatively costly in terms of capacity usage. On the other hand, the 1:1 method assumes that in failure-free conditions traffic is transmitted over the working path, which means extra traffic can be transported along the backup path in failure-free conditions. When a failure occurs along the working path, the extra traffic must be prevented from entering the backup path to enable switching the traffic affected by the failure to the backup path. Consequently, in comparison with 1 + 1, the 1:1 approach requires higher recovery times [5].

In the context of anycast flows, path protection methods such as DPP and SBPP can be used differently to the protection of unicast flows. More specifically, anycasting assumes that the same content/service can be provisioned by several DCs located in the network. In consequence, working and backup paths of the same demand can use different DC nodes. For ease of reference, the DC node used by the working path is denoted as *working DC* and the DC node used by the backup path is referred to as *backup DC*. There are two main reason for this relocation approach using different DC nodes for working and backup paths. Firstly, this concept protects the network against a failure of a single DC node, improving network survivability. Secondly, in some cases a backup path connected to another DC node can provide better performance according to the considered metric compared to a backup path that uses the same DC node as the working path. It should be noted that both associated anycast demands (upstream and downstream) must use the same DC node for working paths and backup paths, respectively [88]. The relocation approach is presented in [70, 89–99].

To illustrate the concept of a backup DC, a simple example is shown in Fig. 2.1. An anycast client is located at node 1 and three DCs are placed in the network at nodes $c, f$ and $g$. The anycast client uses node $c$ as the working DC and node $f$ as the backup DC. Both downstream and upstream connections are shown. Note that if the backup DC is located in the same node as the working DC (node $c$), the shortest backup paths will include at least 4 links (for instance path $(a, d, e, b, c)$), when the backup DC located at node $f$ is used, the backup path is shorter and includes only 3 links (path $(a, d, e, f)$).

## 2.5.1  Formulations

In this section, two optimization models using DPP and SBPP approaches to provide network survivability are formulated and discussed. Both models use the node-link formulation of anycast and unicast flows with some enhancements for addressing additional survivability requirements. The network is protected against a single link failure. The following models can be easily adapted to tackle other types of failures, such as a single node failure or a region failure [70, 98, 99].

**Dedicated Path Protection**

In DPP, two types of flow variables are used for each demand: $x_{ed}$ denoting whether working path of demand $d$ uses link $e$, and $b_{ed}$ denoting whether the working path of demand $d$ uses link $e$. In anycast demands, two types of DC node selection variables are also required, i.e., $z_{vd}$ defining whether DC node $v$ is selected as a working DC for demand $d$ and $w_{vd}$ defining whether DC node $v$ is selected as a backup DC for demand $d$.

---

**CON/AU/FA/DPP/Cost**

**sets**

| | |
|---|---|
| $V$ | nodes |
| $R$ | data center nodes |
| $E$ | links |
| $D$ | anycast demands |
| $\delta^+(v)$ | links leaving node $v$ |
| $\delta^-(v)$ | links entering node $v$ |
| $D$ | demands (anycast and unicast) |
| $D^{UN}$ | unicast demands |
| $D^{AN}$ | anycast demands |
| $D^{DS}$ | anycast downstream demands |
| $D^{US}$ | anycast upstream demands |

**constants**

$h_d$     volume of demand $d$
$c_e$     capacity of link $e$
$\zeta_e$     unit routing cost on link $e$
$s_d$     source node of demand $d$
$t_d$     destination node of demand $d$
$\tau(d)$     index of a demand associated with demand $d$. If $d$ is a downstream demand,
        then $\tau(d)$ must be an upstream demand and vice versa

**variables**

$x_{ed}$     $=1$, if demand $d$ uses link $e$ on working path; 0, otherwise (binary)
$b_{ed}$     $=1$, if demand $d$ uses link $e$ on backup path; 0, otherwise (binary)
$z_{vd}$     $=1$, if DC node $v$ is selected as a working DC for demand $d$; 0, otherwise
        (binary)
$w_{vd}$     $=1$, if DC node $v$ is selected as a backup DC node for demand $d$; 0, otherwise
        (binary)

**objective**

$$\text{minimize} \quad F = \sum_{e \in E} \sum_{d \in D} \zeta_e h_d (x_{ed} + b_{ed}) \tag{2.5.1a}$$

**constraints**

$$\sum_{e \in \delta^+(v)} x_{ed} - \sum_{e \in \delta^-(v)} x_{ed} = \begin{cases} +1 & \text{if } v = s_d \\ -1 & \text{if } v = t_d, \\ 0 & \text{otherwise} \end{cases} \quad v \in V, d \in D^{UN} \tag{2.5.1b}$$

$$\sum_{e \in \delta^+(v)} b_{ed} - \sum_{e \in \delta^-(v)} b_{ed} = \begin{cases} +1 & \text{if } v = s_d \\ -1 & \text{if } v = t_d, \\ 0 & \text{otherwise} \end{cases} \quad v \in V, d \in D^{UN} \tag{2.5.1c}$$

$$\sum_{e \in \delta^+(v)} x_{ed} - \sum_{e \in \delta^-(v)} x_{ed} = \begin{cases} +z_{vd} & \text{if } v \in R \\ -1 & \text{if } v = t_d \\ 0 & \text{otherwise} \end{cases} \quad v \in V, d \in D^{DS} \tag{2.5.1d}$$

$$\sum_{e \in \delta^+(v)} b_{ed} - \sum_{e \in \delta^-(v)} b_{ed} = \begin{cases} +w_{vd} & \text{if } v \in R \\ -1 & \text{if } v = t_d \\ 0 & \text{otherwise} \end{cases} \quad v \in V, d \in D^{DS} \tag{2.5.1e}$$

$$\sum_{e \in \delta^+(v)} x_{ed} - \sum_{e \in \delta^-(v)} x_{ed} = \begin{cases} +1 & \text{if } v = s_d \\ -z_{vd} & \text{if } v \in R \\ 0 & \text{otherwise} \end{cases} \quad v \in V, d \in D^{US} \tag{2.5.1f}$$

$$\sum_{e \in \delta^+(v)} b_{ed} - \sum_{e \in \delta^-(v)} b_{ed} = \begin{cases} +1 & \text{if } v = s_d \\ -w_{vd} & \text{if } v \in R \\ 0 & \text{otherwise} \end{cases} \quad v \in V, d \in D^{US} \tag{2.5.1g}$$

$$z_{vd} = z_{v\tau(d)}, \quad d \in D^{DS}, v \in R \tag{2.5.1h}$$

$$w_{vd} = w_{v\tau(d)}, \quad d \in D^{DS}, v \in R \tag{2.5.1i}$$

$$\sum_{v \in R} z_{vd} = 1, \quad d \in D^{AN} \tag{2.5.1j}$$

$$\sum_{v \in R} w_{vd} = 1, \quad d \in D^{AN} \tag{2.5.1k}$$

$$x_{ed} + b_{ed} \leq 1, \quad e \in E, d \in D \tag{2.5.1l}$$

$$\sum_{d \in D} h_d(x_{ed} + b_{ed}) \leq c_e, \quad e \in E. \tag{2.5.1m}$$

The objective (2.5.1a) is to find the routing of both working and backup paths for all anycast and unicast demands using the DPP approach and minimizing routing cost. Equations (2.5.1b) and (2.5.1c) define the unicast flow conservation constraints for working and backup paths, respectively. Equations (2.5.1d)–(2.5.1g) define the corresponding flow conservation constraints for working and backup paths of downstream and upstream anycast demands. Constraints (2.5.1h) and (2.5.1i) are in the model to ensure that working paths of two associated anycast demands $d$ and $\tau(d)$ use the same working and backup DC node, respectively. Equations (2.5.1j) and (2.5.1k) ensure that each anycast demand is assigned to exactly one working and backup DC node, respectively. Constraint (2.5.1l) expresses the DPP, i.e., it ensures that working and backup path are link disjoint for each demand. Finally, the inequality (2.5.1m) ensures the link capacity constraint taking into account flows of both working and backup paths.

The model (2.5.1a)–(2.5.1m) does not include a coupling between the working and backup DC nodes. Two cases are possible for each anycast demand: (i) working and backup DCs are located in different network nodes and (ii) working and backup DC nodes are located in the same network node. For ease of reference, the above model is referred to as the ADN (Any DC node) model. Some additional constraints on the selection of working and backup DC nodes are introduced and discussed below.

The disjoint DC node (DDN) scenario assumes that the working and backup DC nodes are disjoint for each anycast demand:

$$\sum_{v \in R} (z_{vd} + w_{vd}) \leq 1, \quad d \in D^{DS}. \tag{2.5.1n}$$

DDN model given by (2.5.1a)–(2.5.1n) provides protection against a single DC node failure, since it protects each anycast demand against any single DC node failure including the working DC node.

Let $\kappa_{vd}$ denote a binary constant which is 1 if $v$ is the nearest DC for anycast demand $d$ and 0 otherwise. To find the value of $\kappa_{vd}$, the shortest paths between the client node of anycast demand $d$ and each DC node $v \in R$ are calculated and the DC node with the lowest value is selected. The next model, known as the nearest

DC node (NDN), ensures that both working and backup DC nodes are located in the same network node, which is the closest to the client node of a particular anycast demand:

$$z_{vd} = w_{vd} = \kappa_{vd}, \quad d \in D^{DS}, v \in R. \tag{2.5.1o}$$

The main advantage of the NDN model given by (2.5.1a)–(2.5.1m) and (2.5.1o) is the fact that in many anycasting systems the client is assigned to the nearest DC node by default. A more comprehensive treatment of DPP of anycast flows and various working and backup DC node scenarios is given in [70, 98].

### Shared Backup Path Protection

The next model assumes the SBPP approach applied to protect the network against a single link failure [99]. Since the SBPP model is similar to the DPP model formulated as (2.5.1), we only present additional elements here. To control the sharing of capacity among backup paths three new variables are introduced. First, let binary variable $y_{deg}$ denote whether link $e$ is used for a backup path of demand $d$ in the event of link $g$ failure. Using variable $y_{deg}$, we are able to provide sharing of the backup capacity among demands for which working paths are failure disjoint. Next, variable $y_{eg}$ defines the spare capacity on link $e$ required in the event of link $g$ failure. Finally, variable $y_e$ denotes the maximum amount of spare capacity on link $e$ required in the event of a single link failure, and is calculated simply as the maximum value $y_{eg}$ considering all single link failures one by one.

------

### CON/AU/FA/SBPP/Cost

**variables (additional)**

$y_{deg}$   =1, if in the event of link $g$ failure link $e$ is used as a backup path of demand $d$; 0, otherwise

$y_{eg}$   spare capacity on link $e$ required in the event of a link $g$ failure (integer)

$y_e$   maximum spare capacity on link $e$ required in the event of a single link failure (integer)

**objective**

$$\text{minimize} \quad F = \sum_{e \in E} \sum_{d \in D} \zeta_e h_d x_{ed} + \sum_{e \in E} \zeta_e y_e \tag{2.5.2a}$$

**constraints** (2.5.1a)–(2.5.1k) **and**

$$x_{ed} + b_{gd} \leq 1 + y_{deg}, \quad e, g \in E : e \neq g, d \in D \tag{2.5.2b}$$

$$2y_{deg} \leq x_{ed} + b_{gd}, \quad e, g \in E : e \neq g, d \in D \tag{2.5.2c}$$

$$y_{eg} = \sum_{d \in D} h_d y_{deg}, \quad e, g \in E : e \neq g \tag{2.5.2d}$$

$$y_{eg} \leq y_e, \quad e, g \in E : e \neq g \tag{2.5.2e}$$

$$y_e + \sum_{d \in D} h_d x_{ed} \leq c_e, \quad e \in E. \tag{2.5.2f}$$

The objective function (2.5.2a) aims to minimize the network similarly to function (2.5.1a); however, the cost related to working paths is included directly in the objective function (first term), while the cost related to backup paths is limited to the cost of spare capacity (second term). When compared with the DPP, the SBPP model (2.5.2) includes the same constraints regarding the definition of flows and survivability requirements, and thus these constraints are not repeated in the formulation. There are five new constraints following from the SBPP approach. Constraint (2.5.2b) and (2.5.2c) are used to define variable $y_{deg}$. Constraint (2.5.2b) ensures that if both variables of the left-hand side are 1 (i.e., demand $d$ is affected by the failure of link $g$ and link $e$ is used by the backup path of demand $d$), variable $y_{deg}$ must be 1. In turn, constraint (2.5.2b) guarantees that if at least one of the variables $x_{ed}$ and $b_{gd}$ is 0, then $y_{deg}$ must also be set to 0. Equality (2.5.2d) directly defines variable $y_{eg}$ that denotes the spare capacity required for on link $e$ when link $g$ is broken. Constraint (2.5.2e) is used to find value of $y_e$ defined as the maximum amount of spare capacity on link $e$ taking into account all failure scenarios. Finally, inequality (2.5.2f) imposes the link capacity constraint, i.e., for each link $e \in E$ the flow allocated to working paths and the capacity left for backup paths cannot exceed the link capacity.

## 2.5.2  Numerical Results

This section presents and discusses results of numerical experiments reported in [99]. The main goal of the experiments was to compare the SBPP approach assuming sharing of the backup capacity against the DPP approach, where the backup capacity is not shared in the context of two working and backup DC node scenarios, namely, ADN and NDN. Numerical experiments were performed with the CPLEX solver [100] using optimization models (2.5.1) and (2.5.2) formulated in the previous section.

All simulations were run on the NSF network (Fig. A.4, Table A.3). All links were assigned a capacity equal to 40 Gb/s. Several scenarios referring to 2 and 3 DC nodes available in the network were evaluated. DCs were located at nodes with a relatively high value of the average node degree. To examine the influence of anycast traffic on the analyzed protection mechanisms, we define the *anycast ratio* (AR) parameter. Let $h^{Any}$ and $h^{Uni}$ denote the overall volume of all anycast and unicast demands, respectively. Next, let $h^{All} = h^{Any} + h^{Uni}$ be the overall demand in the network. The *AR* parameter is defined as the volume (capacity) of all anycast demands divided by the volume of all demands in the network, i.e., $AR = h^{Any}/h^{All}$. Eight scenarios of network load were examined in terms of the AR parameter, namely, 0, 10, 20, ..., 80 %. In each case, three sets of demands were generated at random, giving 24 different demand sets in total. The number of unicast demands in each set was selected in the range 7–44, while the corresponding number of anycast demands was in the range 8–28. The demand volume was selected from the range 1–9 Gb/s in order to obtain the particular anycast proportion parameter value. For each set of demands, experiments considered two scenarios (2 and 3 DC nodes) and four models (SBPP-ADN, SBPP-NDN, DPP-ADN, DPP-NDN). This yields the overall number of 192 distinct experiments.

To report the results showing the comparison between the SBPP and DPP approaches, we use a ratio calculated as $F^{SBPP}/F^{DPP}$, where $F^{SBPP}$ and $F^{DPP}$ denote the value of a given performance metric obtained for SBPP and DPP models, respectively. It should be noted that if the value of this ratio is lower than 1, then SBPP provides a lower value of a particular performance metric compared with DPP.

Figure 2.2 shows the $cost^{SBPP}/cost^{DPP}$ ratio as a function of the anycast ratio parameter. In turn, Table 2.1, presents the average values of the ratio between SBPP and DPP (averaged over all cases of the analyzed anycast traffic proportion) for the following performance metrics: cost (objective function of both models), capacity utilization (ratio of network capacity allocated to demands), working and backup path length in km, and hop count for both unicast and anycast demands.

The average value of the $cost^{SBPP}/cost^{DPP}$ ratio (taking into account all experiments) is 0.64, which means that SBPP outperforms DPP by 36 %. The detailed analysis of the results presented in Fig. 2.2 indicates that the ratio between both models becomes less evident with the increase of the anycast ratio. This can be explained by the fact that anycast demands must have one of the end nodes located at the DC node. Let us recall that the capacity sharing used in the SBPP approach assumes that the backup capacity can be shared by demands that have failure-disjoint working paths. However, the majority of anycast traffic concentrates on links adja-

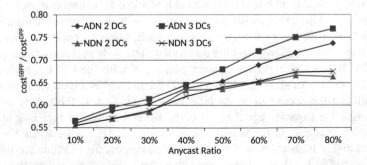

**Fig. 2.2**  Average cost ratio between SBPP and DPP models as a function of the anycast ratio

**Table 2.1**  Average ratio between SBPP and DPP approaches for various performance metrics

| Number of DC nodes | 2 | 3 | 2 | 3 |
|---|---|---|---|---|
| DC node scenarios | ADN | ADN | NDN | NDN |
| Cost | 0.65 | 0.67 | 0.62 | 0.62 |
| Capacity utilization | 0.60 | 0.61 | 0.56 | 0.57 |
| Unicast working path length | 1.01 | 1.01 | 1.01 | 1.05 |
| Unicast backup path length | 1.71 | 1.68 | 1.78 | 1.86 |
| Unicast working path hops | 1.01 | 1.01 | 0.99 | 1.03 |
| Unicast backup path hops | 1.49 | 1.43 | 1.53 | 1.55 |
| Anycast working path length | 1.01 | 1.06 | 1.01 | 1.03 |
| Anycast backup path length | 2.00 | 2.09 | 1.79 | 1.91 |
| Anycast working path hops | 1.00 | 0.98 | 1.01 | 1.02 |
| Anycast backup path hops | 1.54 | 1.60 | 1.43 | 1.56 |

cent to DC nodes. Accordingly, there are fewer opportunities to share the backup capacity in the event of a failure of a link adjacent to a DC node.

Moreover—as shown in Fig. 2.2—the increase of the number of DC nodes (here from 2 to 3) also reduces the gap between SBPP and DPP. As DC nodes are spread over the network, anycast demands use shorter backup paths than unicast demands in terms of the hop count. Accordingly, when the number of DC nodes increases, the average path hop count decreases, which means that once again there are fewer possibilities to share the backup capacity.

As shown in Table 2.1, the DC node location scenario (ADN vs. NDN) does not have a significant impact on the SBPP/DPP ratio. Nevertheless, the ADN approach provides lower values of the cost percentage difference (equivalent to higher values of the SBPP/DPP ratio) compared to the NDN scenario. This is because the ADN scenario is flexible and benefits more from switching anycast demands to another DC node after a network failure. However, in the NDN approach, both working and backup paths of anycast demands are assigned to the same, closest DC node. Therefore, in this case anycast traffic performs similarly to unicast traffic; in consequence—as explained above in the context of cost values (see Fig. 2.2)—the difference between the SBPP and DPP approaches increases.

Another interesting observation is that the capacity utilization metric returns a performance similar to the cost objective, i.e., as the anycast ratio parameter and the number of DC nodes increase, the difference between SBPP and DPP decreases. Note that on average, the SBPP model requires 42 % less capacity than DPP. Moreover, the results show that the SBPP model yields significantly longer backup paths compared to the DPP model. This is because the objective function (2.5.2a) includes the cost of the shared backup capacity, not the cost of every backup path as in the DPP model. Hence, in the SBPP model backup paths are selected in order to share the backup capacity, even if they become relatively long. More information on SBPP protection of anycast and unicast flows and additional results can be found in [99].

## 2.6    Protection Design with Anycast and Unicast Flows

This section continues the discussion on the protection of connection-oriented networks with anycast flows and presents a network design problem. In contrast to the
previous section where the node-link notation was used to model network flows, here
we use the link-path approach. Moreover, a general concept of *failure state* (situation) is applied to model network failures [4]. Simply put, a failure state models
any failure that can occur in the network and it is specified by the availability status
of the network elements (links and nodes). The DPP approach is as the protection
method; the optimization and heuristic can be modified easily to also address the
SBPP approach.

### 2.6.1    Formulation

The optimization model involves the joint optimization of anycast and unicast flows
protected by the DPP method [101–103]. The objective is to minimize the cost of
the network essential to fully protect all flows (unicast and anycast). As in [4], the
notion of failure states is used. Let set $S$ contain all failure states considered in the
network including the special state $s_0$ denoting the normal state of the network when
all network elements are available. Each failure state $s \in S$ is defined by a vector of
binary link availability coefficients $\alpha_s = (\alpha_{1s}, \alpha_{2s}, \ldots, \alpha_{|E|s})$, i.e., in a given failure
state a particular link $e$ is either fully available ($\alpha_{es} = 1$) or it is completely broken
($\alpha_{es} = 0$). This approach, makes it easy to model various failure scenarios. For
instance, if a single link failure scenario is studied, then set $S$ includes $|E| + 1$ failure
states, $s_0$ denotes the normal state with all links available ($\alpha_{es_0} = 1$ for each $e \in E$)
and $s_e$ denotes the situation when link $e$ is broken ($\alpha_{es_e} = 0$).

To provide path protection for each demand $d \in D$, set $P(d)$ including candidate
pairs of the failure situation disjoint paths is given. Each pair of paths is denoted
as $(w_{dp}, b_{dp})$, where $w_{dp}$ refers to the working path and $b_{dp}$ refers to the backup path.
Constant $\delta_{edp}$ defines working path $w_{dp}$ and is 1 if link $e$ belongs to $w_{dp}$. In turn,
constant $\beta_{edp}$ describes the backup path $b_{dp}$ and is 1 if link $e$ belongs to $b_{dp}$. The
working path $w_{dp}$ (used in the normal, failure-free network state) is protected by
the dedicated backup path $b_{dp}$, which is failure-situation disjoint with the working
path $w_{dp}$. This approach ensures that when a particular working path is broken, its
backup path must be available for each failure state $s \in S$. The binary availability
coefficient $\theta_{dps}$ indicates whether the working path $w_{dp}$ is affected by a failure $s$. In
particular, $\theta_{dps} = \prod_{e \in w_{dp}} \alpha_{es}$ since if at least one link $e$ of path $w_{dp}$ is broken in state
$s$ ($\alpha_{es} = 0$), then path $w_{dp}$ is not available ($\theta_{dps} = 0$) and must be restored using
backup path $b_{dp}$.

Because associated anycast connections $d$ and $\tau(d)$ must be connected to the
same DC, there are two possible cases when a failure affects demand $d$ and/or $\tau(d)$.
Firstly, in failure state $s$ only one demand of the pair $(d, \tau(d))$ is broken. Let $d$ denote

the broken connection, i.e., working path $w_{dp}$ selected for demand $d$ containing a link broken in state $s$. Thus, backup path $b_{dp}$ must use the same DC node as the DC included in working path $w_{\tau(d)q}$ of demand $\tau(d)$. In the second case, both associated anycast demands $(d, \tau(d))$ fail due to failure $s$ (i.e., both working paths $w_{dp}$ and $w_{\tau(d)q}$ are broken in situation $s$). Then, to restore demands $d$ and $\tau(d)$, backup paths $b_{dp}$ and $b_{\tau(d)q}$ can use a different DC to the node included in corresponding working paths $w_{dp}$ and $w_{\tau(d)q}$, respectively. This approach following directly from the anycast paradigm should make it possible to reduce network cost, and it is analogous to the backup DC concept described in Sect. 2.5.

---

## CON/AU/ND/DPP/Cost/Link-path

**sets**

$E$      links

$D$      demands (anycast, unicast)

$D^{DS}$      anycast downstream demands

$P(d)$      pairs of failure disjoint candidate paths for flows realizing demand $d$. If $d$ is a unicast demand, the candidate path connects end nodes of the demand. If $d$ is an anycast upstream demand, the candidate path connects client node and DC node. If $d$ is a downstream demand, the candidate path connects the DC node and the client node

$S$      failure states (situations)

**constants**

$\delta_{edp}$      $=1$, if link $e$ belongs to working path $w_{dp}$ realizing demand $d$; 0, otherwise

$\beta_{edp}$      $=1$, if link $e$ belongs to backup path $b_{dp}$ realizing demand $d$; 0, otherwise

$h_d$      volume of demand $d$

$\xi_e$      unit cost of link $e$

$M$      size of the link capacity module

$\tau(d)$      index of a demand associated with demand $d$. If $d$ is a downstream demand, then $\tau(d)$ must be an upstream demand and vice versa

$\theta_{dps}$      binary availability coefficient of working path $w_{dp}$ in state $s$

$o(p)$      origin node of working path $p$

$t(p)$      destination node of working path $p$

$\bar{o}(p)$      origin node of backup path $p$

$\bar{t}(p)$      destination node of backup path $p$

**variables**

$x_{dp}$      $=1$, if pair of paths $(w_{dp}, b_{dp})$ is used to realize demand $d$; 0, otherwise (binary)

$y_e$      capacity of link $e$ as the number of capacity modules (integer)

**objective**

$$\text{minimize} \quad F = \sum_{e \in E} \xi_e y_e \tag{2.6.1a}$$

**constraints**

$$\sum_{p \in P(d)} x_{dp} = 1, \quad d \in D \tag{2.6.1b}$$

$$\sum_{d \in D} \sum_{p \in P(d)} x_{dp} h_d (\delta_{edp} \theta_{dps} + \beta_{edp}(1 - \theta_{dps})) \leq M y_e, \quad e \in E, s \in S \tag{2.6.1c}$$

$$\sum_{p \in P(d)} x_{dp} o(p) = \sum_{p \in P(\tau(d))} x_{\tau(d)p} t(p), \quad d \in D^{DS} \tag{2.6.1d}$$

$$\sum_{p \in P(d)} x_{dp} \bar{o}(p) = \sum_{p \in P(\tau(d))} x_{\tau(d)p} \bar{t}(p), \quad d \in D^{DS}. \tag{2.6.1e}$$

The objective (2.6.1a) is to minimize the cost of network capacity required for working flows and for protection against all failure scenarios included in set $S$. Condition (2.6.1b) is in the model to guarantee that a single pair of paths (working and backup) is selected to realize demand $d$. Inequality (2.6.1c) controls the link capacity constraint. More precisely, for each link $e \in E$ and each possible failure state $s \in S$, the allocated link capacity (right-hand side of (2.6.1c)) must exceed the flow on the link (left-hand side of (2.6.1c)). Note that if working path $w_{dp}$ is not available (i.e., at least one link belonging to $w_{dp}$ is broken is state $s$ and $\theta_{dps} = 0$), then backup path $b_{dp}$ is activated. Therefore, the left-hand side of the link capacity constraint (2.6.1c), includes both working flows transmitted in the event of failure $s$ (term $\sum_{d \in D} \sum_{p \in P(d)} x_{dp} h_d \delta_{edp} \theta_{dps}$) and backup flows activated after failure $s$ (term $\sum_{d \in D} \sum_{p \in P(d)} x_{dp} h_d \beta_{edp}(1 - \theta_{dps})$). Constraint (2.6.1c) assumes the stub-release scenario, i.e., the flow of the broken working path is released in the network and this free capacity can be used for restoration [4]. Constraints (2.6.1d) and (2.6.1e) ensure that working and backup paths of two associated anycast demands connect the same pair of nodes, respectively.

Note that in the problem (2.6.1a)–(2.6.1e) the DC node of an anycast demand can be changed due to the restoration process (backup DC concept). To remove this option, the following constraint must be added to the model:

$$\sum_{p \in P(d)} x_{dp} o(p) = \sum_{p \in P(\tau(d))} x_{\tau(d)p} \bar{o}(p), \quad d \in D^{DS}. \tag{2.6.1f}$$

More discussion on joint optimization of anycast and unicast flows with additional survivability constraints can be found in [101–103].

## 2.6.2 Cut Inequalities

Cut inequalities are used to facilitate the optimization process. More specifically, cut inequalities enable the branching phase of the branch-and-cut algorithm to use additional information included in the cuts in calculations of more effective bounds. A cut-and-branch variant of the branch-and-cut algorithm assumes that cut inequalities are added in the root node of the solution tree only. In consequence, all generated cuts are valid throughout the whole solution tree [104]. The main advantage of this approach is that the cut inequalities are calculated once only, and more time can be spent generating relatively tight bounds compared to the classical scenario, when cuts are generated in each node of the solution tree. For more general information on applications of branch-and-cut algorithms with cut inequalities to various network optimization problems see [4, 104–115].

The first cut inequality proposed for problem (2.6.1) is a version of a partition inequality obtained by separating network nodes into two subsets and analyzing the flow and capacity of links included in the cut between these two sets [108]. The key modifications of the below approach—compared to the classical partition inequality—follow from two specific features of the problem (2.6.1), i.e., anycasting and survivability.

Set $V$ contains all nodes in the considered network. Let $R$ denote the set embracing all nodes that host a data center and let $C = V \backslash R$ denote the set of all other network nodes. For ease of notation, let $o(e)$ and $t(e)$ be the origin node and destination node of link $e$, respectively. Analogously, let $o(d)$ and $t(d)$ denote the origin node and destination node of demand $d$, respectively. Let us recall that in the context of anycast demands, the origin (destination) node of upstream (downstream) connection is the client node. The destination (origin) nodes of upstream (downstream) demand are selected from DC nodes included in set $R$.

Let $\eta(W)$ denote a graph cut induced by $W \subseteq V$, i.e., $\eta(W)$ includes all links with the origin node in set $W$ and the destination node in its complement set $(V \backslash W)$. Note that links are directed, i.e., in the case of $\eta(W)$ links originate in $W$. Moreover, $h(W, W')$ is the overall flow related to all demands that have the origin node included in $W$ and the destination node included in $W'$:

$$h(W, W') = \sum_{d:o(d)\in W, t(d)\in W'} h_d. \tag{2.6.2}$$

For instance, $h(R, C)$ defines the overall flow from client nodes to DC nodes taking into account both unicast and anycast demands. To formulate the DC partition inequality, the following inequalities are defined:

$$\sum_{e\in\eta(R)} My_e \geq h(R, C) \tag{2.6.3a}$$

$$\sum_{e\in\eta(C)} My_e \geq h(C, R). \tag{2.6.3b}$$

Inequality (2.6.3a) is explained as follows. The left-hand side of (2.6.3a) is the overall link capacity necessary to be installed on links included in $\eta(R)$ to satisfy traffic coming from nodes included in $R$ to nodes included in $C$ (right-hand side), i.e., cut $\eta(R)$ must carry the whole anycast downstream traffic as well as unicast traffic related to demands that originate in nodes from $R$. Similarly, inequality (2.6.3b) is formulated for traffic in the opposite direction. Using the mixed-integer rounding (MIR) approach [108, 111], it is possible to make (2.6.3a) and (2.6.3b) stronger as follows:

$$\sum_{e\in\eta(R)} y_e \geq \lceil \frac{h(R,C)}{M} \rceil \tag{2.6.4a}$$

$$\sum_{e\in\eta(C)} y_e \geq \lceil \frac{h(C,R)}{M} \rceil. \tag{2.6.4b}$$

Finally, network survivability can be taken into account to provide more effective cut inequalities. In a nutshell, assuming a single link failure scenario and using a single backup path to protect the network in 100 %, the left-hand side of (2.6.4a) and (2.6.4b) can be modified by removing from a particular cut one by one a single link. At the same time, all other remaining links must provide enough capacity to carry the whole traffic between sets $R$ and $C$ in both directions:

$$\sum_{e\in\eta(R)\backslash e'} y_e \geq \lceil \frac{h(R,C)}{M} \rceil, \quad e' \in \eta(R) \tag{2.6.5a}$$

$$\sum_{e\in\eta(C)\backslash e'} y_e \geq \lceil \frac{h(C,R)}{M} \rceil, \quad e' \in \eta(C). \tag{2.6.5b}$$

Constraints (2.6.5a) and (2.6.5b) are the final versions of the partition inequality.

To formulate the next inequality, the following notation is introduced. First recall that $\delta^+(v)$ is defined as a set of links that originate at node $v$ and $\delta^-(v)$ is a set of all links that terminate in node $v$. Moreover, let $h^+(v) = \sum_{d:o(d)=v} h_d$ and let $h^-(v) = \sum_{d:t(d)=v} h_d$ denote the overall demand flow leaving and entering node $v$, respectively. This cut inequality is again based on the concept of partition inequality and the MIR approach. In fact, the outgoing and incoming demands are analyzed for each node $v \in V$. Since a single backup path protection method is used, similarly to (2.6.5a) and (2.6.5b), the capacity of links in the outgoing (incoming) node $v$ excluding any single link must be sufficient to satisfy the outgoing (incoming) overall demand flow of node $v$. Therefore, the node partition inequality can be formulated as follows:

$$\sum_{e\in\delta^+(v)\backslash e'} y_e \geq \lceil \frac{h^+(v)}{M} \rceil, \quad v \in V, e' \in \delta^+(v) \tag{2.6.6a}$$

$$\sum_{e \in \delta^-(v) \setminus e'} y_e \geq \lceil \frac{h^-(v)}{M} \rceil, \quad v \in V, e' \in \delta^-(v). \tag{2.6.6b}$$

Finally, a rather obvious cut inequality involves using a solution of problem (2.6.1) provided by the heuristic algorithm as an upper bound of the objective function (2.6.1a).

Results showing the effectiveness of cut inequalities proposed above are presented and discussed in [102]. In turn, heuristic algorithms solving problem (2.6.1) can be found in [102, 103].

## 2.7 p-Cycle Protection of Anycast Flows

Sections 2.5 and 2.6 focused on path protection methods. Further interesting concepts that can be used to provide network survivability are methods based on ring topology, e.g., bidirectional line switched rings (BLSRs) or unidirectional path-switched rings (UPSRs). The main advantage of ring-based survivability is the fact that rings use a simple switching mechanism which permits very fast restoration. However, the key disadvantage of ring-based methods is the high consumption of spare capacity required to provide survivability. i.e., a redundancy of at least 100 % is needed. Moreover, in recent years there has been a trend to shift from ring-based to mesh-based networks, therefore the use of protection rings is on the decline [5, 6, 116, 117].

An interesting survivability approach that combines the relatively short restoration time offered in ring-based protection with efficient usage of network capacity provided in mesh protection is the concept of *p-cycles* proposed in the late 1990s. In particular, a p-cycle can be defined as a preconfigured ring created in a mesh network. The main advantage of the p-cycle concept is that as well as providing protection for all on-cycle links (as in traditional ring-based protection), a p-cycle can also protect *straddling* links that are not on the p-cycle but both their end nodes are included in the p-cycle. This additional protection improves the capacity efficiency of p-cycles over traditional ring-based schemes [6, 117, 118].

The majority of earlier research into p-cycles focused on unicast flows, e.g., [6, 118–129]. Moreover, some papers also considered p-cycle protection of multicast flows, e.g., [130–141]. This section presents a novel p-cycle known as the Anycast-Protecting p-Cycle (APpC) proposed to protect anycast flows in connection-oriented networks [142–145].

The APpC approach is used to protect anycast flows related to streaming services provided in backbone networks. More precisely, a set of streaming servers (data centers) is located in some nodes of the network and each server provides the same signal with the same rate, e.g., IPTV service. End users requesting the streaming service are connected to backbone network nodes by access networks. However, we focus on backbone network optimization only, therefore all users connected to the

same backbone network are aggregated and represented as a single anycast client
that must receive the streaming service. In other words, if at least one user connected
to a particular backbone network node requests the streaming service, this node must
be provided with a connection to a streaming server. Transmission in a failure-state
of the network is provided by anycast connections, while protection against single
link failures is provided by p-cycles. We start by describing the concept of Anycast-
Protecting p-Cycles, followed by formulating and discussing the optimization model
and presenting and analyzing numerical results.

## 2.7.1   Anycast-Protecting p-Cycles

The concept of Anycast-Protecting p-Cycles (APpC) was first proposed in [142] and
applied to various optimization problems with anycast flows [143–146].

A classical p-cycle supports the protection of physical links that are on-cycle links
or straddling links. Therefore, if all network links are to be protected, the p-cycles
must be configured such that each network link is an on-cycle link or a straddling link
of at least one p-cycle established in the network [6]. An illustrative example is shown
in Fig. 2.3. The network consists of 10 nodes. There are two streaming servers $r_1$ and
$r_2$ located in nodes $a$ and $c$, respectively. The anycast client $c_1$ is placed at node $j$. A
working path $(a, e, h, j)$ (solid bold line) is established to provide a streaming service
for client $c_1$, which means that client $c_1$ uses server $r_1$. To provide 100 % protection
of the working path $p_1 = (a, e, h, j)$, two classical p-cycles are established (dotted
lines): $q_1 = (a, d, e, b, a)$ and $q_2 = (e, f, i, j, h, g, e)$. More precisely, link $(a, e)$
is a straddling link of p-cycle $q_1$, link $(e, h)$ is a straddling link of p-cycle $q_2$ and
link $(h, j)$ is an on-cycle link of p-cycle $q_2$.

The authors of [124] propose a new idea of *flow* p-cycles. The key innovation
is that protection is provided on the level of flow paths instead of physical links as
in the case of classical p-cycles. Accordingly, instead of a single physical link, a

**Fig. 2.3** Example of
classical p-cycles

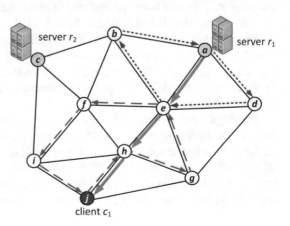

client $c_1$

**Fig. 2.4** Example of an
anycast-protection p-cycles

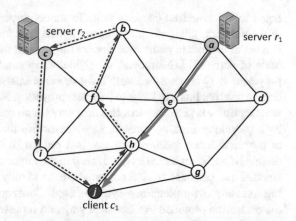

path segment can be considered to straddle the cycle, i.e., two end nodes of the path
segment are included in the cycle to enable protection of the whole path segment. This
results in a more efficient allocation of spare capacity and leads to a decrease in the
capacity usage, while still providing 100 % protection. Moreover, [147] introduces
an extension of flow p-cycles known as *Failure-Independent Path-Protection* (FIPP)
p-cycles. The main assumption behind FIPP p-cycles is that failure-independent paths
are grouped and only one p-cycle is needed to protect each group. This approach
results in an even more efficient allocation of spare capacity in comparison to flow
p-cycles.

The anycast-protecting p-cycle approach proposed in this section combines the
concept of classical and flow p-cycles with additional anycasting properties. We use
the fact that each data center (streaming server) provides the same signal. Therefore,
if a particular p-cycle includes one streaming server, it can protect a working path
of a client connected to another server. This is described using a simple example
shown in Fig. 2.4. The general assumptions are the same as in Fig. 2.3. To protect the
working path established for client $c_1$, only one anycast-protecting p-cycle is required
$q_3 = (b, c, i, j, h, f, b)$. It should be stressed that p-cycle $q_3$ includes streaming server
$r_2$ located at node $c$. Note that only link $(h, j)$ is protected in a classical way, i.e., as an
on-cycle link of p-cycle $q_3$. Two other links of the working path $p_1$, $(a, e)$ and $(e, h)$,
are protected due to the fact that p-cycle $q_3$ includes server $r_2$ and thus if any of these
links fail (meaning that client $c_1$ is disconnected form server $r_1$ located at node $a$),
client $c_1$ can still can receive the streaming signal using p-cycle $q_3$ and server $r_2$.

### 2.7.2   Formulation

Since only streaming servers provide transmission to anycast clients, only downstream anycast demands are considered. Moreover, because all streaming servers
provide the same signal with the same bit-rate, each demand has the same volume

equal to 1 normalized capacity unit. To model anycast flows in a failure-free state of the network, link-path modeling is applied, and for each anycast demand $d$ there is a set of candidate paths $P(d)$ connecting one of the streaming servers and client node of demand $d$. Moreover, set $Q$ including candidate p-cycles is given. Each p-cycle $q \in Q$ is described with a set of constants. Firstly, let binary constant $\beta_{eq}$ denote whether link $e$ belongs to (creates) p-cycle $q$. Next, binary constant $\gamma_{eq}$ defines whether link $e$ is protected in a classical way (as an on-cycle link or a straddling link) by a p-cycle $q$. Finally, constant $\omega_{eqdp}$ describes the anycast-protection properties of p-cycles. More specifically, $\omega_{eqdp}$ is 1 only if link $e$ belongs to path $p$ realizing demand $d$ and is protected by a p-cycle $q$ in the anycast-protection mode. It should be stressed that constant $\omega_{eqdp}$ is activated (set to 1) only if additional protection resulting from anycast-protection p-cycles is used. Consequently, $\omega_{eqdp}$ does not duplicate the protection provided in a classical way and represented by constant $\gamma_{eq}$.

The routing of working paths is controlled by the binary variable $x_{dp}$. Link capacity expressed in normalized units is denoted by the integer variable $y_e$. The selection of p-cycles required to fully protect the network is described by four types of variables. Firstly, binary variable $z_{eqdp}$ denotes whether p-cycle $q$ is used to protect path $p$ realizing demand $d$ in the event of link $e$ failure. Secondly, the binary variable $z_{eqd}$ indicates whether p-cycle $q$ is applied to protect demand $d$ in the event of link $e$ failure. Thirdly, the integer variable $z_{eq}$ represents the number of required copies of p-cycle $q$ in the event of link $e$ failure. Finally, the integer variable $z_q$ is the maximum value over $z_{eq}$ for each network failure, and thus it denotes the total number of required copies of p-cycle $q$.

---

**CON/A/ND/PCycle/Cost**

**sets**

$E$        links
$D$        anycast demands (downstream)
$P(d)$    candidate paths for flows realizing demand $d$, the candidate path originates
           at the DC node and terminates at the client node
$Q$        candidate p-cycles

**constants**

$\delta_{edp}$   =1, if link $e$ belongs to path $p$ realizing demand $d$; 0, otherwise
$\beta_{eq}$    =1, if link $e$ belongs to p-cycle $q$; 0, otherwise
$\gamma_{eq}$    =1, if link $e$ is protected in a classical way (as an on-cycle link or a straddling
           link) by a p-cycle $q$; 0, otherwise
$\omega_{eqdp}$   =1, if link $e$ belongs to path $p$ realizing demand $d$ and it is protected by
           p-cycle $q$ in the anycast-protection mode, but link $e$ is not protected in the
           classical mode by p-cycle $q$ ($\gamma_{eq} = 0$); 0, otherwise
$\xi_e$        unit cost of link $e$

**variables**

$x_{dp}$    =1, if path $p$ is used to realize demand $d$; 0, otherwise (binary)

$y_e$    capacity of link $e$ as the number of normalized units $M$ (integer)

$z_{eqdp}$    =1, if p-cycle $q$ is used to protect path $p$ realizing demand $d$ in the event of link $e$ failure; 0, otherwise (binary)

$z_{eqd}$    =1, if p-cycle $q$ is used to protect of demand $d$ in the event of link $e$ failure; 0, otherwise (binary)

$z_{eq}$    number of required copies of p-cycle $q$ in the event of link $e$ failure (integer)

$z_q$    total number of required copies of p-cycle $q$ (integer)

**objective**

$$\text{minimize} \quad F = \sum_{e \in E} \xi_e y_e \tag{2.7.1a}$$

**constraints**

$$\sum_{p \in P(d)} x_{dp} = 1, \quad d \in D \tag{2.7.1b}$$

$$\sum_{q \in Q} (\gamma_{eq} + \omega_{eqdp}) z_{eqdp} \geq \delta_{edp} x_{dp}, \quad d \in D, p \in P(d), e \in E \tag{2.7.1c}$$

$$z_{eqd} \geq z_{eqdp}, \quad d \in D, p \in P(d), q \in Q, e \in E \tag{2.7.1d}$$

$$z_{eq} = \sum_{d \in D} z_{eqd}, \quad q \in Q, e \in E \tag{2.7.1e}$$

$$z_q \geq z_{eq}, \quad q \in Q, e \in E \tag{2.7.1f}$$

$$\sum_{d \in D} \sum_{p \in P(d)} \delta_{edp} x_{dp} + \sum_{q \in Q} \beta_{eq} y_q \leq y_e, \quad e \in E. \tag{2.7.1g}$$

The goal of the optimization described in (2.7.1a) is to minimize the cost of the capacity required in the network to establish working paths for every anycast demand and to provide 100 % protection using of p-cycles. Equation (2.7.1b) ensures that there is exactly one routing path chosen for each demand. Constraint (2.7.1c) states that if link $e$ is used to realize demand $d$ ($\delta_{edp} = 1$ and $x_{dp} = 1$), then there must be protection against link $e$ failure by at least one p-cycle $q$, either in a classical way ($\gamma_{eq} = 1$) or in the anycast-protecting mode ($\omega_{eqdp} = 1$). Conditions (2.7.1d) and (2.7.1e) define variables $z_{eqd}$ and $z_{eq}$, respectively. Inequality (2.7.1f) states that variable $z_q$ is the maximum of $z_{eq}$ variables taking into consideration every single link failure. The last inequality (2.7.1g) is the link capacity constraint. Note that the first term on the left-hand side of (2.7.1g) denotes the working capacity required on link $e$ to realize all anycast demands, while the second term represents the spare capacity needed for p-cycles.

Model (2.7.1) can be modified so that only classical p-cycles are used by removing constant $\omega_{eqdp}$ from the left-hand side of constraint (2.7.1c). Furthermore, if routing of anycast demands (working paths) is fixed and values of variables $x_{dp}$ are given in advance as constants, model (2.7.1) can be applied to optimize spare capacity usage only. For more details see [142–146].

Moreover, note that an idea similar to anycast-protection of p-cycles can be applied to protect multicast flows as shown in [132, 133, 146]. Let us recall, that in anycast-protection of p-cycles, alternative data (signal) sources were accessible by using another DC included in a particular p-cycle. Considering multicast flows, alternative signal sources are found within the current multicast tree used for streaming. More specifically, if a p-cycle includes any network node connected to the tree and not affected by the network failure (i.e., a node not disconnected from the root node), such a node can provide the signal to nodes of the multicast tree affected by the failure and disconnected from the root node.

## 2.7.3   Numerical Results

To obtain optimal results for the CON/A/ND/PCycle/Cost problem (2.7.1), the Gurobi solver [148] was applied. In [142–146] two heuristic methods of solving several optimization problems with anycast-protecting p-cycles were proposed and analyzed: a greedy algorithm Feedback Functional Efficiency Ratio Algorithm ($F^2$ERH) and the Simulated Annealing (SA) approach. However, since the Gurobi solver provided optimal results in a relatively short execution time, here we focus only on the optimal results. The main goal of the numerical result experiments was to compare the performance of anycast-protecting p-cycles and classical p-cycles.

The following four network topologies included in the SNDlib library [149] were tested: cost266 (37 nodes and 114 links), germany50 (50 nodes and 176 links), giul39 (39 nodes and 172 links), and pioro40 (40 nodes and 178 links). Test scenarios (problem instances) were created at random according to several unique parameters: number and location of streaming servers (data centers), number and location of anycast clients, candidate p-cycles. For working flows three candidate paths using the k-shortest path algorithm were generated. We chose this number of candidate paths to find a good trade-off between the overall problem size resulting directly from the number of candidate paths and precision of the model in terms of routing diversity.

In order to generate candidate p-cycles, two algorithms (p-cycle generators) were used: Straddling Link Algorithm (SLA) known as the Straddling Span Algorithm (SSA) [6, 150] and Expand [151]. We used values 5, 10, 15, 20 and 25 of the p-cycle maximum hop limit to generate candidate p-cycles. The reason for limiting the maximum hop limit for p-cycles is as follows. The length of a p-cycle estimates the transmission delay experienced in a particular p-cycle used to protect the working flows. Minimizing the transmission delay is an important Quality of Service (QoS) requirement expected in networks, especially when multimedia or real-time data is transmitted. Instead of adding an extra constraint to the optimization model, the p-cycle length (number of hops)—and in consequence—transmission delay can be addressed during the phase of generating candidate p-cycles. This approach facilitates solving the optimization model using branch-and-bound algorithms. Moreover, without limiting the p-cycle hop count, some generators yield very high numbers of candidate p-cycles, significantly increasing the time needed to solve the model in an optimal way.

Table 2.2 shows the average percentage gain in the network cost of using anycast-protecting p-cycles (APpC) instead of classical p-cycles (CpC). The results are presented for all tested networks and two cases: spare capacity optimization and joint working and backup capacity optimization. The results are averaged over 75 different tests assuming 20 clients, 4 DCs and the Expand generator with 10 hops limit. It is clear that using classical p-cycles is between 15.2 and 19.3 % more expensive on average than using anycast-protecting p-cycles.

Tables 2.3 and 2.4 show more detailed results using the cost266 network for spare capacity and joint working and spare capacity optimization problems, respectively. Anycast-protecting p-cycles and classical p-cycles are compared in terms of the following performance metrics: average network cost (columns 3–5), average p-cycle length (in hops) selected in the optimization (columns 6–7) and average number of selected p-cycles (columns 8–9). The results are presented separately for seven cases regarding p-cycle generator scenarios (rows): the SLA generator with hop limit 5

**Table 2.2** Anycast-protecting p-cycles versus classical p-cycles—average percentage gap of the network cost for 20 clients, 4 DCs and expand generator with 10 hops path limit

| Network | Spare capacity (%) | Joint capacity (%) |
|---|---|---|
| cost266 | 18.1 | 18.0 |
| germany50 | 18.7 | 17.1 |
| giul39 | 16.8 | 15.2 |
| pioro40 | 19.3 | 18.6 |

**Table 2.3** Anycast-protecting p-cycles versus classical p-cycles for spare capacity optimization—various parameters obtained for cost266 network

| p-cycles | | Av. network cost | | | Av. p-cycle length | | Av. number of p-cycles | |
|---|---|---|---|---|---|---|---|---|
| Generator | Hop limit | CpC | APpC | Gap (%) | CpC | APpC | CpC | APpC |
| SLA | 5 | 606 | 573 | 7.4 | 3.4 | 3.4 | 23.2 | 21.7 |
| SLA | 10 | 492 | 450 | 10.8 | 7.3 | 7.5 | 13.7 | 12.5 |
| Expand | 5 | 591 | 559 | 7.9 | 3.9 | 4.0 | 22.7 | 21.2 |
| Expand | 10 | 536 | 473 | 14.8 | 8.4 | 8.6 | 11.6 | 10.1 |
| Expand | 15 | 501 | 446 | 13.2 | 12.4 | 12.5 | 9.5 | 8.5 |
| Expand | 20 | 493 | 438 | 13.5 | 17.9 | 18.0 | 9.0 | 8.1 |
| Expand | 25 | 493 | 442 | 11.8 | 20.3 | 20.6 | 8.9 | 8.1 |

**Table 2.4**  Anycast-protecting p-cycles versus classical p-cycles for joint working and spare capacity optimization—different parameters obtained for cost266 network

| p-cycles | | Av. network cost | | | Av. p-cycle length | | Av. number of p-cycles | |
|----------|-----------|-----|------|---------|-----|------|-----|------|
| Generator | Hop limit | CpC | APpC | Gap (%) | CpC | APpC | CpC | APpC |
| SLA | 5 | 761 | 689 | 11.7 | 3.7 | 3.4 | 18.0 | 22.5 |
| SLA | 10 | 641 | 576 | 12.1 | 8.0 | 7.5 | 10.0 | 12.6 |
| Expand | 5 | 748 | 674 | 12.2 | 4.0 | 3.7 | 17.3 | 21.9 |
| Expand | 10 | 663 | 585 | 13.4 | 9.0 | 8.7 | 8.0 | 10.1 |
| Expand | 15 | 630 | 565 | 11.6 | 13.0 | 12.9 | 6.9 | 8.2 |
| Expand | 20 | 631 | 568 | 11.1 | 18.6 | 18.4 | 6.7 | 7.7 |
| Expand | 25 | 558 | 505 | 10.6 | 21.7 | 21.5 | 6.0 | 6.7 |

and 10, and the Expand generator with hop limit 5, 10, 15, 20 and 25. Since the SLA algorithm generates relatively short p-cycles, the maximum considered hop count was limited to 10. The results are averaged over 4875 distinct cases different in terms of the anycast client number (2–26), anycast client location, DC number (2–8) and DC location.

The first observation is that these results confirm that using anycast-protecting p-cycles reduces the network cost related to spare and working capacity when compared to classical p-cycles. Tables 2.3 and 2.4 show that the gain of using anycast-protecting p-cycles is lower than that in the results shown in Table 2.2. This is due to the number of anycast client assumed in both cases, i.e., the results in Tables 2.3 and 2.4 are averaged over various number of anycast clients (2–26), while Table 2.2 presents results yielded for 20 anycast clients. Comprehensive analysis of the results clearly reveals that as anycast clients increase, the gap between both types of p-cycles grows. For more details on this issue see [144–146].

The second interesting observation refers to the examination of the p-cycle type and hop limit influence on the results. It is clear that, in general, increasing the hop count limit of p-cycles reduces the network cost for both anycast-protecting p-cycles and classical p-cycles. This is because longer p-cycles protect a large number of working flows. Moreover, for a higher hop count limit, a lower number of p-cycles is required to provide protection for all working flows, making management of the network simpler. However, as mentioned above, longer p-cycles can significantly increase the transmission delay experienced in the event of a network failure.

The average execution time required to solve optimization models using the Gurobi solver was counted in tens of seconds. The key difficulty in solving the models was the memory requirement resulting from the high number of variables and constraints. Accordingly, for larger problem instances the Gurobi solver returned frequently the 'out of memory' error and was unable to provide a feasible solution of the problem. For more results and discussions on the application of anycast-protection p-cycles, see [142–146].

## 2.8  Multi-layer Optimization

All optimization problems presented so far in this book are related to single-layer networks. The concept of *multi-layer* network modeling has been gaining a lot of attention in recent years due to growing demands to provide more precise models that enable efficient optimization and in consequence cost savings in design and operation of computer and communication networks. Many issues related to multi-layer modeling have been addressed in numerous books and papers, e.g., [4, 6, 152–163]. However, multi-layer optimization with anycast flows has only been considered in [49, 164].

It should be stressed that the multi-layer network approach makes it possible to optimize the whole network much more efficiently compared to the single-layer method where each layer is optimized separately, which cannot guarantee the global optimality of the solution. Nevertheless, optimization of multi-layer networks brings additional challenges. In particular, because more network layers are considered, the optimization problem increases; in consequence this triggers the need to develop new heuristic algorithms, since exact solutions given by branch-and-bound and branch-and-cut methods can be obtained for relatively small networks only. A more comprehensive treatment of modeling and optimization of multi-layer networks is given in [4].

### 2.8.1  Formulation

In this section, we present a two-layer network design problem with simultaneous unicast, multicast and anycast flows realized in the upper layer [49, 164]. The network consists of two layers; upper layer links are denoted by set $E$ and lower layer links are represented as set $G$. Anycast, multicast and unicast traffic demands included in set $D$ are defined in the upper layer, since it is assumed that DCs are accessed by protocols or technologies implemented in the upper layer of the network. For each demand $d \in D$, a set of candidate structures $P(d)$ is given. Each candidate structure $p \in P(d)$ is defined as a set of upper layer links. The concept of two-layer network modeling assumes that each link $e \in E$ of the upper layer can be realized using paths $q \in Q(e)$ established in the lower layer, i.e., each path $q \in Q(e)$ is defined as a set of lower layer links. In consequence, links $e \in E$ of the upper layer create a virtual topology used to realize traffic demands. Going down the network hierarchy, upper layer links included in set $E$ represent the demand pattern of the lower layer. Note that to formulate a multi-layer model, this approach is repeated going down the network hierarchy for each subsequent layer.

The connection-oriented approach is applied in both layers, i.e., a single path routing is ensured. The link capacity of both layers is allocated in modules; the upper layer capacity uses modules of size $M$, while the lower layer capacity is allocated with modules of size $N$. Integer variables $y_e$ and $u_g$ denote the capacity (in modules) assigned to upper and lower layers, respectively. Binary variable $x_{dp}$ denotes the routing structure selected to realize demand $d$, while integer variable $z_{eq}$ indicates how many copies of path $q$ are used to realize upper layer link $e$ using lower layer resources. Variable $z_{eq}$ is a non-binary integer, in order to allow capacity $y_e$ assigned to link $e$ to be greater than the size of one capacity module $M$.

To illustrate this idea, the following example of MPLS over a WDM network is presented. The upper layer uses the MPLS protocol, and upper layer demands included in set $D$ and the upper layer capacity are expressed in Mb/s. In turn, the lower layer is based on the WDM technology with 40 Gb/s per one wavelength. Therefore, the upper layer capacity module is $M = 40$ Gb/s. Continuing the example, the lower layer links $g \in G$ are represented as fibers and it is assumed that the lower layer capacity module is $N = 80$, which means that each fiber can support at most 80 wavelengths [4, 49, 164].

---

**CON/AMU/MLN/Cost**

**sets**

$E$       links of upper layer
$D$       demands (anycast, multicast, unicast)
$D^{DS}$   anycast downstream demands
$P(d)$   candidate structures for flows realizing demand $d$. If $d$ is a unicast demand, the candidate structure is a path connecting end nodes of the demand. If $d$ is an anycast upstream demand, the candidate structure is a path connecting the client node and DC node. If $d$ is a downstream demand, candidate structure is a path connecting the DC node and the client node. If $d$ is a multicast demand, the candidate structure is a tree that includes the root node and all receivers of the demand
$Q(e)$   candidate paths in the lower layer for flows realizing the upper layer link $e$
$G$       links of lower layer

**constants**

$\delta_{edp}$   $=1$, if link $e$ belongs to structure $p$ realizing demand $d$; 0, otherwise
$h_d$     volume of demand $d$
$\gamma_{geq}$   $=1$, if link $g$ of lower layer belongs to path $q$ realizing link $e$ of upper layer; 0, otherwise
$\xi_e$      unit cost of link $e$
$\kappa_g$     unit cost of link $g$
$M$      size of the upper layer link capacity module
$N$      size of the lower layer link capacity module
$\tau(d)$   index of a demand associated with demand $d$. If $d$ is a downstream demand, then $\tau(d)$ must be an upstream demand and vice versa
$s(p)$   origin node of path $p$
$t(p)$   destination node of path $p$

**variables**

$x_{dp}$ =1, if structure $p$ is used to realize demand $d$; 0, otherwise (binary)
$y_e$    capacity of upper layer link $e$ as the number of capacity modules (integer)
$z_{eq}$   number of path $q$ instances used to realize the capacity of link $e$ (integer)
$u_g$    capacity of lower layer link $g$ as the number of capacity modules (integer)

**objective**

$$\text{minimize} \quad F = \sum_{e \in E} \xi_e y_e + \sum_{g \in G} \kappa_g u_g \tag{2.8.1a}$$

**constraints**

$$\sum_{p \in P(d)} x_{dp} = 1, \quad d \in D \tag{2.8.1b}$$

$$\sum_{d \in D} \sum_{p \in P(d)} \delta_{edp} x_{dp} h_d \leq M y_e, \quad e \in E \tag{2.8.1c}$$

$$\sum_{p \in P(d)} x_{dp} s(p) = \sum_{p \in P(\tau(d))} x_{\tau(d)p} t(p), \quad d \in D^{DS} \tag{2.8.1d}$$

$$\sum_{q \in Q(e)} z_{eq} = y_e, \quad e \in E \tag{2.8.1e}$$

$$\sum_{e \in E} \sum_{q \in Q(e)} \gamma_{geq} z_{eq} h_d \leq N u_g, \quad g \in G. \tag{2.8.1f}$$

The objective function (2.8.1a) aims to minimize the cost of capacity assigned in both network layers. Constraints (2.8.1b)–(2.8.1d) are the same as in the single layer network design problem formulated in Sect. 2.3.1. Equality (2.8.1e) ensures that each upper layer link is realized by a set of lower layer paths. Condition (2.8.1f) states that flow in each lower layer link cannot exceed its capacity.

Due to the complexity of multi-layer models, heuristic algorithms are required to solve larger problem instances. One approach is to tackle the optimization in all network layers jointly. However, since routing and capacity decision variables of both layers are bound to each other, this strategy may not be efficient. Another approach is to optimize each network layer in a separate phase, i.e., the problem in the upper layer is solved first, and the lower layer is optimized next. Algorithms developed for single layer problems can be used, i.e., methods proposed in Sects. 2.2.2 and 2.3.2. If the algorithms used to optimize each layer are relatively fast, the procedure can be repeated many times and in each subsequent iteration, information obtained in the previous iteration can be used to improve the performance of the overall optimization process.

# References

1. Perros, H.: Connection-Oriented Networks: SONET/SDH, ATM. MPLS and Optical Networks. Wiley, New York (2005)
2. Rosen, E., Viswanathan, A., Callon, R.: Multiprotocol label switching architecture, RFC3013, Internet Engineering Task Force (2001)
3. Kasim, A.: Delivering Carrier Ethernet: Extending Ethernet Beyond the LAN, 1st edn. McGraw-Hill Inc, New York (2008)
4. Pioro, M., Medhi, D.: Routing, Flow, and Capacity Design in Communication and Computer Networks. Morgan Kaufmann, San Francisco (2004)
5. Vasseur, J., Pickavet, M., Demeester, P.: Network Recovery: Protection and Restoration of Optical, SONET-SDH, IP, and MPLS. Morgan Kaufmann Publishers Inc., San Francisco (2004)
6. Grover, Wayne D.: Mesh-based Survivable Transport Networks: Options and Strategies for Optical, MPLS. SONET and ATM Networking. Prentice Hall PTR, Upper Saddle River (2003)
7. Habib, M.F., Tornatore, M., De Leenheer, M., Dikbiyik, F., Mukherjee, B.: Design of disaster-resilient optical datacenter networks. J. Lightwave Technol. **30**(16), 2563–2573 (2012)
8. Ramaswami, R., Sivarajan, K., Sasaki, G.: Optical Networks: A Practical Perspective, 3rd edn. Morgan Kaufmann Publishers Inc, San Francisco (2009)
9. Simmons, J.: Optical Network Design and Planning, 2nd edn. Springer (2014)
10. IEEE. 802.1ah, Provider backbone bridging (2008)
11. IEEE. 802.1ay, Provider backbone bridging traffic engineering (2009)
12. Allan, D., Bragg, N., Mcguire, A., Reid, A.: Ethernet as carrier transport infrastructure. IEEE Commun. Mag. **44**(2), 95–101 (2006)
13. Takacs, A., Green, H., Tremblay, B.: GMPLS controlled ethernet: an emerging packet-oriented transport technology. IEEE Commun. Mag. **46**(9), 118–124 (2008)
14. Walkowiak, K.: Anycasting in connection-oriented computer networks: models, algorithms and results. Int. J. Appl. Math. Comput. Sci. **20**(1), 207–220 (2010)
15. Fratta, L., Gerla, M., Kleinrock, L.: The flow deviation method: an approach to store-and-forward communication network design. Networks **3**(2), 97–133 (1973)
16. Bienstock, D., Raskina, O.: Asymptotic analysis of the flow deviation method for the maximum concurrent flow problem. Math. Program. **91**(3), 479–492 (2002)
17. Burns, J.E., Ott, T., Krzesinski, A.E., Muller, K.: Path selection and bandwidth allocation in MPLS networks. Perform. Eval. **52**(2–3), 133–152 (2003)
18. Duhamel, Ch., Mahey, P.: Multicommodity flow problems with a bounded number of paths: a flow deviation approach. Networks **49**(1), 80–89 (2007)
19. Fratta, L., Gerla, M., Kleinrock, L.: Flow deviation: 40 years of incremental flows for packets, waves, cars and tunnels. Comput. Netw. **66**(0), 18–31 (2014). Leonard Kleinrock Tribute Issue: A Collection of Papers by his Students
20. Gerla, M., Kleinrock, L.: On the topological design of distributed computer networks. IEEE Trans. Commun. **25**(1), 48–60 (1977)
21. LeBlanc, L., Chifflet, J., Mahey, P.: Packet routing in telecommunication networks with path and flow restrictions. INFORMS J. Comput. **11**(2), 188–197 (1999)
22. Murakami, K., Kim, H.S.: Virtual path routing for survivable ATM networks. IEEE/ACM Trans. Netw. **4**(1), 22–39 (1996)
23. Ng, T., Hoang, D.B.: Joint optimization of capacity and flow assignment in a packet-switched communications network. IEEE Trans. Commun. **35**(2), 202–209 (1987)
24. Ouorou, A., Mahey, P.: Vial, J-Ph: A survey of algorithms for convex multicommodity flow problems. Manage. Sci. **46**(1), 126–147 (2000)
25. Walkowiak, K.: A new method of primary routes selection for local restoration. In: Mitrou, N., Kontovasilis, K., Rouskas, G.N., Iliadis, I., Merakos, L. (eds.) Networking 2004. Lecture Notes in Computer Science, vol. 3042, pp. 1024–1035. Springer, Berlin (2004)

26. Walkowiak, K.: Heuristic algorithm for anycast flow assignment in connection-oriented networks. In: Sunderam, V.S., van Albada, G., Sloot, P.M.A., Dongarra, J. (eds.) Computational Science (ICCS 2005). Lecture Notes in Computer Science, vol. 3516, pp. 1092–1095. Springer, Berlin (2005)

27. Walkowiak, K.: A flow deviation algorithm for joint optimization of unicast and anycast flows in connection-oriented networks. Computational Science and Its Applications (ICCSA 2008). Lecture Notes in Computer Science, vol. 5073, pp. 797–807. Springer, Berlin (2008)

28. Dias, R.A., Camponogara, E., Farines, J., Willrich, R. Campestrini, A.: Implementing traffic engineering in MPLS-based IP networks with Lagrangean relaxation. In: Proceedings of the Eighth IEEE International Symposium on Computers and Communication (ISCC 2003), pp. 373–378, June 2003

29. Gavish, B., Hantler, S.: An algorithm for optimal route selection in SNA networks. IEEE Trans. Commun. **31**(10), 1154–1161 (1983)

30. Gavish, B., Neuman, I.: A system for routing and capacity assignment in computer communication networks. IEEE Trans. Commun. **37**(4), 360–366 (1989)

31. Gavish, B., Neuman, I.: Routing in a network with unreliable components. IEEE Trans. Commun. **40**(7), 1248–1258 (1992)

32. Holmberg, K., Yuan, D.: A Lagrangean approach to network design problems. Int. Trans. Oper. Res. **5**(6), 529–539 (1998)

33. Holmberg, K., Yuan, D.: A Lagrangian heuristic based branch-and-bound approach for the capacitated network design problem. Oper. Res. **48**(3), 461–481 (2000)

34. Retvdri, G., Biro, J.J., Cinkler, T.: A novel lagrangian-relaxation to the minimum cost multicommodity flow problem and its application to OSPF traffic engineering. In: Proceedings of the Ninth International Symposium on Computers and Communications (ISCC 2004), vol. 2, pp. 957–962, June 2004

35. Walkowiak, K.: Lagrangean heuristic for anycast flow assignment in connection-oriented networks. Computational Science (ICCS 2006). Lecture Notes in Computer Science, vol. 3991, pp. 626–633. Springer, Berlin (2006)

36. Walkowiak, K.: Lagrangean heuristic for primary routes assignment in survivable connection-oriented networks. Comput. Opt. Appl. **40**(2), 119–141 (2008)

37. Held, M., Wolfe, P., Crowder, H.: Validation of subgradient optimization. Math. Program. **6**(1), 62–88 (1974)

38. Coley, D.: An Introduction to Genetic Algorithms for Scientists and Engineers. World scientific, Singapore (1999)

39. Corne, D., Oates, M., Smith, G. (eds.): Telecommunications Optimization: Heuristic and Adaptive Techniques. Wiley, Chichester (2000)

40. Donoso, Y., Fabregat, R.: Multi-Objective Optimization in Computer Networks Using Metaheuristics. Auerbach Publications, Boston (2007)

41. Goldberg, David E.: Genetic Algorithms in Search Optimization and Machine Learning, 1st edn. Addison-Wesley Longman Publishing Co., Inc, Boston (1989)

42. Yang, X.: Nature-Inspired Optimization Algorithms. Elsevier (2014)

43. Arabas, J., Kozdrowski, S.: Applying an evolutionary algorithm to telecommunication network design. IEEE Trans. Evol. Comput. **5**(4), 309–322 (2001)

44. Banerjee, N., Mehta, V., Pandey, S.: A genetic algorithm approach for solving the routing and wavelength assignment problem in WDM networks. In: Proceedings of the 3rd IEEE/IEE International Conference on Networking (ICN 2004), pp.70–78 (2004)

45. Koppen, M., Yoshida, K., Tsuru, M., Oie, Y.: Evolutionary routing-path selection in congested communication networks. In: Proceedings of the IEEE International Conference on Systems, Man and Cybernetics (SMC 2009), pp. 2155–2160, October 2009

46. Pitsillides, A., Stylianou, G., Pattichis, C.S., Sekercioglu, A., Vasilakos, A.: Aggregated bandwidth allocation: investigation of performance of classical constrained and genetic algorithm based optimisation techniques. Comput. Commun. **25**(16), 1443–1453 (2002)

47. Przewozniczek, M., Walkowiak, K.: Evolutionary algorithm for congestion problem in connection-oriented networks. In: Gervasi, O., Gavrilova, M.L., Kumar, V., Laganá, A., Lee,

H.P., Mun, Y., Taniar, D., Tan, C.J.K. (eds.) Computational Science and Its Applications ICCSA. Lecture Notes in Computer Science, vol. 3483, pp. 802–811. Springer, Berlin (2005)

48. Rubio-Largo, Á., Vega-Rodríguez, M.: Applying MOEAs to solve the static routing and wavelength assignment problem in optical WDM networks. Eng. Appl. Artif. Intell. **26**(5U6), 1602–1619 (2013)

49. Walkowiak, K.: Modeling and Optimization of Computer Networks. Wroclaw University of Technology, Wroclaw (2011)

50. Bhanja, U., Mahapatra, S., Roy, R.: An evolutionary programming algorithm for survivable routing and wavelength assignment in transparent optical networks. Inf. Sci. **222**, 634–647 (2013). Including Special Section on New Trends in Ambient Intelligence and Bio-inspired Systems

51. Bhanja, U., Mahapatra, S.: A metaheuristic approach for optical network optimization problems. Appl. Soft Comput. **13**(2), 981–997 (2013)

52. El-Alfy, E., Mujahid, S., Selim, S.: A pareto-based hybrid multiobjective evolutionary approach for constrained multipath traffic engineering optimization in MPLS/GMPLS networks. J. Netw. Comput. Appl. **36**(4), 1196–1207 (2013)

53. Li, K., Zhou, X., Zhang, W.: An efficient anycast routing algorithm for load-balancing based on evolutionary algorithm. In: Proceedings of the 3rd IEEE International Conference on Broadband Network and Multimedia Technology (IC-BNMT 2010), pp. 48–54, October 2010

54. Przewozniczek, M., Walkowiak, K.: uasi-hierarchical evolutionary algorithm for flow optimization in survivable MPLS networks. Computational Science and Its Applications (ICCSA 2007). Lecture Notes in Computer Science, vol. 4707, pp. 330–342. Springer, Berlin (2007)

55. Glover, F.: Tabu search fundamentals and uses. Technical report, University of Colorado (1995)

56. Kirkpatrick Jr., S., Gelatt, C.D., Vecchi, M.P.: Optimization by simulated annealing. Science **220**, 671–680 (1983)

57. Resende, M.: Greedy randomized adaptive search procedures (grasp). Encycl. optim. **2**, 373–382 (2001)

58. Talbi, El.G.: Metaheuristics - From Design to Implementation. Wiley, New York (2009)

59. Hyytia, E.: Heuristic algorithms for the generalized routing and wavelength assignment problem. In: Proceedings of the 17th Nordic Teletraffic Seminar(NTS-17), pp. 373–386 (2004)

60. Shen, Jian: Fuyong, Xu, Zheng, Peng: A tabu search algorithm for the routing and capacity assignment problem in computer networks. Comput. Oper. Res. **32**(11), 2785–2800 (2005)

61. Bakiras, S., Loukopoulos, T.: Combining replica placement and caching techniques in content distribution networks. Comput. Commun. **28**(9), 1062–1073 (2005)

62. Bektas, T., Oguz, O., Ouveysi, I.: Designing cost-effective content distribution networks. Comput. Oper. Res. **34**(8), 2436–2449 (2007)

63. Krishnan, P., Raz, D., Shavitt, Y.: The cache location problem. IEEE/ACM Trans. Netw. **8**(5), 568–582 (2000)

64. Li, B., Golin, M., Italiano, G., Deng, X., Sohraby, K.: On the optimal placement of web proxies in the internet. In: Proceedings of Annual Joint Conference of the IEEE Computer and Communications (INFOCOM 1999), pp. 1282–1290 (1999)

65. Markowski, M., Kasprzak, A.: The web replica allocation and topology assignment problem in wide area networks: algorithms and computational results. Computational Science and Its Applications (ICCSA 2005). Lecture Notes in Computer Science, vol. 3483, pp. 772–781. Springer, Berlin (2005)

66. Markowski, M., Kasprzak, A.: The three-criteria servers replication and topology assignment problem in wide area networks. Computational Science and Its Applications (ICCSA 2006). Lecture Notes in Computer Science, vol. 3982, pp. 1119–1128. Springer, Berlin (2006)

67. Markowski, M., Kasprzak, A.: An exact algorithm for the servers allocation, capacity and flow assignment problem with cost criterion and delay constraint in wide area networks. Computational Science (ICCS 2007). Lecture Notes in Computer Science, vol. 4487, pp. 442–445. Springer, Berlin (2007)

68. Qiu, L., Padmanabhan, V., Voelker, G.: On the placement of web server replicas. In: Proceedings of Annual Joint Conference of the IEEE Computer and Communications (INFOCOM 2001), pp. 1587–1596 (2001)

69. Thouin, F., Coates, M.: Video-on-demand server selection and placement. In: Mason, L., Drwiega, T., Yan, J. (eds.) Managing Traffic Performance in Converged Networks. Lecture Notes in Computer Science, vol. 4516, pp. 18–29. Springer, Berlin (2007)

70. Walkowiak, K., Rak, J.: Simultaneous optimization of unicast and anycast flows and replica location in survivable optical networks. Telecommun. Syst. **52**(2), 1043–1055 (2013)

71. Wang, Jia: A survey of web caching schemes for the internet. CM SIGCOMM Comput. Commun. Rev. **29**(5), 36–46 (1999)

72. Wang, y., Li, Z., Tyson, G., Uhlig, S., Xie, G.: Optimal cache allocation for content-centric networking. In: Proceedings of the 21st IEEE International Conference on Network Protocols (ICNP 2013), pp. 1–10, October 2013

73. Woeginger, G.: Monge strikes again: optimal placement of web proxies in the internet. Oper. Res. Lett. **27**(3), 93–96 (2000)

74. Baev, I., Rajaraman, R.: Approximation algorithms for data placement in arbitrary networks. In: Proceedings of the Twelfth Annual ACM-SIAM Symposium on Discrete Algorithms, SODA '01. Society for Industrial and Applied Mathematics pp. 661–670, Philadelphia (2001)

75. Fang, W., Wang, Z., Lloret, J., Zhang, D., Yang, Z.: Optimising data placement and traffic routing for energy saving in backbone networks. Trans. Emerg. Telecommun. **25**(9), 914–925 (2014)

76. Araldo, A., Mangili, M., Martignon, F., Rossi, D.: Cost-aware caching: optimizing cache provisioning and object placement in ICN. In: Proceedings of the IEEE Global Communications Conference (GLOBECOM 2014), pp. 1108–1113, December 2014

77. Bektas, T., Cordeau, J., Erkut, E., Laporte, G.: Exact algorithms for the joint object placement and request routing problem in content distribution networks. Comput. Oper. Res. **35**(12), 3860–3884 (2008). Part Special Issue: Telecommunications Network Engineering

78. Cidon, I., Kutten, S., Soffer, R.: Optimal allocation of electronic content. In: Proceedings of Annual Joint Conference of the IEEE Computer and Communications (INFOCOM 2001), vol. 3, pp. 1773–1780 (2001)

79. Kangasharju, J., Roberts, J., Ross, K.: Object replication strategies in content distribution networks. Comput. Commun. **25**(4), 376–383 (2002)

80. Laoutaris, N., Zissimopoulos, V., Stavrakakis, I.: Joint object placement and node dimensioning for internet content distribution. Inf. Process. Lett. **89**(6), 273–279 (2004)

81. Laoutaris, N., Zissimopoulos, V., Stavrakakis, I.: On the optimization of storage capacity allocation for content distribution. Comput. Netw. **47**(3), 409–428 (2005)

82. Abhari, A., Soraya, M.: Workload generation for youtube. Multimed. Tools. Appl. **46**(1), 91–118 (2010)

83. Breslau, L., Cao, P., Fan, L., Phillips, G., Shenker, S.: Web caching and zipf-like distributions: evidence and implications. In: Proceedings of Annual Joint Conference of the IEEE Computer and Communications (INFOCOM 1999), vol. 1, pp. 126–134, March 1999

84. Cho, K., Lee, M., Park, K., Kwon, T.T., Choi, Y., Pack, S.: Wave: popularity-based and collaborative in-network caching for content-oriented networks. In: Proceedings of the IEEE Conference on Computer Communications Workshops (INFOCOM WKSHPS 2012), pp. 316–321, March 2012

85. Gill, P., Arlitt, M., Li, Z., Mahanti, A.: Youtube traffic characterization: a view from the edge. In: Proceedings of the 7th ACM SIGCOMM Conference on Internet Measurement, IMC '07, New York, pp. 15–28. ACM (2007)

86. Hasslinger, G., Hohlfeld, O.: Efficiency of caches for content distribution on the internet. In: Proceedings of the 22nd International Teletraffic Congress (ITC 2010), pp. 1–8, September 2010

87. Janaszka, T., Bursztynowski, D., Dzida, M.: On popularity-based load balancing in content networks. In: Proceedings of the 24th International Teletraffic Congress (ITC 2012), pp. 1–8, September 2012

88. Walkowiak, K.: A unified approach to survivability of connection-oriented networks. Computer and Information Sciences (ISCIS 2005). Lecture Notes in Computer Science, vol. 3733, pp. 3–12. Springer, Berlin (2005)
89. Bui, M., Jaumard, B., Develder, C.: Anycast end-to-end resilience for cloud services over virtual optical networks. In: Proceedings of the 15th International Conference onTransparent Optical Networks (ICTON 2013), pp. 1–7, June 2013
90. Buysse, J., De Leenheer, M., Develder, C., Dhoedt, B.: Exploiting relocation to reduce network dimensions of resilient optical grids. In: Proceedings of the 7th International Workshop on Design of Reliable Communication Networks (DRCN 2009), pp. 100–106, October 2009
91. Buysse, J., De Leenheer, M., Dhoedt, B., Develder, C.: On the impact of relocation on network dimensions in resilient optical grids. In: Proceedings of the 14th Conference on Optical Network Design and Modeling (ONDM 2010), pp. 1–6, February 2010
92. Develder, C., Buysse, J., De Leenheer, M., Jaumard, B., Dhoedt, B.: Resilient network dimensioning for optical grid/clouds using relocation. In: Proceedings of the IEEE International Conference on Communications (ICC 2012), pp. 6262–6267, June 2012
93. Develder, C., Buysse, J., Shaikh, A., Jaumard, B., De Leenheer, M., Dhoedt, B.: Survivable optical grid dimensioning: Anycast routing with server and network failure protection. In: Proceedings of the IEEE International Conference on Communications (ICC 2011), pp. 1–5, June 2011
94. Develder, C., Buysse, J., Dhoedt, B., Jaumard, B.: Joint dimensioning of server and network infrastructure for resilient optical grids/clouds. IEEE/ACM Trans. Netw. **22**(5), 1591–1606 (2014)
95. Goscien, R., Walkowiak, K., Klinkowski, M.: Gains of anycast demand relocation in survivable elastic optical networks. In: Proceedings of the 6th International Workshop on Reliable Networks Design and Modeling (RNDM 2014), pp. 109–115, November 2014
96. Jaumard, B., Buysse, J., Shaikh, A., De Leenheer, M., Develder, C.: Column generation for dimensioning resilient optical grid networks with relocation. In: Proceedings of the IEEE Global Telecommunications Conference (GLOBECOM 2010), pp. 1–6, December 2010
97. Shaikh, A., Buysse, J., Jaumard, B., Develder, C.: Anycast routing for survivable optical grids: Scalable solution methods and the impact of relocation. IEEE/OSA J. Opt. Commun. Netw. **3**(9), 767–779 (2011)
98. Walkowiak, K., Rak, J.: Joint optimization of anycast and unicast flows in survivable optical networks. In: Proceedings of the 14th International Telecommunications Network Strategy and Planning Symposium (NETWORKS 2010), pp. 1–6, September 2010
99. Walkowiak, K., Rak, J.: Shared backup path protection for anycast and unicast flows using the node-link notation. In: Proceedings of the 2011 IEEE International Conference on Communications (ICC 2011), pp. 1–6, June 2011
100. IBM. ILOG CPLEX optimizer. http://www.ibm.com
101. Gladysz, J., Walkowiak, K.: Modeling of survivable network design problems with simultaneous unicast and anycast flows. In: Proceedings of the 2nd International Logistics and Industrial Informatics (LINDI 2009), pp. 1–6, September 2009
102. Gladysz, J., Walkowiak, K.: Optimization of survivable networks with simultaneous unicast and anycast flows. In: Proceedings of the International Conference on Ultra Modern Telecommunications Workshops (ICUMT 2009), pp. 1–6, October 2009
103. Gladysz, J., Walkowiak, K.: Tabu search algorithm for survivable network design problem with simultaneous unicast and anycast flows. Int. J. Electr. Telecommun. **56**(1), 39–45 (2010)
104. Mitchell, J.: Branch-and-cut methods for combinatorial optimization problems. In: Resende, M., Pardalos, P. (eds.) Handbook of Applied Optimization. Oxford University Press, Oxford (2002)
105. Barnhart, C., Hane, Ch., Vance, P.: Using branch-and-price-and-cut to solve origin-destination integer multicommodity flow problems. Oper. Res. **48**(2), 318–326 (2000)
106. Bienstock, D., Muratore, G.: Strong inequalities for capacitated survivable network design problems. Math. Program. **89**(1), 127–147 (2000)

107. Caccetta, L., Hill, S.P.: A branch and cut method for the degree-constrained minimum spanning tree problem. Networks **37**(2), 74–83 (2001)
108. Gunluk, O.: A branch-and-cut algorithm for capacitated network design problems. Math. Program. **86**, 17–39 (1998)
109. Klinkowski, M., Pioro, M., Zotkiewicz, M., Ruiz, M., Velasco, L.: Valid inequalities for the routing and spectrum allocation problem in elastic optical networks. In: Proceedings of the 16th International Conference on Transparent Optical Networks (ICTON 2014), pp. 1–5, July 2014
110. Koster, A., Kutschka, M., Raack, Ch.: Robust network design: formulations, valid inequalities, and computations. Networks **61**(2), 128–149 (2013)
111. Marchand, H., Wolsey, L.: Aggregation and mixed integer rounding to solve MIPs. Oper. Res. **49**(3), 363–371 (2001)
112. Minoux, M.: Discrete cost multicommodity network optimization problems and exact solution methods. Ann. Oper. Res. **106**(1–4), 19–46 (2001)
113. Ortega, F., Wolsey, L.: A branch-and-cut algorithm for the single-commodity, uncapacitated, fixed-charge network flow problem. Networks **41**(3), 143–158 (2003)
114. Raack, Ch., Koster, A., Orlowski, S., Wessäly, R.: On cut-based inequalities for capacitated network design polyhedra. Networks **57**(2), 141–156 (2011)
115. Tomaszewski, A., Pioro, M., Dzida, M., Mycek, M., Zagozdzon, M.: Valid inequalities for a shortest-path routing optimization problem. In: Proceedings of the 3rd International Network Optimization Conference (INOC 2007) (2007)
116. Chow, T.Y., Chudak, F., Ffrench, A.M.: Fast optical layer mesh protection using pre-cross-connected trails. IEEE/ACM Trans. Netw. **12**(3), 539–548 (2004)
117. Stamatelakis, D., Grover, W.D.: IP layer restoration and network planning based on virtual protection cycles. IEEE J. Sel. Areas Commun. **18**(10), 1938–1949 (2000)
118. Stamatelakis, D., Grover, W.D.: Theoretical underpinnings for the efficiency of restorable networks using preconfigured cycles (ldquo;p-cycles rdquo;). IEEE Trans. Commun. **48**(8), 1262–1265 (2000)
119. Gruber, C.G.: Resilient networks with non-simple p-cycles. In: Proceedings of the 10th International Conference on Telecommunications, ICT 2003, vol. 2, pp. 1027–1032, February 2003
120. Li, W., Doucette, J., Zuo, M.: p-cycle network design for specified minimum dual-failure restorability. In: Proceedings of the IEEE International Conference on Communications, ICC '07, pp. 2204–2210, June 2007
121. Mukherjee, D.S., Assi, C., Agarwal, A.: Alternate strategies for dual failure restoration using p-cycles. In: IEEE International Conference onCommunications, ICC '06, vol. 6, pp. 2477–2482, June 2006
122. Oliveira, H.M.N.S., da Fonseca, N.L.S.: Algorithm for FIPP p-cycle path protection in flexgrid networks. In: Proceedings of the Global Communications Conference (GLOBECOM), pp. 1278–1283. IEEE, December 2014
123. Rocha, C., Jaumard, B., Bougue, P.: Directed vs. undirected p-cycles and FIPP p-cycles. In: Proceedings of the International Network Optimization Conference, INOC 2009 (2009)
124. Shen, G., Grover, W.D.: Extending the p-cycle concept to path segment protection for span and node failure recovery. IEEE J. Sel. Areas Commun. **21**(8), 1306–1319 (2003)
125. Szigeti, J., Cinkler, T.: Evaluation and estimation of the availability of p-cycle protected connections. Telecommun. Syst. **52**(2), 767–782 (2013)
126. Wei, Y., Xu, K., Jiang, Y., Zhao, H., Shen, G.: Optimal design for p-cycle-protected elastic optical networks. Photonic Netw. Commun. **29**, 1–12 (2015)
127. Y. Wei, K. Xu, H. Zhao, and G. Shen. Applying p-cycle technique to elastic optical networks. In Proceedings of the 2014 International Conference on Optical Network Design and Modeling (ONDM 2014), pages 1–6, May 2014
128. Wu, J., Liu, Y., Yu, C., Wu, Y.: Survivable routing and spectrum allocation algorithm based on p-cycle protection in elastic optical networks. Optik - Int. J. Light. Electron Opt. **125**(16), 4446–4451 (2014)

129. Zhang, F., Zhong, W.: A novel path-protecting p-cycle heuristic algorithm. In: Proceedings of the International Conference on Transparent Optical Networks (ICTON 2006), vol. 3, pp. 203–206, June 2006

130. Feng, T., Ruan, L., Zhang, W.: Intelligent p-cycle protection for dynamic multicast sessions in WDM networks. IEEE/OSA J. Opt. Commun. Netw. **2**(7), 389–399 (2010)

131. Frikha, A., Cousin, B., Lahoud, S.: Extending node protection concept of p-cycles for an efficient resource utilization in multicast traffic. In: Proceedings of the 2011 IEEE 36th Conference on Local Computer Networks (LCN), pp. 175–178, October 2011

132. Smutnicki, A., Walkowiak, K.: p-cycle based multicast protection - a new ILP formulation. In: Proceedings of the 2nd Baltic Congress on Future Internet Communications (BCFIC 2012), pp. 268–274, April 2012

133. Smutnicki, A., Walkowiak, K.: A new approach to optimization of p-cycle protected multicast optical networks. In: Proceedings of the 5th International Congress on Ultra Modern Telecommunications and Control Systems and Workshops (ICUMT 2013), pp. 74–81, September 2013

134. Zhang, F., Zhong, W.: Applying p-cycles in dynamic provisioning of survivable multicast sessions in optical WDM networks. In: Proceedings of the 2007 Conference on Optical Fiber Communication and the National Fiber Optic Engineers Conference (OFC/NFOEC 2007), pp. 1–3, March 2007

135. Zhang, F., Zhong, W.: Optimized design of node-and-link protecting p-cycle with restorability constraints for optical multicast traffic protection. In: Proceedings of the 14th OptoElectronics and Communications Conference (OECC 2009), pp. 1–2, July 2009

136. Zhang, F., Zhong, W.: p-cycle based tree protection of optical multicast traffic for combined link and node failure recovery in WDM mesh networks. IEEE Commun. Lett. **13**(1), 40–42 (2009)

137. Zhang, F., Zhong, W.: Performance evaluation of optical multicast protection approaches for combined node and link failure recovery. J. Lightwave Technol. **27**(18), 4017–4025 (2009)

138. Zhang, F., Zhong, W.: Extending p-cycles to source failure recovery for optical multicast media traffic. IEEE/OSA J. Opt. Commun. Netw. **2**(10), 831–840 (2010)

139. Zhong, W., Zhang, F.: An overview of p-cycle based optical multicast protection approaches in mesh wdm networks. Opt. Switch. Netw. **8**(4), 259–274 (2011). Optical network architectures and management

140. Zhong, W., Zhang, F.: p-cycle based optical multicast protection approaches for combined node and link failure recovery. In: Proceedings of the Asia Communications and Photonics Conference and Exhibition (ACP 2010), pp. 367–368, December 2010

141. Zhong, W., Zhang, F.: Source failure recovery for optical multicast traffic in WDM networks. In: Proceedings of the 13th International Conference on Transparent Optical Networks (ICTON 2011), pp. 1–4, June 2011

142. Smutnicki, A., Walkowiak, K: Optimization of p-cycles for survivable anycasting streaming. In: Proceedings of the 7th International Workshop on Design of Reliable Communication Networks (DRCN 2009), pp. 227–234, October 2009

143. Smutnicki, A., Walkowiak, K.: Optimal results for anycast-protecting p-cycles problem. In: Proceedings of the 2010 International Congress on Ultra Modern Telecommunications and Control Systems and Workshops (ICUMT 2010), pp. 511–517, October 2010

144. Smutnicki, A., Walkowiak, K.: Joint working and spare capacity assignment for anycast streaming in survivable networks protected by p-cycles. In: Proceedings of the 3rd International Congress on Ultra Modern Telecommunications and Control Systems and Workshops (ICUMT 2011), pp. 1–6, October 2011

145. Smutnicki, A., Walkowiak, K.: A heuristic approach to working and spare capacity optimization for survivable anycast streaming protected by p-cycles. Telecommun. Syst. **56**(1), 141–156 (2014)

146. Smutnicki, A.: Algorithms for flow and capacity assignment in p-cycle protected computer networks (in Polish). Ph.D. thesis, Wroclaw University of Technology, Department of Systems and Computer Networks, Report PRE008, 2012. Supervisor: K. Walkowiak

147. Kodian, A., Grover, W.D., Doucette, J.: A disjoint route-sets approach to design of path-protecting p-cycle networks. In: Proceedings. 5th International Workshop on Design of Reliable Communication Networks, (DRCN 2005), p. 8, October 2005
148. Inc., Gurobi Optimization. Gurobi optimizer reference manual
149. Orlowski, S., Pióro, M., Tomaszewski, A., Wessäly, R.: SNDlib 1.0 – Survivable network design library. In: Proceedings of the 3rd International Network Optimization Conference (INOC 2007), Spa, April 2007. http://sndlib.zib.de
150. Zhang, H., Yang, O.: Finding protection cycles in DWDM networks. In: Proceedings of the IEEE International Conference on Communications (ICC 2002), vol. 5, pp. 2756–2760 (2002)
151. Doucette, J., He, D., Grover, W.D., Yang, O.:. Algorithmic approaches for efficient enumeration of candidate p-cycles and capacitated p-cycle network design. In: Proceedings of the Fourth International Workshop on Design of Reliable Communication Networks, DRCN 2003), pp. 212–220, October 2003
152. Belotti, P., Capone, A., Carello, G., Malucelli, F.: Multi-layer MPLS network design: the impact of statistical multiplexing. Comput. Netw. 52(6), 1291–1307 (2008)
153. Capone, A., Carello, G., Matera, R.: Multi-layer network design with multicast traffic and statistical multiplexing. In: Proceedings of the IEEE Global Telecommunications Conference (GLOBECOM 2007), pp. 2565–2570, November 2007
154. Gerstel, O., Filsfils, C., Telkamp, T., Gunkel, M., Horneffer, M., Lopez, V., Mayoral, A.: Multi-layer capacity planning for IP-optical networks. IEEE Commun. Mag. 52(1), 44–51 (2014)
155. Katib, I., Medhi, D.: A network optimization model for multi-layer IP/MPLS over OTN/DWDN networks. In: Nunzi, Giorgio, Scoglio, Caterina, Li, Xing (eds.) IP Operations and Management. Lecture Notes in Computer Science, vol. 5843, pp. 180–185. Springer, Berlin (2009)
156. Katib, I., Medhi, D.: Optimizing node capacity in multilayer networks. IEEE Commun. Lett. 15(5), 581–583 (2011)
157. Katib, I., Medhi, D.: IP/MPLS-over-OTN-over-DWDM multilayer networks: an integrated three-layer capacity optimization model, a heuristic, and a study. IEEE Trans. Netw. Serv. Manage. 9(3), 240–253 (2012)
158. Liu, Y., Tipper, D., Vajanapoom, K.: Spare capacity allocation in two-layer networks. IEEE J. Sel. Areas Commun. 25(5), 974–986 (2007)
159. Mikoshi, T., Takenaka, T., Sugiyama, R., Masuda, A., Shiomoto, K., Hiramatsu, A.: High-speed calculation method for large-scale multi-layer network design problem. In: Proceedings of the XVth International Telecommunications Network Strategy and Planning Symposium (NETWORKS 2012), pp. 1–6, October 2012
160. Pedrola, O., Ruiz, M., Velasco, L., Careglio, D., de Dios, O.G., Comellas, J.: A grasp with path-relinking heuristic for the survivable IP/MPLS-over-WSON multi-layer network optimization problem. Comput. Oper. Res. 40(12), 3174–3187 (2013)
161. Ruiz, M., Pedrola, O., Velasco, L., Careglio, D., Fernaì-Palacios, J., Junyent, G.: Survivable IP/MPLS-over-WSON multilayer network optimization. IEEE/OSA J. Opt. Commun. Netw. 3(8), 629–640 (2011)
162. Schnitter, S., Barth, A., Schnitter, O., Horneffer, M.: Benefits from 2-layer traffic engineering for IP/MPLS networks. In: Proceedings of the the the 13th International Telecommunications Network Strategy and Planning Symposium (Networks 2008), volume Supplement, pp. 1–6, September 2008
163. Zhang, Xiaoning: Shen, Feng, Wang, Li, Wang, Sheng, Li, Lemin, Luo, Hongbin: Two-layer mesh network optimization based on inter-layer decomposition. Photonic Netw. Commun. 21(3), 310–320 (2011)
164. Gladysz, J., Walkowiak, K.: Design of MPLS over DWDM architecture for unicast and anycast flows. In: Balandin, S., Roman, D., Yevgeni, K. (eds.) Smart Spaces and Next Generation Wired/Wireless Networking. Lecture Notes in Computer Science, pp. 172–183. Springer, Berlin (2010)

# Chapter 3
# Elastic Optical Networks

This chapter presents issues related to the optimization of Elastic Optical Networks (EONs). More specifically, several optimization problems that arise in EONs in the context of cloud computing and content-oriented networking requirements are formulated and discussed. As in the previous chapter, the problems assume optimization of unicast, anycast and multicast network flows in several different combinations. As well as formulating the problems as ILP models, we also provide and analyze algorithms and results of numerical experiments for selected problems.

## 3.1 Introduction

Elastic Optical Networks (EONs) are a new concept in spectrally-efficient and flexible optical transport networks to have been developed in recent years. The development of EON has been mainly driven by business requirements. More specifically, the rapid network traffic growth has brought existing systems (Wavelength Switched Optical Networks (WSONs) implemented with the Wavelength Division Multiplexing (WDM) technology) to the capacity limit. One potential solution to this capacity problem is simply lighting new fiber pairs to expand the bandwidth of existing network links. However, this approach is not highly scalable and does not provide economies of scale. Therefore, since more cost-effective solutions are needed by telecoms, the EON concept addresses a number of business requirements which are currently gaining in importance [1–3].

It should be noted that the technology behind EONs is a spectrum-sliced elastic optical path network (SLICE) architecture—an innovative and promising solution for new generation optical networks enabling provisioning beyond 100 Gb/s connections. The key advantage of the SLICE architecture is its support of sub-wavelength, superwavelength and multiple-rate data traffic accommodation in a spectrum-efficient way. The concept of SLICE was first proposed by Jinno et al. [4] and described further in [1, 2, 5–10].

© Springer International Publishing Switzerland 2016
K. Walkowiak, *Modeling and Optimization of Cloud-Ready
and Content-Oriented Networks*, Studies in Systems, Decision
and Control 56, DOI 10.1007/978-3-319-30309-3_3

The main requirement for new optical technologies such as EONs is spectral efficiency. Historically, WDM systems have been addressing traffic growth by increasing the number of wavelengths in a single fiber, and by increasing the bit-rate provided by each wavelength starting with 16 wavelengths of 2.5 Gb/s in the late 1990s to $80 \times 100$ Gb/s systems in 2012. However, in practice the fixed grid of 50 GHz used in WDM limits the maximum wavelength bit-rate to 100 Gb/s. New network services such as CDNs and cloud computing, and the growing popularity of centralized data processing in dedicated data centers, have driven the need to provision connections of 400 Gb/s or even 1 Tb/s, which cannot be implemented directly in WDM systems. Therefore, a *flexible* grid architecture is adapted in EONs and a finer grid granularity is used. The ITU-T standard [11] issued in 2012 permits any combination of wavelength spacing using 6.25 GHz slices (spectrum slots), and a bandwidth assigned to each lightpath equal to integral multiples of 12.5 GHz (even number of 6.25 GHz slots). In addition, the Internet Engineering Task Force (IETF) plans to support flexible grids in Generalized Multi-Protocol Label Switching (GMPLS) [12]. Figure 3.1 illustrates the fixed grid of WDM systems and flexible grid provided in EONs. In the example, the fixed grid provisions three lightpaths on three adjacent wavelengths. Irrespective of the requested bit-rate in the range up to 100 Gb/s, each WDM lightpath consumes 50 GHz of spectrum. In contrast, the flexible grid allocation has a higher granularity of the spectrum starting from small channels of 12.5 GHz established to support relatively low bit-rates, up to lightpaths with eight or more channels supporting bit-rates up to 400 Gb/s or even 1 Tb/s. One of the steps towards a bit-rate exceeding 100 Gb/s bit-rate, is the concept of *superchannels* [3, 13], which have a bandwidth of more than 50 GHz channels, e.g., 100 GHz. Expending the optical spectrum makes it possible to obtain high bit-rates, although this approach is not as spectral-efficient as EONs. For more discussions on the evolution of optical networks towards elastic optical networks see [1–3, 14–20].

The basic optimization problem arising in the context of EONs is the Routing and Spectrum Allocation (RSA) problem, also known as the Routing and Spectrum Assignment [1–3, 21]. The first ILP formulations of the RSA problem were presented

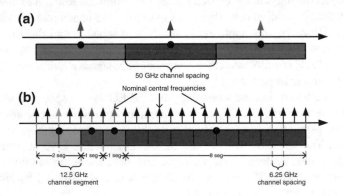

**Fig. 3.1** Optical channel assignment under **a** fixed and **b** flexible grid

in [22–29]. The RSA problem assumes that for each demand to be provisioned in the EON, a routing path connecting end nodes of the demand must be selected, and a spectrum must be allocated along the links included in the routing path. The spectrum allocated to the demand consists of multiple slices (spectrum slots). Due to the EON architecture, the allocated spectral resources must fulfill two important constraints. Firstly, the *continuity constraint* states that in an absence of spectrum converters, the demand must use exactly the same spectrum slots (optical corridor) in all links included in the routing path. Secondly, the *contiguity constraint* requires that slices assigned to a particular demand must be adjacent (contiguous). The RSA problem has been shown to be $\mathcal{NP}$-complete [23, 29].

Figure 3.2 presents an example of the RSA problem. The traffic pattern (shown in the upper figure) includes five demands with bit-rates from 50 to 400 Gb/s established between four nodes in the network, namely, nodes $a$, $b$, $c$ and $d$. The EON network encompasses five nodes ($a$, $b$, $c$, $d$ and $e$). The demands are provisioned on routing paths and spectra (ranges of slices) as illustrated in the lower figure satisfying the continuity and contiguity constraints.

Note that the RSA problem is similar to the Routing and Wavelength Assignment (RWA) problem formulated for WDM networks [3, 18]. In summary, the RWA problem involves selecting a single routing path and one wavelength (spectrum slot) for each demand. In RWA, the wavelength continuity constraint must be met, i.e., the same wavelength must be used for a single demand in all links included in the routing path, except when the use of wavelength converters is allowed. The key difference between RSA and RWA is that the former must satisfy an additional contiguity constraint, since a single demand can be composed of multiple spectrum slices. Moreover, as the RSA problem uses a finer spectrum granularity than RWA (6.25 GHz vs. 50 GHz), a greater number of individual spectrum slots is controlled during optimization, which results in a higher problem complexity. For more information on various aspects of EONs, see [1–3, 14–21].

**Fig. 3.2** Illustration of the Routing and Spectrum Allocation (RSA) problem

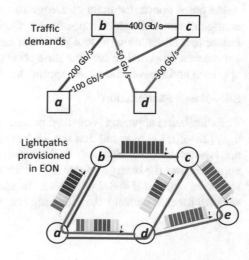

### 3.1.1   Modeling

The EON considered here is modeled as a directed graph $G = (V, E)$, where $V$ is a set of network nodes and $E$ is a set of fiber links that connect pairs of network nodes. On each network link $e \in E$, available spectrum resources are divided into frequency slices included in set $S$. By grouping a number of frequency slices $s \in S$, optical channels of different width can be created. EON operates within a flexible ITU-T grid of 6.25 GHz granularity [11]. Note that in this book, the term *slice* is used to denote a single unit of spectrum used in EONs (frequency range) with a usual size of 6.25 or 12.5 GHz. However, it should be stressed that in some papers on EONs, the authors use *slot* instead of slice.

Demands are included in set $D$. Below, we consider anycast, unicast and multicast demands (flows). The introduction to this section focuses on unicast flows, since the adaptation to other types of flows is straightforward. If the link-path notation is used, a set of candidate paths $P(d)$ is given for each demand $d \in D$. Each candidate path $p \in P(d)$ originates in the source node of demand $d$ and terminates in the destination node of demand $d$. Moreover, each demand $d$ is described by bit-rate $h_d$. Let constant $n_{dp}$ denote the number of slices required for demand $d$ with bit-rate $h_d$ realized on candidate path $p$. For a comprehensive discussion on how to calculate $n_{dp}$ see to Sect. 3.1.2. It should be noted that in this chapter it is usually assumed that $n_{dp}$ is calculated to include the guard bands required in optical networks [3, 20]. Consistent with the definition of the RSA problem, demand $d$ must be assigned with a routing path $p$ and set of $n_{dp}$ adjacent slices allocated in all links included in path $p$.

A very common assumption in optimization of optical networks is that demands between two nodes are bidirectionally symmetric, i.e., if there is a demand from node $a$ to node $b$, then there is equivalent demand from node $b$ to node $a$ and both demands have the same bit-rate [3]. However, all optimization models presented in this chapter assume unidirectional traffic, since this book focuses on services such as cloud-computing and content distribution, which are mostly asymmetric.

As noted above, the main challenge in mathematical modeling of EONs is the contiguity constraint. More specifically, the ILP model of the RSA problem must ensure that slices assigned to a particular demand are adjacent (contiguous). Two approaches proposed in the literature to cope with this issue are the *slice-based* approach and the *channel-based* approach.

**Slice-Based Formulation**

The slice-based approach is covered in numerous papers on optimization of EONs, e.g., [22–39]. Let us recall that the RSA problem involves selecting a routing path and spectrum for each demand. To denote selection of a routing path, the slice-based approach uses the binary variable $x_{dp}$ which is 1 if candidate path $p \in P(d)$ is used to realize demand $d$ and 0 otherwise. To ensure that precisely one routing path is selected for each demand, the following constraint is used in slice-based modeling:

$$\sum_{p \in P(d)} x_{dp} = 1, \quad d \in D. \tag{3.1.1a}$$

Using variable $x_{dp}$, it is easy to calculate links that are used by demand $d$ according to the selected path $p$. To this end, let binary variable $y_{ed}$ denote whether demand $d$ uses link $e$:

$$\delta_{edp} x_{dp} \leq y_{ed}, \quad d \in D, p \in P(d), e \in E. \tag{3.1.1b}$$

The main idea behind slice-based modeling is to directly define spectrum slices assigned to each demand. More specifically, for each demand $d \in D$ the integer variable $w_d$ denotes the starting slice used for demand $d$ and the integer variable $y_d$ indicates the ending slice used for demand $d$. Since variables $w_d$ and $y_d$ are defined regardless of particular network links, the continuity constraint is guaranteed, i.e., the demand uses the same spectrum slices in all links included in the routing path. In turn, to ensure the contiguity constraint resulting in all slices assigned to a particular demand being adjacent, the following inequality must be satisfied:

$$y_d - w_d + 1 \geq \sum_{p \in P(d)} x_{dp} n_{dp}, \quad d \in D. \tag{3.1.1c}$$

To avoid spectrum overlapping, an additional variable is required. Let $o_{di}$ be 1 if the starting slice of demand $d$ is smaller than that of demand $i$ and 0 otherwise. Since only one demand can use the smaller starting slice, the following equality holds:

$$o_{di} + o_{id} = 1, \quad d, i \in D : d \neq i. \tag{3.1.1d}$$

However, not all pairs of demands need to be controlled against spectrum overlapping. Accordingly, the binary variable $c_{di}$ is used to denote whether demands $d$ and $i$ use common link(s). The following constraint defines $c_{di}$ according to values of variables $y_{ed}$ and $y_{ei}$:

$$c_{di} \geq y_{ed} + y_{ei} - 1, \quad e \in E, d, i \in D : d \neq i. \tag{3.1.1e}$$

To guarantee the non-overlapping constraint, which states that a slice on a particular link can be allocated to at most one demand, two additional constraints are needed:

$$y_i - w_d + 1 \leq |S|(1 + o_{di} - c_{di}), \quad d, i \in D : d \neq i \tag{3.1.1f}$$

$$y_d - w_i + 1 \leq |S|(2 - o_{di} - c_{di}), \quad d, i \in D : d \neq i. \tag{3.1.1g}$$

It is clear that constraint (3.1.1f) is only active if both demands $d$ and $i$ share at least one link ($c_{di} = 1$) and the starting slice of demand $d$ is larger or equal than that of demand $i$ ($o_{di} = 0$). Then, the right-hand side of (3.1.1f) equals 0, which ensures that $y_i + 1 \leq w_d$, i.e., demands $d$ and $i$ do not use common slices and the spectrum is not overlapped. The next constraint (3.1.1g) is activated only if both demands $d$ and $i$ use the same link ($c_{di} = 1$) and the starting slice of demand $d$ is smaller than that of demand $i$ ($o_{di} = 1$). Analogously, if the right-hand side of (3.1.1g) is set to 0, then $y_d + 1 \leq w_i$, which once again guarantees that a slice on a particular link is allocated to at most one of $d$ and $i$ demands.

Model (3.1.1) represents one possible slice-based formulation, since other similar approaches have also been proposed in the literature. However, according to results reported in [26], model (3.1.1) outperforms other formulations.

**Channel-Based Formulation**

The channel-based approach was first proposed in [26] and it has been widely applied to formulate various optimization problems related to EONs, e.g., [35, 40–59]. A *channel* can be defined as a pre-computed set of spectrum contiguous frequency slices of a particular size (number of slices). Figure 3.3 shows an illustrative example of the channel-approach. There are eight slices available in the network denoted as $s_1, s_2, \ldots, s_8$. Accordingly, seven channels with a size of two slices ($n = 2$) can be created, namely, channels $(s_1, s_2), (s_2, s_3), \ldots, (s_7, s_8)$. In turn, assuming four slices ($n = 4$), we can create five channels: $(s_1, s_2, s_3, s_4), (s_2, s_3, s_4, s_5), \ldots, (s_5, s_6, s_7, s_8)$. Next, three channels with six slices ($n = 6$) are available: $(s_1, s_2, s_3, s_4, s_5, s_6)$, $(s_2, s_3, s_4, s_5, s_6, s_7)$ and $(s_3, s_4, s_5, s_6, s_7, s_8)$. Finally, only one channel with eight slices can be created.

Let $C_n$ denote a set of all channels with a size of $n$ slices. Note that if there are $|S|$ slices in the EON, set $C_n$ includes $|S| - n + 1$ channels. Let $C(d) = C_{n_d}$ denote

**Fig. 3.3** Example channels created for flexible grid with eight available slices

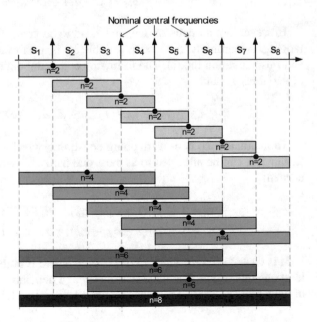

a set of candidate channels for demand $d$ when the number of slices required for demand $d$ is fixed and given by constant $n_d$. However, due to the fact that EONs can support various modulation formats, it is possible that the same demand $d$ can require different numbers of slices depending on the routing path selected to realize the demand (for more details see Sect. 3.1.2). Therefore, we define $C(d, p)$ as a set of candidate channels for demand $d$ using path $p$. To define channel $c$ in a more formal way, constant $\gamma_{dpcs}$ is used. In particular, $\gamma_{dpcs}$ is 1, if channel $c$ associated with demand $d$ on path $p$ (i.e., included in set $C(d, p)$) uses slice $s$ and 0 otherwise.

Having defined sets of candidate channels, the spectrum can be allocated to the RSA problem by using a binary variable which directly denotes which channel is selected for a particular demand. To combine the routing and spectrum (channel) allocation in a one decision variable, let $x_{dpc}$ denote a binary variable which is 1 if channel $c$ on candidate path $p$ is used to realize demand $d$ and 0 otherwise. Note that to satisfy continuity and contiguity constraints for each demand $d \in D$ only one path and one channel can be selected, i.e., the following equality must be satisfied:

$$\sum_{p \in P(d)} \sum_{c \in C(d,p)} x_{dpc} = 1, \quad d \in D. \tag{3.1.2a}$$

Moreover, to ensure that slice $s$ in link $e$ is used for one demand only, i.e., to avoid spectrum overlapping, the following constraint is formulated:

$$\sum_{d \in D} \sum_{p \in P(d)} \sum_{c \in C(d,p)} \gamma_{dpcs} \delta_{edp} x_{dpc} \leq 1, \quad e \in E, s \in S. \tag{3.1.2b}$$

The channel-based formulation (3.1.2) compact at only two sets of constraints, which may facilitate the optimization process. However, before conducting a comprehensive performance comparison of the slice-based and channel-based approaches in the context of different types of flows, shown below, we should point out a potential drawback of the channel-based approach. When the bit-rate of a demand is not fixed (for instance the bit-rate is determined in the optimization process) and as a result the value of $n_{dp}$ is also not fixed, set of candidate channels $C(d, p)$ cannot be limited to channels of one size (i.e., $n_{dp}$ slices) and must include channels of different size. Thus, the size of set $C(d, p)$ increase significantly, which in turn increases the size of the problem (number of variables). For a comprehensive comparison and discussion of slice-based and channel-based spectrum modeling in EONs refer to [26].

### 3.1.2 Distance Adaptive Transmission

One of the key innovations of EONs is that they enable distance-adaptive transmission. For the purpose of our discussion, we assume that various modulation formats included in set $M$ are available in the EON. Here, we use six modulation formats, namely BPSK, QPSK, and x-QAM, where x belongs to 8, 16, 32, 64, with spectral efficiency equal to 1, 2, ..., 6 [b/s/Hz], respectively. However, the discussion and models presented below are generic and can be adapted to other scenarios in terms of the number of types and modulation formats.

Note that modulation formats provide a trade-off between spectrum efficiency and transmission range, i.e., more spectrum-efficient modulation formats provide shorter transmission ranges while requiring fewer spectrum slices. Therefore, if spectrum usage is the only optimization objective, more effective modulation formats are selected albeit with shorter transmission ranges, increasing the number of regenerators required in the network and increasing network cost. As such, a reasonable strategy—known as *distance-adaptive transmission* (DAT)—is to preselect a modulation format for a particular demand based on the transmission distance only. More precisely, all available modulation formats are analyzed according to the transmission distance and the most effective format not exceeding the transmission range is selected. Thus, additional regenerators are not required to regenerate the signal, while spectrum usage is kept on a reasonable level. Another advantage of this approach is that instead of a RMSA (Routing, Modulation and Spectrum Allocation) problem, we solve a simpler RSA problem as the modulation is selected in advance as described above.

To illustrate the DAT approach, the following example is considered. Table 3.1 shows spectrum requirements and transmission ranges of modulation formats calculated according to [60] for bit-rate 200 Gb/s. It is assumed that the transmission distance (i.e., length of a path selected to realize the demand) is 1100 km; the modulation format 8-QAM is selected according to the DAT rule, since the transmission range of 8-QAM (1325 km) is greater than the distance considered. Consequently, only eight slices and no regenerators are required to realize the demand. Note that less spectrum is needed (six slices) if the modulation format 16-QAM is chosen; however, an extra regenerator is required, since the transmission range of 16-QAM (1031 km) is lower than 1100 km.

It should be noted that the DAT rule is commonly used in optimization of EONs in the context of RSA problems, e.g., [8, 32, 36, 44–47, 54–59, 61]

**Table 3.1** Distance-adaptive modulation formats for bit-rate 200 Gb/s

|                              | BPSK | QPSK | 8-QAM | 16-QAM | 32-QAM | 64-QAM |
|------------------------------|------|------|-------|--------|--------|--------|
| Required number of slices    | 18   | 10   | 8     | 6      | 6      | 6      |
| Maximum range (km)           | 1912 | 1618 | 1325  | 1031   | 738    | 444    |

## 3.2 Routing and Spectrum Allocation for Anycast Flows

The Routing and Spectrum Allocation problem is the basic optimization problem encountered in EONs (see Sect. 3.1). This section presents the RSA problem in the context of anycast flows. Three formulations are presented and examined: node-link, slice-based and channel-based. The number of slices required for each demand is calculated according to the DAT rule described in Sect. 3.1.2.

### 3.2.1 Formulations

It is assumed that there are $r$ data centers (DCs) located in the EON at some network nodes (we do not focus here on the DC/replica location problem). Moreover, each DC provides the same service/content and there is no limit on the number of served clients. Consequently, each DC can provision the requested service/content for every anycast demand. Each DC is connected to the backbone EON network by a local access link with a large bandwidth and hence capacity of this link is not included as a constraint in the optimization model.

In this section, all demands to be provisioned in the EON are anycast. More precisely, anycast demand $d$ is defined by a client node and bit-rate $h_d$. There are two types of anycast demands: upstream (from the client node to the DC node server) and downstream (from the DC node to the client node). Two demands (downstream and upstream) realizing the same anycast request are known as *associated*. Let $\tau(d)$ denote the associated demand of demand $d$. Both associated demands $d$ and $\tau(d)$ must be connected to the same DC node. For more details on anycast modeling refer to Sect. 1.2.2.

#### Node-Link Formulation

The first model is based on the node-link formulation of anycast flows. However, following the basic formulation shown in Sect. 1.2.2, the model must be modified, since the bandwidth requirement is not known in advance due to the DAT rule. More specifically, as described in Sect. 3.1.2, the number of slices required for a particular demand is a function of the transmission distance. Moreover, it should be stressed that in some physical transmission models (e.g., [60]), the transmission reach of an optical signal is not only a function of the selected modulation format but it also depends on the transported bit-rate. Therefore, it is assumed that the number of slices required for a particular demand can depend on both the transmission distance and the bit-rate. Since the bit-rate is defined for each demand, instead of referring directly to its value, the demand index denoted as $d$ is used. The slice-based approach is used to model spectrum usage. For each modulation format $m$ and demand $d$, constant $b_{dm}$ denotes the maximum supported distance range for modulation format $m$ and bit-rate $h_d$ defined for demand $d$. Furthermore, $n_{dm}$ denotes the number of slices required when modulation $m$ is used for the bit-rate of demand $d$. Set $M$ includes all available modulation formats (i.e., 64-QAM, 32-QAM, 16-QAM, 8-QAM, QPSK and BPSK) sorted according to the increasing values of the transmission range and number of required slices. Consequently, $m = 1$ denotes 64-QAM, $m = 2$ denotes 32-QAM, etc., which ensures that $b_{d(m+1)} \geq b_{dm}$ and $n_{d(m+1)} \leq n_{dm}$. Moreover, let $a_{dm}$ denote a lower bound of the distance range supported by modulation format $m$ and demand $d$. For $m = 1$, $a_{d1} = 0$, but if $m > 1$, then $a_{dm} = b_{d(m-1)} + 1$.

The main challenge of the following formulation is to include the option of calculating the number of slices required for each demand according to the distance of a routing path selected for a particular demand. Let binary variable $x_{ed}$ denote the selected routing path of demand $d$. $x_{ed}$ is 1 if demand $d$ uses link $e$. Using values of $x_{ed}$, we can calculate the length of the routing path selected for demand $d$ as $x_d = \sum_{e \in e} l_e x_{ed}$, where $l_e$ denotes the length of link $e$. Next, having the path length given by variable $x_d$, a modulation format can be selected according to the DAT rule. For this purpose, the path length $x_d$ is compared to the lower bound of the transmission range $a_{dm}$ of all modulation formats $m \in M$. An auxiliary binary variable $u_{dm}$ is used to check whether any modulation format $i \leq m$ can be applied for demand $d$ according to the DAT rule. Consistently with the values of $n_{dm}$, constant $h_{dm} = n_{dm} - n_{d(m-1)}$ can be defined as the number of additional slices required for demand $d$, if modulation format $m$ is applied instead of modulation format $m - 1$. Finally, the number of slices required for demand $d$ the selected routing path is given by the formula $u_d = \sum_{m \in M} u_{dm} h_{dm}$.

---

### EON/A/RSA/Spectrum/Node-link/Slice-based

**sets**

| | |
|---|---|
| $V$ | nodes |
| $E$ | links |
| $R$ | DC nodes |
| $\delta^+(v)$ | links leaving node $v$ |
| $\delta^-(v)$ | links entering node $v$ |
| $D$ | anycast demands |
| $D^{DS}$ | anycast downstream demands |
| $D^{US}$ | anycast upstream demands |
| $M$ | modulation formats |

**constants**

| | |
|---|---|
| $h_d$ | volume (requested bit-rate) of demand $d$ (Gb/s) |
| $b_{dm}$ | maximum distance range supported for modulation format $m$ and demand $d$ (km) |
| $a_{dm}$ | lower bound of the distance range supported for modulation format $m$ and demand $d$. If $m = 1$, then $a_{d1} = 0$. If $m > 1$, then $a_{dm} = b_{d(m-1)} + 1$ (km) |
| $n_{dm}$ | number of slices required for demand $d$ (with bit-rate $h_d$) using modulation format $m$ |

$h_{dm}$ number of additional slices required for demand $d$ if modulation format $m$ is applied instead of modulation format $m-1$, $h_{dm} = n_{dm} - n_{d(m-1)}$

$l_e$ length of link $e$ (km)

$l^{max}$ maximum distance of a path in the network (km)

$n$ number of slices available on each fiber link

$s_d$ source node of demand $d$

$t_d$ destination node of demand $d$

$\tau(d)$ index of a demand associated with demand $d$. If $d$ is a downstream demand, then $\tau(d)$ must be an upstream demand and vice versa

**variables**

$x_{ed}$ =1, if demand $d$ uses link $e$; 0, otherwise (binary)

$x_d$ length of path created for demand $d$ (continuous)

$u_{dm}$ =1, if any modulation format $i \leq m$ can be applied for demand $d$ according to DAT; 0, otherwise (binary)

$u_d$ number of slices required for demand $d$ (integer)

$o_{di}$ =1, if the starting slice of demand $d$ is smaller than that of demand $i$; 0, otherwise (binary)

$c_{di}$ =1, if demands $d$ and $i$ use common link(s); 0, otherwise (binary)

$w_d$ indicates the starting slice used for demand $d$ (integer)

$y_d$ indicates the ending slice used for demand $d$ (integer)

$y$ indicates the maximum slice used in the network (integer)

$z_{vd}$ =1, if DC located at node $v$ is selected to realize anycast demand $d$; 0, otherwise (binary)

**objective**

$$\text{minimize} \quad F = y \tag{3.2.1a}$$

**constraints**

$$\sum_{e \in \delta^+(v)} x_{ed} - \sum_{e \in \delta^-(v)} x_{ed} = \begin{cases} +z_{vd} & \text{if } v \in R \\ -1 & \text{if } v = t_d, \\ 0 & \text{otherwise} \end{cases} \quad v \in V, d \in D^{DS} \tag{3.2.1b}$$

$$\sum_{e \in \delta^+(v)} x_{ed} - \sum_{e \in \delta^-(v)} x_{ed} = \begin{cases} +1 & \text{if } v = s_d \\ -z_{vd} & \text{if } v \in R, \\ 0 & \text{otherwise} \end{cases} \quad v \in V, d \in D^{US} \tag{3.2.1c}$$

$$x_d \geq \sum_{e \in e} l_e x_{ed}, \quad d \in D \tag{3.2.1d}$$

$$l^{max} u_{dm} \geq (x_d - a_{dm}), \quad d \in D, m \in M \tag{3.2.1e}$$

$$u_d \geq \sum_{m \in M} u_{dm} h_{dm}, \quad d \in D \tag{3.2.1f}$$

$$c_{di} \geq x_{ed} + x_{ei} - 1, \quad e \in E, d, i \in D : d \neq i \tag{3.2.1g}$$

$$o_{di} + o_{id} = 1, \quad d, i \in D : d \neq i \tag{3.2.1h}$$

$$y_i - w_d + 1 \leq n(1 + o_{di} - c_{di}), \quad d, i \in D : d \neq i \tag{3.2.1i}$$

$$y_d - w_i + 1 \leq n(2 - o_{di} - c_{di}), \quad d, i \in D : d \neq i \tag{3.2.1j}$$

$$y_d - w_d + 1 \geq u_d, \quad d \in D \tag{3.2.1k}$$

$$y \geq y_d, \quad d \in D \tag{3.2.1l}$$

$$z_{vd} = z_{v\tau(d)}, \quad d \in D, v \in R \tag{3.2.1m}$$

$$\sum_{v \in R} z_{vd} = 1, \quad d \in D. \tag{3.2.1n}$$

The goal of optimization (3.2.1a) is to minimize the overall spectrum usage defined as the maximum number of slices allocated in the network to provision the demands. Equations (3.2.1b) and (3.2.1c) are flow conservation constraints defined using the node-link formulation for downstream and upstream anycast flows, respectively. More specifically, the left-hand side of each equation defines the difference of flows entering and leaving node each $v \in V$. The right-hand side of the equations depends on the demand type. For downstream demands, the flow difference of node $v$ is 1 if node $v$ is the DC node selected for demand $d$ according to variable $z_{vd}$, $-1$ if node $v$ is the destination (client) node of demand $d$, and 0 in all other cases. On the other hand, for upstream demands, the flow difference of node $v$ is: 1 if node $v$ is the source (client) node demand $d$; $-1$ if node $v$ is the DC node of demand $d$ according to variable $z_{vd}$; and 0 otherwise.

Constraints (3.2.1d)–(3.2.1f) are used to obtain the number of slices required for demand $d$ according to the DAT rule. More specifically, constraint (3.2.1d) defines the length of a routing path selected for demand $d$ according to the values of variables $x_{ed}$ and lengths of links $l_e$. Inequality (3.2.1e) ensures the correct value of variable $u_{dm}$ which denotes whether any modulation format $i \leq m$ can be applied to demand $d$. Finally, condition (3.2.1f) defines the number of slices required for demand $d$ (variable $u_d$).

The next group of constraints (3.2.1g)–(3.2.1k) controls the spectrum usage. More specifically, these conditions ensure the non-overlapping constraint, which states that a slice on a particular link can be allocated to at most one demand. Condition (3.2.1g) checks whether two different demands $d$ and $i$ use the same link $e$. Inequality (3.2.1h) guarantees that either the starting slice of demand $d$ is smaller than that of demand $i$ or the opposite. Constraint (3.2.1i) is activated only if both analyzed demands $d$ and $i$ use the same link ($c_{di} = 1$) and the starting slice of demand $d$ is larger than or equal to demand $i$ ($o_{di} = 0$). In such a case, the right-hand side of (3.2.1i) is 0, which ensures the correct relationship between variables $y_i$ and $w_d$ denoting the ending slice used for demand $i$ and the starting slice used for demand $d$, respectively. In turn, constraint (3.2.1j) operates only if both analyzed demands $d$ and $i$ use the same link ($c_{di} = 1$) and the starting slice of demand $d$ is smaller than that of demand $i$ ($o_{di} = 1$). Similarly, this means that the right-hand side of (3.2.1j) is set to 0 which guarantees that $y_d + 1 \leq w_i$. Note that constraints (3.2.1h)–(3.2.1j) are in the model to avoid spectrum overlapping of two different demands $d$ and $i$. Equation (3.2.1k) ensures that the requested number of slices for each demand is satisfied. Constraint (3.2.1l) defines the maximum slice index used in the network.

The last two constraints (3.2.1m)–(3.2.1n) are related to anycast flows. The former expresses the rule that both associated demands $d$ and $\tau(d)$ must use the same DC node, while the latter states that exactly one node can be selected as the DC node for each demand.

### Link-Path and Slice-Based Formulation

The next model uses the link-path formulation explained in Sects. 1.2.1 and 1.2.2. The main difference from the node-link formulation is that the routing process is conducted by selecting a path from candidate paths that are pre-calculated for each demand. A binary decision variable $x_{dp}$ is applied to denote whether path $p$ is selected to realize demand $d$. For more details on link-path modeling of anycast flows refer to Sect. 1.2.2. To model spectrum usage, we use the same approach as above, therefore the model is known as slice-based.

---

### EON/A/RSA/Spectrum/Link-path/Slice-based

**sets**

| | |
|---|---|
| $E$ | links |
| $S$ | slices |
| $D$ | anycast demands |
| $D^{DS}$ | anycast downstream demands |
| $P(d)$ | candidate paths for flows realizing demand $d$. If $d$ is an anycast upstream demand, the candidate path connects the client node and the DC node. If $d$ is a downstream demand, the candidate path connects the DC node and the client node |

**constants**

| | |
|---|---|
| $\delta_{edp}$ | $=1$, if link $e$ belongs to path $p$ realizing demand $d$; 0, otherwise |
| $n_{dp}$ | requested number of slices for demand $d$ on path $p$ |
| $n$ | number of slices available on each fiber link |
| $o(p)$ | origin node of path $p$ |
| $t(p)$ | destination node of path $p$ |
| $\tau(d)$ | index of a demand associated with demand $d$. If $d$ is a downstream demand, then $\tau(d)$ must be an upstream demand and vice versa |

**variables**

| | |
|---|---|
| $x_{dp}$ | $=1$, if candidate path $p$ is used to realize demand $d$; 0, otherwise (binary) |
| $y_{ed}$ | $=1$, if demand $d$ uses link $e$; 0, otherwise (binary) |
| $u_d$ | number of slices required for demand $d$ (integer) |

$o_{di}$     =1, if the starting slice of demand $d$ is smaller than that of demand $i$; 0, otherwise (binary)

$c_{di}$     =1, if demands $d$ and $i$ use common link(s); 0, otherwise (binary)

$w_d$     indicates the starting slice used for demand $d$ (integer)

$y_d$     indicates the ending slice used for demand $d$ (integer)

$y$      indicates the maximum slice used in the network (integer)

**objective**

$$\text{minimize} \quad F = y \tag{3.2.2a}$$

**constraints**

$$\sum_{p\in P(d)} x_{dp} = 1, \quad d \in D \tag{3.2.2b}$$

$$\sum_{p\in P(d)} \delta_{edp}x_{dp} \leq y_{ed}, \quad d \in D, e \in E \tag{3.2.2c}$$

$$\sum_{p\in P(d)} n_{dp}x_{dp} \leq u_d, \quad d \in D \tag{3.2.2d}$$

$$c_{di} \geq y_{ed} + y_{ei} - 1, \quad e \in E, d, i \in D : d \neq i \tag{3.2.2e}$$

$$o_{di} + o_{id} = 1, \quad d, i \in D : d \neq i \tag{3.2.2f}$$

$$y_i - w_d + 1 \leq n(1 + o_{di} - c_{di}), \quad d, i \in D : d \neq i \tag{3.2.2g}$$

$$y_d - w_i + 1 \leq n(2 - o_{di} - c_{di}), \quad d, i \in D : d \neq i \tag{3.2.2h}$$

$$y_d - w_d + 1 \geq u_d, \quad d \in D \tag{3.2.2i}$$

$$y \geq y_d, \quad d \in D \tag{3.2.2j}$$

$$\sum_{p\in P(d)} o(p)x_{dp} = \sum_{p\in P(\tau(d))} t(p)x_{\tau(d)p}, \quad d \in D^{DS}. \tag{3.2.2k}$$

New constraints of the above model—compared to model (3.2.1)—follow from the link-path formulation. In particular, Eq. (3.2.2b) ensures that exactly one candidate path is selected to realize each demand. Condition (3.2.2c) is in the model to obtain values of variables $y_{ed}$ denoting whether demand $d$ uses link $e$. Inequality (3.2.2d) is used to define variable $u_d$, which denotes the number of slices required for demand $d$. Constraints (3.2.2e)–(3.2.2j) which control spectrum usage are the same as in the previous model (3.2.1). The last equality (3.2.2k) guarantees the anycast constraint required in the model to ensure that two associated anycast demands use the same DC node.

## Link-Path and Channel-Based Formulation

The third model proposed to formulate the RSA problem with anycast flows is based on the link-path formulation and the channel-based approach to spectrum control.

---

### EON/A/RSA/Spectrum/Link-path/Channel-based

**sets**

| | |
|---|---|
| $E$ | links |
| $S$ | slices |
| $D$ | anycast demands |
| $D^{DS}$ | anycast downstream demands |
| $P(d)$ | candidate paths for flows realizing demand $d$. If $d$ is an anycast upstream demand, the candidate path connects the client node and the DC node. If $d$ is a downstream demand, the candidate path connects the DC node and the client node |
| $C(d, p)$ | candidate channels for demand $d$ allocated on path $p$ |

**constants**

| | |
|---|---|
| $\delta_{edp}$ | $=1$, if link $e$ belongs to path $p$ realizing demand $d$; 0, otherwise |
| $n_{dp}$ | requested number of slices for demand $d$ on path $p$ |
| $\gamma_{dpcs}$ | $=1$, if channel $c$ associated with demand $d$ on path $p$ uses slice $s$; 0, otherwise |
| $\tau(d)$ | index of a demand associated with demand $d$. If $d$ is a downstream demand, then $\tau(d)$ must be an upstream demand and vice versa |
| $o(p)$ | origin node of path $p$ |
| $t(p)$ | destination node of path $p$ |

**variables**

| | |
|---|---|
| $x_{dpc}$ | $=1$, if channel $c$ on candidate path $p$ is used to realize demand $d$; 0, otherwise (binary) |
| $y_{es}$ | $=1$, if slice $s$ is occupied on link $e$; 0, otherwise (binary) |
| $y_s$ | $=1$, if slice $s$ is occupied on any network link; 0, otherwise (binary) |
| $y$ | indicates the maximum slice used in the network (integer) |

**objective**

$$\text{minimize} \quad F = \sum_{s \in S} y_s \tag{3.2.3a}$$

**constraints**

$$\sum_{p\in P(d)}\sum_{c\in C(d,p)} x_{dpc} = 1, \quad d \in D \tag{3.2.3b}$$

$$\sum_{d\in D}\sum_{p\in P(d)}\sum_{c\in C(d,p)} \gamma_{dpcs}\delta_{edp}x_{dpc} \leq y_{es}, \quad e \in E, s \in S \tag{3.2.3c}$$

$$\sum_{e\in E} y_{es} \leq |E| \, y_s, \quad s \in S \tag{3.2.3d}$$

$$\sum_{p\in P(d)}\sum_{c\in C(d,p)} o(p)x_{dpc} = \sum_{p\in P(\tau(d))}\sum_{c\in C(\tau(d),p)} t(p)x_{\tau(d)pc}, \quad d \in D^{DS}. \tag{3.2.3e}$$

As in both previous models, the objective function (3.2.3a) represents spectrum usage in the network. However, in this model the value of the objective function is calculated as a sum of binary variables $y_s$, which denote whether slice $s$ is used in the network to realize a demand. The first constraint given by (3.2.3b) is slightly different from the previous model (3.2.2), since it makes the routing decision (selecting a routing path $p$), as well as choosing the spectrum channel $c$. Spectrum control is provisioned by inequality (3.2.3c), which imposes that a slice $s$ on a particular link $e$ can be used by at most one demand. Constraint (3.2.3d) defines variable $y_s$. The last equation (3.2.3e) represents the anycast constraint.

### 3.2.2  Numerical Results

This section examines the performance of formulations proposed for the RSA problem with anycast flows: the node-link model formulated in (3.2.1) and denoted in short as NL, the link-path and slice-based model formulated in (3.2.2) and referred to as SB, and the link-path and channel-based model formulated in (3.2.3) and denoted in short as CB. Let us recall that the last two models use the link-path formulation with a set of candidate paths for each demand. Different values of candidate paths defined for each node pair in the network were used in the experiments, namely, $k = 2, 3, 5, 10$. The candidate paths for each node pair were calculated using the k-shortest path algorithm with a link metric defined as the link length given in kilometers. Since anycast demands can use various DCs, the number of candidate paths for each anycast demand was $kr$, where $r$ is the number of DCs in the network.

Two network topologies were tested in two experiments: the German national network DT14 with 14 nodes and 46 directed links (Fig. A.2, Table A.2), and the pan-European network Euro16 with 16 nodes and 48 directed links (Fig. A.5, Table A.4). Ten sets of anycast demands were generated for each network with $|D| = 10, 12, 14, 16, 18, 20, 22, 24$ and 26. The bit-rate of each demand was randomly selected from 10 to 400 Gb/s with 10 Gb/s granularity. There were two nodes with a data center for each network. The CPLEX solver was used to compare the models [62]. A time limit of 1 h was set for solving each problem instance, and all other solver settings were left as default. Experiments were run on a PC with IntelCore i7-2620M CPU and 4GB RAM.

In the experiments, it was assumed that EON uses BV-Ts (Bandwidth-Variable Transponders) supporting the PDM-OFDM technology with modulation formats of BPSK, QPSK, and x-QAM, where x belongs to 8, 16, 32, 64. The spectral efficiency of the modulation formats is equal to 1, 2, ..., 6 [b/s/Hz], respectively. PDM stands for Polarization Division Multiplexing, which doubles the spectral efficiency. EON operates within a flexible ITU-T grid of a 6.25 GHz granularity [11]. The BV-Ts enable a bit-rate adaptability with a 10 Gb/s granularity. The transmission model proposed in [60] is used. The model estimates the transmission reach of an optical signal as a function of the selected modulation level and transported bit-rate. A 12.5 GHz guard band between neighboring connections is introduced. In all scenarios, the transmission reach is extended using regenerators, which are applied whenever necessary. To select the modulation format, the DAT rule is applied as described in Sect. 3.1.2.

In preliminary experiments, we noted that the execution time of the channel-based model is highly sensitive to the number of slices available on each fiber link (parameter $n$). More specifically, this parameter directly defines the number of available channels. For instance, if $n = 40$, then one can create 39 channels with a width of two slices, 37 channels with a width of four slices, etc. Consequently, the value of $n$ has a major impact on the problem complexity expressed in the number of variables, since the number of the main decision variable $x_{dpc}$ depends linearly on the number of channels.

To study this issue in more detail, we conducted the following experiment. Using a randomly selected traffic pattern of DT14 with 26 demands, the optimal result (i.e., the minimum number of required slices) was determined first. Next, the channel-based model was run starting with $n$ equal to the optimal value; we then increased $n$ by two slices up to 20 additional slices. Figure 3.4 shows results obtained for different numbers of candidate paths, namely $k = 2, 3, 5, 10$. The execution time of the channel-based model grows almost exponentially as the number of additional

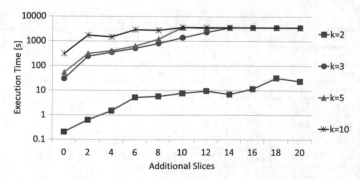

**Fig. 3.4** Execution time of the channel-based model as a function of the number of additional slices for the DT14 network and different number of candidate paths

slices increases (the y-axis is in the logarithmic scale), and finally reaches the 3600 s limit for $k = 3, 5, 10$. Moreover, the execution time strongly depends on the number of candidate paths used in the optimization.

We introduced the following procedure to improve the performance of the channel-based model. Firstly, each problem instance was solved by the heuristic algorithm AFA (for more details on the algorithm see Sect. 3.3.2). Next, the heuristic solution was set as parameter $n$ for the channel-based model. The execution time of the AFA algorithm for small networks was negligibly small and had no impact on the execution time of the channel-based model. For the slice-based model, we set $n = 40$ in all experiments, since this model was not dependent on the number of slices available.

Three performance metrics are reported in tables below: spectrum usage, optimality gap of results, and execution time. The first metric denotes the number of slices required, calculated according to the objective function used in models being compared. The second metric shows the percentage distance between the best result and the lower bound provided by CPLEX. This metric is due to the fact that the CPLEX solver is run with a 1 h time limit, and if the optimum result is not yielded within the hour, an optimality gap shows the maximum possible gap between the actual result and the optimum. Finally, the last reported metric denotes the execution time of the CPLEX solver in seconds. The results of each performance metric shown in the following tables are averaged over ten demand sets created for each examined value of $|D|$. More precisely, there are three tables related to each reported metric for each network topology. Tables 3.2, 3.4 and 3.6 show results related to the DT14 network, while Tables 3.3, 3.5 and 3.7 include results obtained for the Euro16 network. Each table contains ten columns. The first column shows the number of demands in the traffic pattern. The remaining columns contain results of the formulations analyzed in the node-link (NL), slice-based (SB) with different values of $k$ and channel-based (CB) with different values of $k$ experiments.

**Table 3.2** Comparison of ILP models for the RSA problem with anycast flows—average number of slices required for the DT14 network

| $|D|$ | NL | SB | | | | CB | | | |
|---|---|---|---|---|---|---|---|---|---|
| | | $k = 2$ | $k = 3$ | $k = 5$ | $k = 10$ | $k = 2$ | $k = 3$ | $k = 5$ | $k = 10$ |
| 10 | 8.0 | 8.2 | 8.0 | 8.0 | 8.0 | 8.2 | 8.0 | 8.0 | 8.0 |
| 12 | 8.2 | 8.6 | 8.2 | 8.2 | 8.2 | 8.6 | 8.2 | 8.2 | 8.2 |
| 14 | 8.4 | 9.2 | 8.4 | 8.4 | 8.4 | 9.2 | 8.4 | 8.4 | 8.4 |
| 16 | 12.4 | 12.0 | 9.6 | 9.6 | 9.6 | 12.0 | 9.6 | 9.6 | 9.6 |
| 18 | 13.1 | 11.4 | 9.4 | 9.4 | 9.4 | 11.4 | 9.4 | 9.4 | 9.4 |
| 20 | 19.0 | 11.6 | 9.6 | 9.6 | 9.6 | 11.6 | 9.6 | 9.6 | 9.6 |
| 22 | 24.8 | 13.0 | 12.0 | 12.0 | 12.0 | 13.0 | 12.0 | 12.0 | 12.0 |
| 24 | – | 14.4 | 12.0 | 12.0 | 12.0 | 14.4 | 12.0 | 12.0 | 12.0 |
| 26 | – | 13.8 | 12.2 | 12.2 | 12.2 | 13.8 | 12.2 | 12.2 | 12.2 |

**Table 3.3** Comparison of ILP models for the RSA problem with anycast flows—average number of slices required for the Euro16 network

| $|D|$ | NL | SB | | | | CB | | | |
|---|---|---|---|---|---|---|---|---|---|
| | | $k=2$ | $k=3$ | $k=5$ | $k=10$ | $k=2$ | $k=3$ | $k=5$ | $k=10$ |
| 10 | 9.8 | 10.2 | 9.8 | 9.8 | 9.8 | 10.2 | 9.8 | 9.8 | 9.8 |
| 12 | 11.8 | 12.6 | 11.8 | 11.8 | 11.8 | 12.6 | 11.8 | 11.8 | 11.8 |
| 14 | 13.0 | 10.8 | 10.6 | 10.6 | 10.6 | 10.8 | 10.6 | 10.6 | 10.6 |
| 16 | 12.2 | 11.8 | 11.0 | 10.8 | 10.8 | 11.8 | 11.0 | 10.8 | 10.8 |
| 18 | 12.0 | 11.0 | 10.8 | 10.8 | 10.8 | 11.0 | 10.8 | 10.8 | 10.8 |
| 20 | – | 13.8 | 12.8 | 12.8 | 12.6 | 13.8 | 12.8 | 12.8 | 12.6 |
| 22 | – | 14.6 | 13.4 | 12.8 | 12.8 | 14.6 | 13.4 | 12.8 | 12.8 |
| 24 | – | 15.4 | 13.6 | 13.4 | 13.4 | 15.4 | 13.6 | 13.4 | 13.4 |
| 26 | – | 14.4 | 14.4 | 14.4 | 14.4 | 14.4 | 14.4 | 14.4 | 14.2 |

**Table 3.4** Comparison of ILP models for the RSA problem with anycast flows—average optimality gap for the DT14 network

| $|D|$ | NL (%) | SB | | | | CB | | | |
|---|---|---|---|---|---|---|---|---|---|
| | | $k=2$ (%) | $k=3$ (%) | $k=5$ (%) | $k=10$ (%) | $k=2$ (%) | $k=3$ (%) | $k=5$ (%) | $k=10$ (%) |
| 10 | 0.00 | 0.00 | 0.00 | 0.00 | 0.00 | 0.00 | 0.00 | 0.00 | 0.00 |
| 12 | 1.00 | 0.00 | 0.00 | 0.00 | 0.00 | 0.00 | 0.00 | 0.00 | 0.00 |
| 14 | 0.00 | 0.00 | 0.00 | 0.00 | 0.00 | 0.00 | 0.00 | 0.00 | 0.00 |
| 16 | 15.00 | 0.00 | 0.00 | 0.00 | 0.00 | 0.00 | 0.00 | 0.00 | 0.00 |
| 18 | 14.05 | 0.00 | 0.00 | 0.83 | 1.67 | 0.00 | 0.00 | 0.00 | 0.00 |
| 20 | 25.77 | 0.00 | 0.00 | 0.00 | 0.00 | 0.00 | 0.00 | 0.00 | 0.00 |
| 22 | 51.40 | 0.00 | 1.67 | 4.10 | 5.10 | 0.00 | 0.00 | 0.00 | 0.00 |
| 24 | – | 1.25 | 1.83 | 6.79 | 8.83 | 0.00 | 0.83 | 0.00 | 0.83 |
| 26 | – | 1.25 | 2.28 | 2.50 | 2.50 | 0.00 | 0.63 | 0.00 | 0.63 |

Let us recall that the NL model (3.2.1) uses all candidate paths for each demand, since it is based on the node-link formulation of multicommodity flows. In contrast, SB (3.2.2) and CB (3.2.3) models are based on the link-path formulation and consequently they only consider a subset of candidate paths (parameter $k$). Therefore, the NL model should provide the best results or at least not worse results, when compared with the SB and CB models. This can be observed in Tables 3.2 and 3.3 for smaller values of $|D|$. However, starting from $|D| = 16$, the NL models is outperformed by SB and CB models even with $k = 2$. This is due to the time limit of 1 h used in CPLEX, as according to the complexity of the NL model, the solver is not able to yield optimal results in an hour for larger values of $|D|$. This is confirmed by the results shown in Tables 3.4 and 3.5, where we report optimality gaps. Moreover, for large values of $|D|$, the NL model is not able to provide any feasible results in 1 h.

**Table 3.5** Comparison of ILP models for the RSA problem with anycast flows—average optimality gap for the Euro16 network

| $|D|$ | NL (%) | SB | | | | CB | | | |
|---|---|---|---|---|---|---|---|---|---|
| | | $k = 2$ (%) | $k = 3$ (%) | $k = 5$ (%) | $k = 10$ (%) | $k = 2$ (%) | $k = 3$ (%) | $k = 5$ (%) | $k = 10$ (%) |
| 10 | 0.00 | 0.00 | 0.00 | 0.00 | 0.00 | 0.00 | 0.00 | 0.00 | 0.00 |
| 12 | 4.52 | 0.00 | 0.00 | 0.00 | 0.00 | 0.00 | 0.00 | 0.00 | 0.00 |
| 14 | 7.50 | 0.00 | 0.00 | 0.00 | 0.00 | 0.00 | 0.00 | 0.00 | 0.00 |
| 16 | 10.60 | 0.00 | 0.00 | 0.00 | 0.00 | 0.00 | 0.00 | 0.00 | 0.00 |
| 18 | 5.83 | 0.00 | 0.00 | 0.00 | 0.00 | 0.00 | 0.00 | 0.00 | 0.00 |
| 20 | – | 0.00 | 0.00 | 0.00 | 1.00 | 0.00 | 0.00 | 0.00 | 0.00 |
| 22 | – | 0.00 | 3.10 | 0.00 | 0.00 | 0.00 | 0.00 | 0.00 | 0.00 |
| 24 | – | 0.00 | 0.00 | 0.00 | 0.00 | 0.00 | 0.00 | 0.00 | 0.00 |
| 26 | – | 0.00 | 1.43 | 3.57 | 10.16 | 0.00 | 0.00 | 0.00 | 0.00 |

**Table 3.6** Comparison of ILP models for the RSA problem with anycast flows—average execution time is seconds for the DT14 network

| $|D|$ | NL | SB | | | | CB | | | |
|---|---|---|---|---|---|---|---|---|---|
| | | $k = 2$ | $k = 3$ | $k = 5$ | $k = 10$ | $k = 2$ | $k = 3$ | $k = 5$ | $k = 10$ |
| 10 | 0.2 | 0.0 | 0.1 | 0.2 | 0.6 | 0.1 | 0.0 | 0.0 | 0.0 |
| 12 | 360.1 | 0.1 | 0.1 | 0.2 | 0.9 | 0.1 | 0.2 | 0.1 | 0.4 |
| 14 | 209.7 | 0.1 | 0.2 | 0.3 | 3.5 | 0.3 | 0.1 | 0.3 | 0.4 |
| 16 | 1421.7 | 6.9 | 0.9 | 1.6 | 124.3 | 2.3 | 0.7 | 1.6 | 6.2 |
| 18 | 1710.8 | 0.4 | 94.5 | 363.0 | 371.5 | 1.3 | 0.9 | 13.3 | 21.7 |
| 20 | 2215.2 | 0.8 | 1.8 | 8.9 | 36.4 | 2.0 | 6.4 | 7.3 | 97.5 |
| 22 | 3410.8 | 291.3 | 397.9 | 1132.0 | 1136.5 | 3.6 | 8.4 | 18.9 | 131.3 |
| 24 | – | 377.3 | 825.8 | 1673.9 | 2006.8 | 17.3 | 370.0 | 224.3 | 389.1 |
| 26 | – | 367.1 | 419.6 | 398.6 | 790.6 | 379.3 | 426.1 | 280.1 | 453.8 |

More specifically, this applies for $|D| > 22$ in the DT14 network and for $|D| > 18$ in the Euro16 network. This indicates that the NL model based on the node-link formulation does not scale well with the size of the problem instance.

Another interesting observation focuses on the impact of the number of candidate paths (parameter $k$) in the SB and CB models. In particular, spectrum usage decreases as $k$ increases, since more routing paths can be selected. However, the differences between results obtained for different values of $k$ are relatively small. This is mainly due to the relatively small size of both networks.

Moreover, comparing results in Tables 3.2 and 3.3 shows that spectrum usage is slightly smaller for the DT14 network than for the Euro16 network. This is because the DT14 network is significantly smaller than the Euro16 network and in conse-

**Table 3.7** Comparison of ILP models for the RSA problem with anycast flows—average execution time is seconds for the Euro16 network

| $|D|$ | NL | SB | | | | CB | | | |
|---|---|---|---|---|---|---|---|---|---|
| | | $k = 2$ | $k = 3$ | $k = 5$ | $k = 10$ | $k = 2$ | $k = 3$ | $k = 5$ | $k = 10$ |
| 10 | 2.3 | 0.1 | 0.1 | 0.2 | 0.8 | 0.0 | 0.0 | 0.2 | 0.2 |
| 12 | 844.2 | 0.2 | 0.3 | 1.5 | 2.9 | 0.2 | 0.4 | 0.3 | 0.5 |
| 14 | 635.9 | 0.2 | 0.4 | 0.9 | 8.6 | 0.2 | 0.3 | 0.6 | 1.1 |
| 16 | 1447.5 | 0.5 | 28.3 | 6.3 | 19.2 | 2.4 | 5.0 | 14.2 | 76.5 |
| 18 | 757.0 | 0.2 | 0.4 | 6.6 | 32.3 | 0.1 | 0.3 | 0.4 | 1.4 |
| 20 | – | 0.5 | 1.9 | 13.2 | 384.7 | 1.2 | 9.8 | 28.2 | 66.7 |
| 22 | – | 1.6 | 724.1 | 93.1 | 230.6 | 9.3 | 14.3 | 33.5 | 77.9 |
| 24 | – | 51.0 | 175.9 | 165.5 | 182.5 | 53.3 | 7.8 | 26.4 | 331.3 |
| 26 | – | 13.3 | 797.2 | 1120.9 | 2056.3 | 10.2 | 207.0 | 281.3 | 67.5 |

quence more spectral efficient modulation formats are selected according to the DAT rule, resulting in lower values of the requested number of slices for each demand.

The last performance metric is the average execution time of all ILP models analyzed (Tables 3.6 and 3.7). The first obvious observation is that, generally speaking, the execution time for each model increases with problem complexity expressed by the number of demands. Moreover, for the SB and CB models, as the number of candidate paths increases, the running time also increases. Perhaps unsurprisingly, the NL model showed the poorest performance. The differences between the SB and CB models are not clear, although the channel-based model outperformed the slice-based approach in most cases.

In conclusion, the channel-based model defined in (3.2.3) provided the best performance in terms of result quality and running time. If the analyzed problem instance is relatively small, the node-link model (3.2.1) can yield the best results in a reasonable time, although for larger problem instances this model experienced scalability problems.

## 3.3 Routing and Spectrum Allocation for Anycast and Unicast Flows

This section focuses on the RSA problem with joint optimization of anycast and unicast flows [57]. According to the results reported above, the channel-based approach is used to model spectrum usage. Firstly, we present the spectrum, average spectrum and cost formulations. Next, we describe several heuristic and metaheuristic algorithms. Finally, we report results of extensive numerical experiments.

### 3.3.1   Formulations

We use the same assumptions as in Sect. 3.2 to modeling the EON. The main difference is that two types of flows are provisioned in the network, namely anycast flows and unicast flows.

**Spectrum Objective**

The first model aims to minimize spectrum usage defined as the maximum index of the slice required in the network to realize all demands included in the traffic matrix.

---

**EON/AU/RSA/Spectrum/Link-path/Channel-based**

**sets**

| | |
|---|---|
| $E$ | links |
| $S$ | slices |
| $D$ | demands (anycast and unicast) |
| $D^{DS}$ | anycast downstream demands |
| $P(d)$ | candidate paths for flows realizing demand $d$. If $d$ is a unicast demand, the candidate path connects the end nodes of the demand. If $d$ is an anycast upstream demand, the candidate path connects the client node and the DC node. If $d$ is a downstream demand, the candidate path connects the DC node and the client node |
| $C(d, p)$ | candidate channels for demand $d$ allocated on path $p$ |

**constants**

| | |
|---|---|
| $\delta_{edp}$ | $=1$, if link $e$ belongs to path $p$ realizing demand $d$; 0, otherwise |
| $n_{dp}$ | requested number of slices for demand $d$ on path $p$ |
| $\gamma_{dpcs}$ | $=1$, if channel $c$ associated with demand $d$ on path $p$ uses slice $s$; 0, otherwise |
| $\tau(d)$ | index of a demand associated with demand $d$. If $d$ is a downstream demand, then $\tau(d)$ must be an upstream demand and vice versa |
| $o(p)$ | origin node of path $p$ |
| $t(p)$ | destination node of path $p$ |

**variables**

| | |
|---|---|
| $x_{dpc}$ | $=1$, if channel $c$ on candidate path $p$ is used to realize demand $d$; 0, otherwise (binary) |
| $y_{es}$ | $=1$, if slice $s$ is occupied on link $e$; 0, otherwise (binary) |
| $y_s$ | $=1$, if slice $s$ is occupied on any network link; 0, otherwise (binary) |

**objective**

$$\text{minimize} \quad F = \sum_{s \in S} y_s \tag{3.3.1a}$$

**constraints**

$$\sum_{p \in P(d)} \sum_{c \in C(d,p)} x_{dpc} = 1, \quad d \in D \tag{3.3.1b}$$

$$\sum_{d \in D} \sum_{p \in P(d)} \sum_{c \in C(d,p)} \gamma_{dpcs} \delta_{edp} x_{dpc} \leq y_{es}, \quad e \in E, s \in S \tag{3.3.1c}$$

$$\sum_{e \in E} y_{es} \leq \mid E \mid y_s, \quad s \in S \tag{3.3.1d}$$

$$\sum_{p \in P(d)} \sum_{c \in C(d,p)} o(p) x_{dpc} = \sum_{p \in P(\tau(d))} \sum_{c \in C(\tau(d),p)} t(p) x_{\tau(d)pc}, \quad d \in D^{DS}. \tag{3.3.1e}$$

The objective of optimization (3.3.1a) is to minimize spectrum usage in terms of the number of slices required in the network to provision the whole traffic matrix. More precisely, variable $y_{es}$ checks whether slice $s$ is occupied on link $e$, i.e., if there is a demand using slice $s$ in link $e$. Next, variable $y_s$ denotes whether slice $s$ is allocated in at least one network link. Therefore, the objective function (3.3.1a) is formulated as $\sum_{s \in S} y_s$. Equation (3.3.1b) imposes that each demand $d$ uses exactly one candidate path and exactly one candidate channel. To ensure the non-overlapping constraints, i.e., that a slice on a particular link can be allocated to at most one demand, we use Eq. (3.3.1c). Note that the non-overlapping constraint follows directly from the definition of optical channels included in set $C(d, p)$. Constraint (3.3.1d) states that slice $s$ is used in the network ($y_s = 1$) only when there is at least one link $e \in E$ for which slice $s$ is allocated to realize a demand. Finally, the last Eq. (3.3.1e) guarantees that both associated anycast demands use candidate paths connected to the same DC node.

**Average Spectrum Objective**

It is worth noting that the objective function used in model (3.3.1) denotes maximum spectrum usage in the most congested link in the network. In the context of anycast flows served by DCs, links adjacent to nodes hosting DCs are usually highly loaded in comparison with other links in the network located far from DCs. Therefore, the spectrum usage defined as function (3.3.1a) is dominated by highly loaded links and does not clearly present the whole situation in terms of spectrum usage in all network links. Therefore, we present a model using the *average spectrum* function. As in model (3.3.1), variable $y_{es}$ denotes whether slice $s$ is allocated in link $e$. Next, using these variables, an integer variable $y_e$ is determined denoting the largest occupied slice for each link $e \in E$. Finally, the objective function defined as the average value of $y_e$ takes into account all network links.

To illustrate the idea of the average spectrum function in comparison with the spectrum function defined in (3.3.1a), we consider the following example. Let set $E$ include four links. Following the routing and spectrum allocation procedure, a solution is obtained where the largest occupied slice index is 8, 6, 4 and 2 for each link, respectively. In consequence, the spectrum usage (defined as the maximum number of slices allocated in the network) is 8, while the average spectrum usage is $5 = (8 + 6 + 4 + 2)/4$.

---

**EON/AU/RSA/Average Spectrum/Link-path/Channel-based**

**sets**

| | |
|---|---|
| $E$ | links |
| $S$ | slices |
| $D$ | demands (anycast and unicast) |
| $D^{DS}$ | anycast downstream demands |
| $P(d)$ | candidate paths for flows realizing demand $d$. If $d$ is a unicast demand, the candidate path connects the end nodes of the demand. If $d$ is an anycast upstream demand, the candidate path connects client the node and the DC node. If $d$ is a downstream demand, the candidate path connects the DC node and the client node |
| $C(d, p)$ | candidate channels for demand $d$ allocated on path $p$ |

**constants**

| | |
|---|---|
| $\delta_{edp}$ | $=1$, if link $e$ belongs to path $p$ realizing demand $d$; 0, otherwise |
| $n_{dp}$ | requested number of slices for demand $d$ on path $p$ |
| $\gamma_{dpcs}$ | $=1$, if channel $c$ associated with demand $d$ on path $p$ uses slice $s$; 0, otherwise |
| $\tau(d)$ | index of a demand associated with demand $d$. If $d$ is a downstream demand, then $\tau(d)$ must be an upstream demand and vice versa |
| $o(p)$ | origin node of path $p$ |
| $t(p)$ | destination node of path $p$ |

**variables**

| | |
|---|---|
| $x_{dpc}$ | $=1$, if channel $c$ on candidate path $p$ is used to realize demand $d$; 0, otherwise (binary) |
| $y_{es}$ | $=1$, if slice $s$ is occupied on link $e$; 0, otherwise (binary) |
| $y_e$ | the largest index of an allocated slice in link $e$ (integer) |

**objective**

$$\text{minimize} \quad F = \frac{1}{|E|} \sum_{e \in E} y_e \tag{3.3.2a}$$

**constraints**

$$\sum_{p\in P(d)}\sum_{c\in C(d,p)} x_{dpc} = 1, \quad d \in D \tag{3.3.2b}$$

$$\sum_{d\in D}\sum_{p\in P(d)}\sum_{c\in C(d,p)} \gamma_{dpcs}\delta_{edp}x_{dpc} \leq y_{es}, \quad e \in E, s \in S \tag{3.3.2c}$$

$$sy_{es} \leq y_e, \quad e \in E, s \in S \tag{3.3.2d}$$

$$\sum_{p\in P(d)}\sum_{c\in C(d,p)} o(p)x_{dpc} = \sum_{p\in P(\tau(d))}\sum_{c\in C(\tau(d),p)} t(p)x_{\tau(d)pc}, \quad d \in D^{DS}. \tag{3.3.2e}$$

The above model concerning the average spectrum function differs from the previous (3.3.1) model only in the objective function (3.3.2a) and constraint (3.3.2d). More precisely, objective function (3.3.2a) denotes the average value of variable $y_e$. In turn, inequality (3.3.2d) is used to define variable $y_e$. Since constraint (3.3.2d) is checked for each $s \in S$, the term $sy_{es}$ on the left-hand side of (3.3.2d) denotes the value of the largest slice allocated ($y_{es} = 1$) in link $e$.

### Cost Objective

The next optimization model aims to minimize the cost of EON. The formulation is generic and can model different types of cost associated with EONs: CAPEX cost, OPEX cost, cost of power consumption, etc. We use constant $\xi_{dp}$ to model cost. More specifically, constant $\xi_{dp}$ denotes the cost of realizing demand $d$ on path $p$. One possible interpretation of $\xi_{dp}$ is the CAPEX cost related to establishing demand $d$ on path $p$ which includes the costs of all elements required to provision demand $d$: transponders, regenerators which are necessary if the length of path $p$ exceeds the transmission range of the modulation format selected, fiber leasing, etc. Additionally, $\xi_{dp}$ denotes the power consumption of demand $d$ realized on path $p$ obtained according to a sum of the power requirements of all transponders and regenerators. Since the value of $\xi_{dp}$ is the input data for the ILP model, the formulation shown below is independent of how constant $\xi_{dp}$ is determined. Thus, various cost models and assumptions can be used to obtain values of $\xi_{dp}$ without having to modify the formulations or the heuristic algorithms presented below.

---

### EON/AU/RSA/Cost/Link-path/Channel-based

**sets**

| | |
|---|---|
| $E$ | links |
| $S$ | slices |
| $D$ | demands (anycast and unicast) |
| $D^{DS}$ | anycast downstream demands |
| $P(d)$ | candidate paths for flows realizing demand $d$. If $d$ is a unicast demand, the candidate path connects the end nodes of the demand. If $d$ is an anycast upstream demand, the candidate path connects the client node and the DC |

node. If $d$ is a downstream demand, the candidate path connects the DC node and the client node

$C(d,p)$    candidate channels for demand $d$ allocated on path $p$

**constants**

$\delta_{edp}$    =1, if link $e$ belongs to path $p$ realizing demand $d$; 0, otherwise

$n_{dp}$    requested number of slices for demand $d$ on path $p$

$\gamma_{dpcs}$    =1, if channel $c$ associated with demand $d$ on path $p$ uses slice $s$; 0, otherwise

$\xi_{dp}$    cost of realizing demand $d$ on path $p$

$\tau(d)$    index of a demand associated with demand $d$. If $d$ is a downstream demand, then $\tau(d)$ must be an upstream demand and vice versa

$o(p)$    origin node of path $p$

$t(p)$    destination node of path $p$

**variables**

$x_{dpc}$    =1, if channel $c$ on candidate path $p$ is used to realize demand $d$; 0, otherwise (binary)

**objective**

$$\text{minimize}\quad F = \sum_{d\in D}\sum_{p\in P(d)}\sum_{c\in C(d,p)} x_{dpc}\xi_{dp} \tag{3.3.3a}$$

**constraints**

$$\sum_{p\in P(d)}\sum_{c\in C(d,p)} x_{dpc} = 1, \quad d\in D \tag{3.3.3b}$$

$$\sum_{d\in D}\sum_{p\in P(d)}\sum_{c\in C(d,p)} \gamma_{dpcs}\delta_{edp}x_{dpc} \leq 1, \quad e\in E, s\in S \tag{3.3.3c}$$

$$\sum_{p\in P(d)}\sum_{c\in C(d,p)} o(p)x_{dpc} = \sum_{p\in P(\tau(d))}\sum_{c\in C(\tau(d),p)} t(p)x_{\tau(d)pc}, \quad d\in D^{DS}. \tag{3.3.3d}$$

As in the previous model, the above formulation is very similar to model (3.3.1). The key difference is the objective function (3.3.3a) which is calculated as the sum of costs generated by each demand established in the network. Therefore, according to the definition of variable $x_{dpc}$, in (3.3.3a) we sum demands, candidate paths and candidate channels to calculate the overall cost resulting from the assignment of demands to candidate paths. The second modification is in constraint (3.3.3c), which ensures that a particular slice $s$ can be used on link $e$ by at most one demand.

### 3.3.2   Algorithms

This section presents several heuristic and metaheuristic algorithms designed to solve the RSA problem with joint optimization of anycast and unicast flows. The algorithms are generally based on the greedy approach.

#### First Fit (FF) Algorithm

The First Fit (FF) procedure for spectrum allocation of a single demand $d$ on a particular routing path $p$ is introduced first. The demand is represented as a set of candidate channels denoted as $C$. The approach is based on a simple method which finds the first channel in the network enabling the allocation of demand $d$ on path $p$ without spectrum overlapping [1]. More specifically, the channel ensures that demand $d$ is allocated on path $p$ starting from the lowest possible slice. The algorithm uses the current state of the network as the input, with information on slices already occupied by previously allocated demands. For each pair link $e$ and channel $c$, we can use function $IsEmptyChannel(e, c)$ to check whether all slices included in $c$ are free (not allocated) on link $e$ or not. Moreover, let $a := Member(A, i)$ return an element included in set $A$ on position $i$.

The concept behind the spectrum allocation approach is presented in Algorithm 3.1. The main loop of the algorithm (lines 2–15) searches channels included in set $C$ to find the first feasible allocation. Note that the *test* flag set in line 4 is used to check whether path $p$ provides the free spectrum in channel $c$ on all links included in path $p$. Inside the main loop, there is an additional loop checking subsequent links included in path $p$ (lines 6–10). In particular, function $IsEmptyChannel(e, c)$ checks whether channel $c$ provides free spectrum on link $e$ for each link $e \in E(p)$. If this is not the case, the loop defined for links $e$ is broken by setting $j := |E(p)| + 1$ (line 8). Moreover, the *test* flag is set to 0 to indicate that channel $c$ does not provide free spectrum on the whole path $p$. If after processing the loop defined in lines 6–10 the *test* flag is still 1, then channel $c$ provides a free spectrum on path $p$ and in consequence $c$ is returned as the selected channel. If the function completes the whole main loop (lines 2–15) without finding a free channel, then in line 16 the function returns an empty set denoting that there is no feasible channel included in set $C$ for the considered demand and routing path. Algorithm 3.1 is formulated as a separate function, since it will be used in subsequent algorithms for the spectrum allocation subproblem. The complexity of this algorithm is bound by $O(|C| |E(p)|)$, where $|C|$ denotes the number of candidate channels available and $|E(p)|$ is the number of links in the considered path $p$.

The FF/AU/RSA method (shown in Algorithm 3.2) is a very simple greedy algorithm based on the First Fit approach and proposed for the RSA problem. As the input data, the FF algorithm needs the network topology represented as a set of links $E$, set of demands $D$ (both anycast and unicast), shortest paths for each demand included in set $P(d)$, and set of candidate channels $C(d, p)$ for each demand $d \in D$ and path $p \in P(d)$. The size of each channel $c \in C(d, p)$ is equal to the spectrum requirement $n_{dp}$ (i.e., the number of slices included in the channel) calculated

---

**Algorithm 3.1** FF_SA (First Fit - Spectrum Allocation)

---

**Require:** set of edges $E$ with current status showing allocation of slices, set of candidate channels
    $C$, set of links $E(p)$ included in path $p$
**Ensure:** selected channel
1: **function** $FF\_SA(C, p)$
2: **for** $i := 1$ **to** $|C|$ **do**
3:    $c := Member(C, i)$
4:    $test := 1$
5:    **for** $j := 1$ **to** $|E(p)|$ **do**
6:      $e := Member(E(p), j)$
7:      **if** $IsEmptyChannel(e, c) = FALSE$ **then**
8:        $j := |E(p)| + 1$
9:        $test := 0$
10:     **end if**
11:     **if** $test = 1$ **then**
12:        **return** $c$
13:     **end if**
14:    **end for**
15: **end for**
16: **return** $\emptyset$
17: **end function**

---

for demand $d$ using path $p$ according to a selected physical EON model. The FF algorithm aims to process all demands in a single run without any special ordering of the demands. The main loop of the FF algorithm (lines 2–7) is defined to process all demands. The routing path for demand $d$ is selected as a shortest path included in set $P(d)$ with index 1 (line 4). To find the spectrum channel on the lowest possible spectrum range, function FF_SA is applied (line 5). Next, function *Allocate_Demand*() ensures that a particular demand is allocated on the selected path and channel (line 6), i.e., all slices included in the selected path and channel are marked as occupied to avoid spectrum overlapping for other demands. Finally, having determined routing paths and spectrum channels for each demand, the value of the objective function is calculated according to the objective function defined in the problem (line 8). Since, the FF/AU/RSA algorithm executes the FF_SA function for every demand included in set $D$, the computational complexity of the FF algorithm can be estimated as $O(|D| \, |C| \, |E|)$, where $|D|$ denotes the number of demands, $|C|$ is the number of channels and $|E|$ is the number of links.

Note that the FF algorithm processes anycast and unicast demands the same way. This is because the network graph is directed and symmetric in terms of the link distances in opposite directions. In consequence, for two associated anycast demands $d$ and $\tau(d)$ the same routing path is selected as the shortest path (albeit in opposite directions), and so both associated demands are assigned to the same DC node, guaranteeing the anycast constraint defined in Eq. (3.3.1e). The FF algorithm is designed directly for model (3.3.1) to minimize spectrum usage. This heuristic can also be used to solve other RSA problems with average spectrum and cost objectives.

---

**Algorithm 3.2** FF/AU/RSA (First Fit for AU/RSA problem)

---

**Require:** set of edges $E$, set of anycast and unicast demands $D$, sets $P(d)$ with a shortest path for
    each demand $d \in D$, candidate channels $C(d, p)$ for each demand $d \in D$ and path $p \in P(d)$

**Ensure:** routing and spectrum allocation for each demand $d \in D$ included in vectors *path* and
    *channel*, value of objective function

1: **procedure** $FF/AU/RSA(D, P(d), C(d, p))$
2:   **for** $i := 1$ **to** $|D|$ **do**
3:     $d := Member(D, i)$
4:     $path[d] := Member(P(d), 1)$
5:     $channel[d] := FF\_SA(C(d, path[d]), path[d])$
6:     $Allocate\_Demand(d, path[d], channel[d])$
7:   **end for**
8:   $Find\_Objective\_Function(path[], channel[])$
9: **end procedure**

---

### Longest Path First (LPF) and Most Subcarriers First (MSF) Algorithms

The FF algorithm usually provides low quality results. Two strategies can be used
to improve the algorithm's performance. Firstly, instead of always using the shortest
path, more candidate paths can be analyzed for each demand. Secondly, the demands
can be processed in a particular order which can provide better allocation of demands
in terms of spectrum usage. Two algorithms that exemplify these two approaches are
described below: and LPF (Longest Path First) and MSF (Most Subcarriers First).
Both algorithms were originally proposed for the RSA problem with unicast flows
only [23]; however, we modified them to enable us to also process anycast demands.

The auxiliary function *FPCSpectrum*() is first defined in Algorithm 3.3, selecting
a path and a channel for a demand aiming to guarantee the best result in terms of
spectrum usage, which is consistent with the spectrum objective function defined in
model (3.3.1). The *FPCSpectrum*() function requires information on the current state
of the network showing slice occupancy as the input. Moreover, the algorithm uses
set $P$ with candidate paths and set $C(p)$ with candidate channels for each path $p \in P$
defined for the demand considered. The method checks all possible candidate paths
(lines 3–10) to find a path which provides the lowest value of the objective function in
terms of spectrum usage. More specifically, for each path $p \in P$ the $FF\_SA(C(p), p)$
function is run to find spectrum channel $c$ according to the First Fit approach (line
5). Next, the *Last_Slice*($c$) function returns the index of the largest slice included in
channel $c$. If the current allocation defined by path $p$ and channel $c$ yields a better
result than that found earlier, the new allocation is saved as the best one (lines 7–
9). The complexity of function *FPCSpectrum*() can be estimated as $O(|P| |C| |E|)$,
where $|P|$ denotes the number of candidate paths, $|C|$ is the number of channels and
$|E|$ estimates the upper bound of the number of links included in the longest path.

Note that the *FPCSpectrum*() function can be modified easily to optimize other
objective functions. In the following, Algorithm 3.6 shows an analogous function
designed to optimize the network cost according to optimization model (3.3.3).

In Algorithm 3.4, we present the pseudocode of algorithm LPF/AU/RSA. The
aim of this method is to process the demands in a certain order. Here, the demand

---

**Algorithm 3.3** FPCSpectrum (Find Path and Channel for Spectrum Objective)

---

**Require:** set of edges $E$ with current status showing allocation of slices, set of candidate paths $P$,
   set of candidate channels for $C(p)$ for each path $p \in P$
**Ensure:** selected path and channel with the lowest index of allocated spectrum slice
1: **function** $FPCSpectrum(P, C(p))$
2:   $s_{min} := \infty$
3:   **for** $i := 1$ **to** $|P|$ **do**
4:     $p := Member(P, i)$
5:     $c := FF\_SA(C(p), p)$
6:     $s := Last\_Slice(c)$
7:     **if** $s < s_{min}$ **then**
8:       $p^\star := p, c^\star := c, s_{min} := s$
9:     **end if**
10:   **end for**
11:   **return** $p^\star$ and $c^\star$
12: **end function**

---

path length is applied as a sorting metric. More precisely, for each demand $d$ metric
$l_d$ denotes the hop count of the shortest path included in set $P(d)$. The demands are
processed in decreasing order of $l_d$ values. The aim of this approach is to process
demands that use longer paths (with more hops), since they consume more spectrum.
Therefore, the demands are sorted using the *Sort_Demands_LPF*() function (line 2).
Next, the demands are processed in a single run using this ordering in the main loop
of the algorithm (lines 3–16). However, anycast and unicast demands are processed
using a different method, which accounts for the anycast constraint (3.3.1e) active
only for anycast demands. More precisely, if demand $d$ is an anycast demand two
cases are considered. Firstly, if the associated demand $\tau(d)$ is not already established
in the network, the path and channel are selected with $FPCSpectrum(P(d), C(d, p))$,
using all candidate paths included in set $P(d)$ (lines 6–8). However, if the associated
demand $\tau(d)$ is already established in the network, demand $d$ must be connected to the
same DC node as $\tau(d)$. Therefore, in this case the path and channel are selected with
$FPCSpectrum(P(d, r), C(d, p))$, but using candidate paths with the same DC node as
demand $\tau(d)$, i.e., the path included in set $P(d, r)$ (lines 9–12). For unicast demands,
the processing is straightforward (line 14). The *Allocate_Demand*() function is run
to allocate the analyzed demand on the selected path and channel (line 15). Note that
the complexity of algorithm LPF/AU/RSA is bound by $O(|D| |P| |C| |E|)$.

The scheme of the LPF heuristic shown in Algorithm 3.4 can be modified to use
a different sorting approach by changing the function in line 2. For instance, the
MSF/AU/RSA algorithm can be obtained by using a function that sorts the demands
in decreasing order of number of slices required for the shortest path allocation.

**Adaptive Frequency Assignment (AFA) Algorithm**

This section presents an Adaptive Frequency Assignment (AFA) algorithm for the
RSA problem with joint optimization of anycast and unicast flows [57]. Note that
the basic version of the AFA algorithm was proposed in [24] in the context of the
RSA problem with unicast flows only.

---

**Algorithm 3.4** LPF/AU/RSA (Longest Path First for AU/RSA problem)

---

**Require:** set of edges $E$, set of anycast and unicast demands $D$, sets $P(d)$ with candidates paths for
   each demand $d \in D$, candidate channels $C(d, p)$ for each demand $d \in D$ and path $p \in P(d)$
**Ensure:** routing and spectrum allocation for each demand $d \in D$ included in vectors *path* and
   *channel*, value of objective function
1: **procedure** *LPF/AU/RSA(D, P(d), C(d, p))*
2:   $D := Sort\_Demands\_LPF(D)$
3:   **for** $i := 1$ **to** $|D|$ **do**
4:     $d := Member(D, i)$
5:     **if** $Type(d) = ANYCAST$ **then**
6:       **if** $Established(\tau(d)) = FALSE$ **then**
7:         $\{path[d], channel[d]\} := FPCSpectrum(P(d), C(d, p))$
8:         $server[d] := Server\_Node(path[d])$
9:       **else**
10:        $r := server[\tau(d)]$
11:        $\{path[d], channel[d]\} := FPCSpectrum(P(d, r), C(d, p))$
12:      **end if**
13:    **end if**
14:    **if** $Type(d) = UNICAST$ **then** $\{path[d], channel[d]\} := FPCSpectrum(P(d), C(d, p))$
15:    $Allocate\_Demand(d, path[d], channel[d])$
16:  **end for**
17:  $Find\_Objective\_Function(path[], channel[])$
18: **end procedure**

---

The main aim of the AFA/AU/RSA method shown in Algorithm 3.5 is to adaptively select a sequence of processed demands in order to minimize spectrum usage
defined in objective function (3.3.1a). The first important modification of AFA, in
comparison to greedy methods shown above (e.g., the LPF method), is that AFA
processes the demands in several separate loops. More precisely, set $D$ including all
demands is divided into several subsets and each subset is processed in a separate
loop (lines 3–8). For this purpose, each demand is assigned with a special metric $n_d$
which is equal to the minimum value of the requested number of slices required for
demand $d$, i.e., $n_d = \min_{p \in P(d)} \{n_{dp}\}$. All demands with the value of this metric equal
to $n$ are included in set $D(n)$. In the main loop of the algorithm (lines 9–30), sets
$D(n)$ are processed one by one according to decreasing value of $n$. The way each
set $D(n)$ is analyzed is the second main modification of AFA to the greedy methods
shown above. To find a demand to be allocated, all not established demands still
included in set $D(n)$ are checked to find the best possible allocation. The outer loop
(lines 10–29) is repeated until all demands in set $D(n)$ are allocated. The inner loop
(lines 11–26) is responsible for finding the best possible allocation from demands
included in set $D(n)$, but not established in the network, which is checked in line
13. The procedure shown in lines 14–24 is analogous to the LPF method shown in
Algorithm 3.4, i.e., the *FPCSpectrum()* function is called to find the path and channel
guaranteeing allocation in the lowest spectrum range. It should be stressed that when
the path and channel are found for the demand, there is no allocation of the demand
in the network. The rationale for this is that the algorithm in lines 14–24 checks

the value of the objective function one-by-one if a particular demand is allocated. Therefore, in line 27, the *Best_Allocation*($D(n)$, *path*, *channel*) function is called to find the best demand $d^*$ for the next allocation.

The *Best_Allocation*() function uses the value of the largest slice index which will be allocated if a particular demand $d$ is allocated according to *path*[$d$] and *channel*[$d$] as the main criterion. However, a special collision metric is applied in the AFA algorithm if there is a tie, i.e., more than one demand yields the same lowest value of the slice index. For each link $e \in E$ we define $c_e = \sum_{d \in D} \sum_{p \in P(d)} \delta_{edp} n_d$. Note that metric $c_e$ approximates the number of slices that may be allocated to link $e$ taking into account all candidate paths for each demand. As the value of $c_e$ increases, potentially more slices can be allocated on link $e$. Consequently, if shorter paths are chosen (according to metric $c_e$), then less congested links are selected and the objective function is decreased. Next, $l_p = \sum_{e \in p} c_e$ is defined as a length of path $p$ calculated according to metric $c_e$. Finally, metric $l_d$ denotes the collision metric of each demand calculated as $l_d = \frac{1}{|P(d)|} \sum_{p \in P(d)} l_p$. It is clear that $l_d$ is the average length of candidate paths of demand $d$ in terms of metric $c_e$. The aim of this collision metric is to avoid potential collisions on popular links which are likely to be used by many demands. In other words, the algorithm promotes demands with candidate paths which do not include links which can potentially be selected in a large number of demands. Note that the complexity of algorithm AFA/AU/RSA is given by $O(|D|^2 |P| |C| |E|)$, since the general processing of AFA/AU/RSA is similar to the LPF/AU/RSA algorithm; however, in the worst case, each demand included in set $D$ is checked $|D|$ times.

Algorithm 3.5 presents a basic version of the AFA method developed in the context of spectrum optimization. However, the AFA method can be easily adapted to other optimization objectives. As an example, we focus on the cost objective defined in model (3.3.3). The main modification is required in the function which selects the most suitable path and channel for a particular demand. The *FPCSpectrum*() function must be substituted (in lines 16, 19 and 23) with another function, which looks for the best allocation in terms of the cost function instead of spectrum usage. Accordingly, Algorithm 3.6 shows a pseudocode of a *FPCCost*() function which returns the path and channel guaranteeing the lowest cost for a particular demand. The main difference when compared to *FPCSpectrum*() function defined in Algorithm 3.3 is that each combination of path and channel is evaluated in terms of the cost which will result if the demand is allocated to the particular path and channel (line 6). If the combination of path and channel results in a reduction of the cost, it is saved as the best solution. Moreover, function *Best_Allocation*() (called in line 27 of the Algorithm 3.5) must be modified in order to find a demand in the set of considered demands which returns the lowest value of the cost function. Using these two modifications, we can obtain an AFA algorithm designed to minimize cost. Similarly, the AFA method shown in Algorithm 3.5 can be modified for the average spectrum objective used in model (3.3.2). The complexity of function *FPCCost* is the same as in the case of function *FPCSpectrum*, namely, $O(|P| |C| |E|)$.

**Algorithm 3.5** AFA/AU/RSA (Adaptive Frequency Assignment for AU/RSA problem)

---

**Require:** set of edges $E$, set of anycast and unicast demands $D$, sets $P(d)$ with candidates paths for each demand $d \in D$, candidate channels $C(d, p)$ for each demand $d \in D$ and path $p \in P(d)$

**Ensure:** routing and spectrum allocation for each demand $d \in D$ included in vectors *path* and *channel*, value of objective function

1: **procedure** *AFA/AU/RSA*$(D, P(d), C(d, p))$
2: $\quad n_{max} := 0$
3: $\quad$ **for** $i := 1$ **to** $|D|$ **do**
4: $\quad\quad d := Member(D, i)$
5: $\quad\quad n_d := \min_{p \in P(d)} \{n_{dp}\}$
6: $\quad\quad D(n_d) := D(n_d) \cup \{d\}$
7: $\quad\quad$ **if** $n_d > n_{max}$ **then** $n_{max} := n_d$
8: $\quad$ **end for**
9: $\quad$ **for** $n := n_{max}$ **to** 1 **do**
10: $\quad\quad$ **while** $D(n) \neq \emptyset$ **do**
11: $\quad\quad\quad$ **for** $i := 1$ **to** $|D(n)|$ **do**
12: $\quad\quad\quad\quad d := Member(D, i)$
13: $\quad\quad\quad\quad$ **if** $Established(d) = FALSE$ **then**
14: $\quad\quad\quad\quad\quad$ **if** $Type(d) = ANYCAST$ **then**
15: $\quad\quad\quad\quad\quad\quad$ **if** $Established(\tau(d)) = FALSE$ **then**
16: $\quad\quad\quad\quad\quad\quad\quad \{path[d], channel[d]\} := FPCSpectrum(P(d), C(d, p))$
17: $\quad\quad\quad\quad\quad\quad$ **else**
18: $\quad\quad\quad\quad\quad\quad\quad r := server[\tau(d)]$
19: $\quad\quad\quad\quad\quad\quad\quad \{path[d], channel[d]\} := FPCSpectrum(P(d, r), C(d, p))$
20: $\quad\quad\quad\quad\quad\quad$ **end if**
21: $\quad\quad\quad\quad\quad$ **end if**
22: $\quad\quad\quad\quad\quad$ **if** $Type(d) = UNICAST$ **then**
23: $\quad\quad\quad\quad\quad\quad \{path[d], channel[d]\} := FPCSpectrum(P(d), C(d, p))$
24: $\quad\quad\quad\quad\quad$ **end if**
25: $\quad\quad\quad\quad$ **end if**
26: $\quad\quad\quad$ **end for**
27: $\quad\quad\quad d^\star := Best\_Allocation(D(n), path, channel)$
28: $\quad\quad\quad Allocate\_Demand(d^\star, path[d^\star], channel[d^\star])$
29: $\quad\quad$ **end while**
30: $\quad$ **end for**
31: $\quad Find\_Objective\_Function(path[], channel[])$
32: **end procedure**

---

## Solution Encoding in Metaheuristic Algorithms

In general, metaheuristic algorithms, including Tabu Search, Simulated Annealing and evolutionary algorithms, can provide an efficient method of solving various types of optimization problems. However, the main challenge is to adapt the basic framework of a particular metaheuristic algorithm according to specific requirements of the optimization problem and develop an efficient solution encoding.

Let us recall that metaheuristic methods usually analyze wide sets of solutions generated by certain disruptions to the current solution. However, in the case of the RSA problem even a small change to a feasible solution can lead to loss of feasibility

---

**Algorithm 3.6** FPCCost (Find Path and Channel for Cost Objective)

---
**Require:** set of edges $E$ with current status showing allocation of slices, set of candidate paths $P$,
    set of candidate channels for $C(p)$ for each path $p \in P$
**Ensure:** selected path and channel guaranteeing the lowest cost
1: **function** *FPCCost*$(P, C(p))$
2:   $cost_{min} := \infty$
3:   **for** $i := 1$ **to** $|P|$ **do**
4:     $p := Member(P, i)$
5:     $c := FF\_SA(C(p), p)$
6:     $cost := Find\_Cost(p, c)$
7:     **if** $c < cost_{min}$ **then**
8:       $p^\star := p, c^\star := c, cost_{min} := cost$
9:     **end if**
10: **end for**
11: **return** $p^\star$ and $c^\star$
12: **end function**

---

since certain constraints of the optimization problem are not fulfilled. On the other hand, metaheuristic algorithms are designed to solve problems without constraints. Therefore, special techniques need to be used to adapt particular metaheuristics to highly constrained problems such as RSA.

The control of constraint fulfilment is first addressed by a suitable solution encoding. The solution representation uses aggregated and simplified information which needs to be processed in order to obtain the complete solution; this processing controls the problem constraints. Consequently, from the perspective of the metaheuristic method, each solution is feasible in terms of the constraint which enables a direct application of the metaheuristic. Moreover, this strategy significantly simplifies the design of the metaheuristic algorithm. A potential drawback is that the representation does not provide an objective function value of the solution, and we need to run an additional process to evaluate the solution. If the calculation process is complex, the running time of the metaheuristic algorithm can increase significantly, since a metaheuristic generally analyzes several solutions.

The second approach which can be applied to control constraints is the penalty function. More specifically, the algorithm accepts solutions which do not satisfy all problem constraints; however, the objective function assigned to these solutions is increased (penalized) with an extra value resulting from overruns of particular constraints. As such, the non-feasible solution is less attractive for the algorithm than feasible solutions. In consequence, such a strategy should lead to a situation when after a number of iterations, the algorithm finds a feasible, high-quality solution. However, if the number of constraints to be controlled by the penalty method is high and the feasible solution space is reduced significantly, this approach can result in an instability of the algorithm in terms of convergence. More precisely, the algorithm is unable to find any feasible solutions except the initial solution. Furthermore, the value of the penalty assigned to constraint overruns should be tuned to provide the best algorithm performance. However, when dealing with problems with many complex constraints, such as RSA problems, the tuning process can be challenging and time-consuming.

Thirdly, a special repair procedure can be applied to correct solutions which break some problem constraints. More precisely, if the solution does not fulfill all constraints, a heuristic algorithm is run to fix it in such a way that its elements which cause the infeasibility are changed. However, as in the case of the penalty function, this approach only works well for relatively simple optimization problems with few constraints. In the context of RSA problems, the repair procedure may be more complex than the basic algorithm. In consequence, the running time of the algorithm using repair procedures may be significantly longer than the basic approaches.

Following the analysis presented above, our experience in the application of meta-heuristics, features of the RSA problem and some preliminary experiments, we use the solution encoding approach as a way of coping with constraints of the RSA problems. We describe our approach in more detail below.

The key feature of the RSA problem is that the overall solution space is very large, while the number of feasible solutions in this space is extremely small. This is due to the construct of the RSA problem, namely, continuity and contiguity constraints. A solution of the RSA problem is defined by two elements selected for each demand: routing path and spectrum channel (set of adjacent spectrum slices). The most intuitive encoding approach is to simply assign two values denoted as $p_d$ and $c_d$ to each demand. Thus, the solution is denoted as follows:

$$X = [(p_1, c_1), (p_2, c_2), \ldots, (p_{|D|}, c_{|D|})]. \tag{3.3.4}$$

To evaluate the solution defined by vector $X$ (3.3.4), all demands must be allocated to routing paths and spectrum channels included in $X$, which yields the usage of spectrum slices in the network. However, the metaheuristic methods usually search the solution space by analyzing a number of solutions generated from the current solution by using operators such as crossover, mutation and local search. It is clear that a very small change in vector $X$ (3.3.4) can lead to the solution being infeasible. For instance, if we only change a routing path for demand $d = 1$ without changing channels, it is likely that demand $d = 1$ will cause spectrum overlapping with other demands on the new path. Similarly, if we only change the spectrum channel for demand $d = 1$, in the majority of cases the new solution will also break the constraint related to spectrum overlapping. In conclusion, the solution encoding given by vector $X$ (3.3.4) does not guarantee that applying metaheuristic operators will lead to feasible solutions. Accordingly, special procedures such as the penalty function or repair procedures may be required, which may significantly decrease the effectiveness of the algorithm or drastically increase the algorithm running time.

Therefore, we propose two encoding approaches for RSA problems that do not experience the problems encountered in encoding defined in (3.3.4). Let $seq_d$ denote a sequence number of demand $d$. This value is used to order all demands for processing., i.e., the demands are processed one by one using the sequence defined by $seq_d$.

The next encoding assumes that we are given a tuple $(p_d, seq_d)$ for each demand $d$ and the solution is defined as a following vector:

$$X = [(p_1, seq_1), (p_2, seq_2), \ldots, (p_{|D|}, seq_{|D|})]. \tag{3.3.5}$$

It should be noted that the encoding shown in (3.3.5) does not include direct information on spectrum allocation. To determine a spectrum channel for each demand according to the solution defined in (3.3.5), the procedure shown in Algorithm 3.7 is applied. Firstly, all demands are sorted according to values of $seq_d$ defined in solution $X$ given as the input for the procedure (line 2). Next, a spectrum channel is selected for each demand using the FF approach (line 6); and the demand is allocated in the network on path $p_d$ defined in solution $X$ and on a spectrum channel obtained in the previous step. The complexity of this algorithm is bound by $O(|D| |C| |E|)$.

---

**Algorithm 3.7** Sol_Eval1 (Solution Evalution 1)

---

**Require:** set of edges $E$, set of demands $D$, sets $P(d)$ with candidates paths for each demand $d \in D$, candidate channels $C(d, p)$ for each demand $d \in D$ and path $p \in P(d)$ solution described as a vector $X = [(p_1, seq_1), (p_2, seq_2), \ldots, (p_{|D|}, seq_{|D|})]$
**Ensure:** routing and spectrum allocation for each demand $d \in D$ included in vectors *path* and *channel*, value of objective function
1: **procedure** $Sol\_Eval1(D, P(d), X)$
2: $\quad D := Sort\_Demands(X)$
3: $\quad$ **for** $i := 1$ **to** $|D|$ **do**
4: $\quad\quad d := Member(D, i)$
5: $\quad\quad path[d] := p_d$
6: $\quad\quad channel[d] := FF\_SA(C(p, d), p_d)$
7: $\quad\quad Allocate\_Demand(d, path[d], channel[d])$
8: $\quad$ **end for**
9: $\quad$ **return** $Find\_Objective\_Function(path[], channel[])$
10: **end procedure**

---

In contrast to the LPF method shown in Algorithm 3.4 and the AFA method presented in Algorithm 3.5, the Sol_Eval1 procedure does not control the anycast constraint defined in Eq. (3.3.1e). This is because routing paths given in solution $X$ satisfy the anycast constraint and thus the procedure does not have to control this issue.

Having defined the Sol_Eval1 method shown in Algorithm 3.7, we can point out the main advantage of the solution encoding (3.3.5): this encoding always provides a feasible solution of the RSA problem, since all problem constraints are fulfilled by using the Sol_Eval1 algorithm. Consequently, there is no need to use penalty functions or repair procedures. However, the main drawback of this concept is that evaluation of a single solution described in (3.3.5) is time consuming.

The next solution encoding approach is a simplified version of the concept used in (3.3.5), namely, for each demand $d$ we only know the sequence number $seq_d$. Thus, the solution vector is defined as:

$$X = [seq_1, seq_2, \ldots, seq_{|D|}]. \tag{3.3.6}$$

As in the previous case, the encoding defined in (3.3.6) does not include all information required to evaluate the solution (find the value of the spectrum usage), since both routing paths and spectrum channels are not known. Again, to resolve the candidate path and spectrum channel for each demand using the solution defined as in (3.3.6), a special a procedure shown in Algorithm 3.8 is applied. The main aim of the Sol_Eval2 algorithm is similar to the LPF method shown in Algorithm 3.4. The key difference is that the demands are processed in an order defined by the solution vector $X$ (line 2). The Sol_Eval2 procedure ensures that a feasible routing path and spectrum channel is selected for each demand. Moreover, the anycast constraint (3.3.1e) is guaranteed, since anycast demands are processed separately from unicast demands. The complexity of Sol_Eval2 can be estimated as $O(|D| \, |P| \, |C| \, |E|)$.

---

**Algorithm 3.8** Sol_Eval2 (Solution Evaluation 2)

---

**Require:** set of edges $E$, set of demands $D$, sets $P(d)$ with candidates paths for each demand $d \in D$, candidate channels $C(d, p)$ for each demand $d \in D$ and path $p \in P(d)$, solution described as a vector $X = [seq_1, seq_2, \ldots, seq_{|D|}]$
**Ensure:** routing and spectrum allocation for each demand $d \in D$ included in vectors *path* and *channel*, value of objective function
1: **procedure** *Sol_Eval2*$(D, P(d), X)$
2: $\quad D := Sort\_Demands(X)$
3: **for** $i := 1$ **to** $|D|$ **do**
4: $\quad d := Member(D, i)$
5: $\quad$ **if** $Type(d) = ANYCAST$ **then**
6: $\quad\quad$ **if** $Established(\tau(d)) = FALSE$ **then**
7: $\quad\quad\quad \{path[d], channel[d]\} := FPCSpectrum(P(d), C(d, p))$
8: $\quad\quad\quad server[d] := Server\_Node(path[d])$
9: $\quad\quad$ **else**
10: $\quad\quad\quad r := server[\tau(d)]$
11: $\quad\quad\quad \{path[d], channel[d]\} := FPCSpectrum(P(d, r), C(d, p))$
12: $\quad\quad$ **end if**
13: $\quad$ **end if**
14: $\quad$ **if** $Type(d) = UNICAST$ **then** $\{path[d], channel[d]\} := FPCSpectrum(P(d), C(d, p))$
15: $\quad Allocate\_Demand(d, path[d], channel[d])$
16: **end for**
17: **return** *Find_Objective_Function*$(path[], channel[])$
18: **end procedure**

---

The solution encoding defined in (3.3.6) with the Sol_Eval2 procedure has the same benefits as the solution encoding (3.3.5), i.e., it is guaranteed that the solution of the RSA problem is feasible. Moreover, there is no need to use penalty functions or repair procedures. Again, the main disadvantage of this encoding is the relatively high execution time needed to evaluate a single solution defined in (3.3.6).

**Tabu Search Algorithm**

The Tabu Search (TS) optimization method TS/AU/RSA designed to solve the RSA problem [43, 54, 63] is based on the classical framework of the TS method described in [64–66]. A similar TS method was introduced in the context of RMSA problems in [47]. As such, this section only presents the key elements of the TS designed for RSA with a special focus on the main differences from the classical framework of the TS method.

The TS/AU/RSA algorithm uses the solution representation defined in (3.3.5), i.e., the solution is encoded by a demand allocation order (demands are allocated one-by-one according to this order) and routing paths. To evaluate a particular solution, Algorithm 3.7 is applied. More precisely, the demands are allocated according to the defined sequence on routing paths included in the solution representation using the First Fit approach, which ensures that the obtained solution is feasible. As the input, TS/AU/RSA uses information defining the RSA problem instance, namely, network topology, set of demands, candidate paths and candidate channels. Moreover, initial solution $X^{init}$ and values of tuning parameters are necessary to start the algorithm. The initial solution required in the TS/AU/RSA method can be provided by any RSA algorithm. The TS/AU/RSA algorithm uses the following variables: current solution denoted as $X$, best solution represented as $X^{best}$, new solution generated from the current one denoted as $X^{new}$, iteration index $i$, no improvement parameter *noImpr* which denotes the number of subsequent iterations without improvemening the objective function, tabu list $TL$ and used moves $UM$. Algorithm 3.9 presents the pseudocode of the TS/AU/RSA method.

The key element of TS is a *move* operation which is required to generate a new solution from the existing solution. The aim is to change some solution attributes to enable the neighborhood search in order to find improvements to the current solution. The tabu list TL includes a search history of the algorithm defined as recent moves which cannot be applied in the subsequent iterations. In the case of TS/AU/RSA, the tabu list is a set of moves which lead to certain improvement of the solution in previous iterations. The tabu list is simply defined as a FIFO queue with the size determined as one of the tuning parameters of the algorithm. Consequently, when the tabu list if full and a new move needs to be added to the list, the oldest move is removed from the list. The tabu list size is an important tuning parameter of the algorithm. If the tabu list is too small, it may lead to a situation where neighborhood solutions are not examined with a sufficient accuracy. In contrast, if this parameter is too large, we may find a local optimum only. The second memory structure applied in TS/AU/RSA is a list of used moves UM embracing recent moves which have not led to any new improvements since the last recorded improvement. The size of the used moves list is not limited, and the list is emptied when an improvement is encountered or during diversification processes. Both lists (tabu and used moves) are used in the move operation, i.e., only moves not included in either list are feasible and can be selected to obtain a new solution.

In the case of the TS/AU/RSA algorithm, three types of the move operation are defined: demand order swap, DC node swap, and path swap. The demand order swap

---

**Algorithm 3.9** TS/AU/RSA (Tabu Search for AU/RSA problem)

---

**Require:** set of edges $E$, set of demands $D$, sets $P(d)$ with candidates paths for each demand $d \in D$, candidate channels $C(d, p)$ for each demand $d \in D$ and path $p \in P(d)$, initial solution described as a vector $X^{init} = [(p_1, seq_1), (p_2, seq_2), \ldots, (p_{|D|}, seq_{|D|})]$, tuning parameters: worsening factor $\beta$, number of iterations $i^{max}$, no improvement limit, length of tabu list

**Ensure:** routing and spectrum allocation for each demand $d \in D$ included in vectors *path* and *channel*, value of objective function

1: **procedure** $TS/AU/RSA(D, P(d), X)$
2:   $F^{init} := Sol\_Eval1(X^{init})$
3:   $X := X^{init}, F := F^{init}$
4:   $X^{best} := X^{init}, F^{best} := F^{init}$
5:   $TL := \emptyset, UM := \emptyset, noImpr := 0$
6:   $i := 1$
7:   **while** $i < i^{max}$ **do**
8:     $X^{new} := Find\_Neighborhood(X), F^{new} := Sol\_Eval1(X^{new})$
9:     **if** $F^{new} < F$ **then**
10:       $X^{old} := X, X := X^{new}$
11:       **if** $F^{new} < F^{best}$ **then**
12:         $X^{best} := X^{new}, F^{best} := F^{new}$
13:         $UM := \emptyset, noImpr := 0$
14:         $TL := TL \cup \{(X^{old}, X^{new})\}$
15:       **else**
16:         $UM := UM \cup \{(X^{old}, X^{new})\}, noImpr := noImpr + 1$
17:       **end if**
18:     **else**
19:       $UM := UM \cup \{(X^{old}, X^{new})\}, noImpr := noImpr + 1$
20:     **end if**
21:     **if** $(noImpr < noImprovmentLimit * (|D| + kr))$ **then** $F := F(1 + \beta)$
22:     $i := i + 1$
23:   **end while**
24:   **return** $Sol\_Eval1(X^{best})$
25: **end procedure**

---

move chooses two demands $d_i$ and $d_j$, and next it swaps the positions of these demands in the allocation order, i.e., $seq_i = seq_j$ and $seq_j = seq_i$. The DC node swap operation can only be used in the context of anycast demands. The main aim of this move is to change a DC node selected for a pair of associated anycast demands. After a change of the DC node, both affected anycast demands use the same local index of the selected path. Finally, the path swap move modifies the selected routing path for one demand. It should be noted that in the case of an anycast demand $d$ already assigned to a DC node $r$ this operation can select a routing path using a set $P(d, r)$ which includes a limited subset of all available paths for demand $d$ including DC node $r$. This procedure ensures that two associated anycast demands are connected to the same DC node and consequently, the anycast constraint defined in Eq. (3.3.1e) is satisfied. The move operations are generally selected at random, although an additional mechanism is applied to prioritize moves which can yield a good solution.

The following major tuning parameters are applied in the TS/AU/RSA algorithm: tabu list size, no improvement limit, worsening factor $\beta$, and number of iterations

$i^{max}$. Algorithm 3.9 shows the main steps of the TS/AU/RSA method. To start, an initial solution included in $X^{init}$ is evaluated and saved as the current and best solution (lines 2–4). In addition, the key variables of the algorithm are initialized (lines 5–6). The main loop of the algorithm repeated $i^{max}$ times is presented in lines 7–23. A new neighborhood solution $X^{new}$ of the current solution $X$ is found by applying one of the available move operations, i.e., demand order swap, DC node swap, path swap (line 8). If the new solution provides an improvement to the current solution, it is saved as the current solution (lines 9–10). Moreover, if the new solution is better than the best one, a new best solution is set, and the last move operation is saved in the tabu list, while used moves list and the no improvement index are reset (lines 11–14). Otherwise, when the new solution does not outperform the best solution or the new solution is worse than the current solution, the last move operation is added to the used moves list and the no improvement index is incremented (lines 15–19). Next, when the number of iterations without improvement of the solution exceeds a predefined threshold calculated as $noImprovmentLimit * (|D| + kr)$, a diversification procedure is run (line 21). For instance, if the no improvement threshold is equal to 30 %, $|D| = 180$, number of candidate paths between each pair of network nodes is $k = 10$ and $r = 2$ data centers are available, then the diversification process is run after 60 iterations without improvement. The diversification mechanism multiplies the value of the objective function $F$ related to the current solution by $1 + \beta$, where $\beta$ is an input parameter to the algorithm. This mechanism is used to expand the neighborhood used in the search process. When the no improvement limit is reached, the neighborhood of the current solution is expanded, which allows the algorithm to leave the potential local optimum. For more details on the TS method refer to [43, 47, 54, 63].

## Simulated Annealing Algorithm

A Simulated Annealing (SA) algorithm for the RSA problem with joint optimization of anycast and unicast flows was proposed in [40]. The SA/AU/RSA method is based on the classical SA concept, which is a generic probabilistic heuristic for the global optimization problems [65–67].

The solution of the optimization problem used in the SA/AU/RSA algorithm is represented as a sequence (order) of demands to be allocated in the network as defined in (3.3.6). Consequently, to calculate the objective function of a particular solution, the demands are allocated in the network one by one according to the particular sequence of demands using Algorithm 3.8.

The SA/AU/RSA algorithm uses the same information as input as the greedy algorithms discussed above, with an additional initial solution $X^{init}$ and values of tuning parameters. Any RSA algorithm can be applied to calculate the initial solution. The main variables used in the algorithm are the current solution denoted as $X$, the best solution represented as $X^{best}$, the new solution generated from the current solution denoted as $X^{new}$, the iteration index $i$, and the current temperature $T$. The value of each solution is denoted using $F$, for instance, the value of the best solution is represented as $F^{best}$.

Three tuning parameters are applied in SA/AU/RSA: initial temperature calculated using parameter $m$, number of iterations $i^{max}$, and cooling factor $j$. The pseudocode of the SA/AU/RSA is shown in Algorithm 3.10. Firstly, the initial solution is evaluated and saved as the current and best solution (lines 2–4). Next, the initial temperature $T$ is calculated using an innovative approach, i.e., we take the value of initial solution $F^{init}$ (number of slices required to establish all demands according to solution encoding $X^{init}$) and multiply it by $m$ tuning parameter (line 6). This approach automates the initial temperature selection process according to the size of the problem represented as the spectrum usage. The main loop of the algorithm (lines 7–20) is processed while the conditions shown in line 7 are satisfied. More specifically, the number of iterations is lower than the maximum number $i^{max}$ and the temperature $T$ has not reached the absolute value which is equal to 0.01. To generate a neighbor of the current state (solution $X$), two randomly selected demands are swapped in the sequence and thus a new solution $X^{new}$ is obtained and evaluated (lines 8–10). Next, the algorithm calculates parameter $\Delta := F^{new} - F$ (line 11) to perform the *Metropolis test* in order to accept a move from $X$ to $X^{new}$ or not (lines 12–17) [66, 68]. If $\Delta \leq 0$ (new solution is not worse than the current solution), this new solution is saved as the current solution (line 13). Moreover, the new solution is compared against the existing best solution, and if any improvement is observed, a new best solution is set (line 14). If there is not improvement of the current solution ($\Delta > 0$), then the new solution $X^{new}$ is accepted with a probability $e^{-\frac{\Delta}{T}}$ (line 16), even though this increases the current solution value. Finally, temperature $T$ is reduced using cooling factor $j$ (line 18).

### 3.3.3 Comparison of Algorithms—Numerical Results

This section focuses on comparisons of algorithms in the context of the AU/RSA problem with the objective of minimizing spectrum usage (model (3.3.1)); the algorithms are the branch and bound method implemented in the CPLEX solver, FF, MSF, LPF, AFA, TS and SA. The aim of the numerical experiments is to compare the results of all algorithms. However, due to complexity of the RSA problem, it was only possible to find optimal solution using the CPLEX solver for relatively few problem instances.

**Simulation Setup**

We examined four networks topologies: NSF15 (Fig. A.4, Table A.3), Euro16 (Fig. A.5, Table A.4), UBN24 (Fig. A.6, Table A.5) and Euro28 (Fig. A.7, Table A.6). We also analyzed several scenarios referring to different numbers of DC nodes (servers). For smaller networks (NSF15 and Euro16) we used one, two or three DCs, while for larger networks (UBN24 and Euro 28) we used between and four DCs. For each number of DCs, four different DC locations were examined. Consequently, 12 (3 × 4) and 16 (4 × 4) different DC scenarios were investigated for smaller and larger topologies, respectively. Sets of unicast and anycast demands were generated at random (end nodes and demand volume). The volume (bit-rate) of a unicast demand was selected in the range 10–100 Gb/s. Anycast traffic was asymmetric

---

**Algorithm 3.10** SA/AU/RSA (Simulated Annealing for AU/RSA problem)

---

**Require:** set of edges $E$, set of demands $D$, sets $P(d)$ with candidates paths for each demand $d \in D$,
candidate channels $C(d, p)$ for each demand $d \in D$ and path $p \in P(d)$, initial solution described
as a vector $X^{init} = [seq_1, seq_2, \ldots, seq_{|D|}]$, tuning parameters: initial temperature parameter $m$,
number of iterations $i^{max}$, cooling factor $j$

**Ensure:** routing and spectrum allocation for each demand $d \in D$ included in vectors *path* and
*channel*, value of objective function

1: **procedure** $SA/AU/RSA(D, P(d), X)$
2:   $F^{init} := Sol\_Eval2(X^{init})$
3:   $X := X^{init}, F := F^{init}$
4:   $X^{best} := X^{init}, F^{best} := F^{init}$
5:   $i := 1$
6:   $T := m * F^{init}$
7:   **while** $i < i^{max}$ and $T > 0.01$ **do**
8:      $d_1 := Rand(D), d_2 := Rand(D)$
9:      $X^{new} := Swap\_Demands(X, d_1, d_2)$
10:     $F^{new} := Sol\_Eval2(X^{new})$
11:     $\Delta := F^{new} - F$
12:     **if** $\Delta \leq 0$ **then**
13:        $X := X^{new}$
14:        **if** $F^{new} < F^{best}$ **then** $X^{best} := X^{new}, F^{best} := F^{new}$
15:     **else**
16:        **if** $Random(0, 1) < e^{-\frac{\Delta}{T}}$ **then** $X := X^{new}, F := F^{new}$
17:     **end if**
18:     $T := T * j$
19:     $i := i + 1$
20: **end while**
21: **return** $Sol\_Eval2(X^{best})$
22: **end procedure**

---

(as for Content Deliver Networks), the downstream demand volume was selected
from 40 to 400 Gb/s, and the upstream anycast demand volume always equaled to
10 Gb/s. Let $h^{Any}$ and $h^{Uni}$ denote the overall volume of all anycast and unicast
demands, respectively. Next, let $h^{All} = h^{Any} + h^{Uni}$ be the overall demand in the network. To examine the impact of anycast traffic on the objective function, we define
the *anycast ratio (AR)* parameter as the volume (capacity) of all anycast demands
divided by the volume of all demands in the network, i.e., $AR = h^{Any}/h^{All}$. Six scenarios of network load were analyzed in terms of the AR parameter, namely, 0, 20, 40,
60, 80 and 100 %. Note that the first case (i.e., $AR = 0\%$) denotes a scenario in which
there is only unicast traffic in the network, while the last case (i.e., $AR = 100\%$)
means that all traffic in the network is anycast.

The physical model of the EON is the half distance law described in [23, 57] for
selecting modulation levels for lightpath connections. More specifically, the modulation level $m_p$ selected for path $p$ depends on the path length $l_p$ and it is equal
to 1, 2, 3, and 4, respectively, for $l_p$ exceeding 1500, 750, 375, and below 375 km.
The requested spectrum for demand $d$ using path $p$ was calculated as $h_d/I_p$, where
$I_p = 2m_p$ [bit/s/Hz] is the transponder spectral efficiency. Moreover, we assumed
that the transponders operate with polarization division multiplexing, which doubles

the spectral efficiency. We did not consider the guard bands separating adjacent spectrum connections. Finally, we used the ITU flexgrid definition [11], which requires the spectrum to be allocated symmetrically around a central frequency and in which the frequency slice width is set to $\Delta_s = 6.25$ GHz. Accordingly, the requested number of slices for demand $d$ realized on path $p$ was calculated using the formula $n_{dp} = 2\lceil h_d/(4m_p\Delta_s)\rceil$. For more details on the simulation setup see [57].

For information on tuning the TS/AU/RSA algorithm see [43, 47, 54, 63]. In turn, results of tuning the SA/AU/RSA algorithm are included in [40].

## Comparison of Algorithms

All algorithms including CPLEX were first executed for smaller networks NSF15 and Euro16, with overall traffic of 2.5 Tb/s introduced to the network. The number of candidate paths was $k = 2$. For each value of the *AR* parameter, three demand sets were tested for 12 different scenarios of the number of DCs (i.e., 1, 2, 3). This gives the overall number of $216 = 3 \times 6 \times 12$ separate experiments. Table 3.8 shows the average results in terms of the optimality gap (distance to optimal results yielded by CPLEX), corresponding lengths of 95 % confidence intervals and average execution time in seconds. Note that for the TS and SA methods, the results are the minimum values obtained over ten repetitions of the algorithms for each individual problem instance. The AFA algorithm provides the best results from greedy methods. However, all greedy algorithms are outperformed by metaheuristic methods, which provide a similar performance with TS showing a minor advantage. Nevertheless, the metaheuristics need significantly more time than greedy methods.

Next, the heuristics were executed for larger problem instances taking into account the UBN24 network with 40 Tb/s of traffic and the Euro28 network with 50 Tb/s of traffic and the number of candidate paths $k = 2, 3, 5, 10$ and 30. For each network and value of $k$, 480 separate problem instances were considered, unique in terms of traffic pattern, number and location of DCs, and value of parameter *AR*. Table 3.9

**Table 3.8** Comparison of optimization algorithms for the RSA problem with anycast and unicast flows for NSF15 and Euro16 networks—average optimality gap, lengths of 95 % confidence intervals and average execution time

|  | CPLEX | FF | MSF | LSF | AFA | TS | SA |
|---|---|---|---|---|---|---|---|
| *Average optimality gap* | | | | | | | |
| NSF15 | – | 45.1 % | 13.1 % | 18.1 % | 7.8 % | 2.7 % | 3.8 % |
| Euro16 | – | 48.6 % | 11.5 % | 14.3 % | 6.9 % | 4.0 % | 4.3 % |
| *Lengths of 95 % confidence intervals* | | | | | | | |
| NSF15 | – | 2.09 % | 1.51 % | 1.78 % | 1.20 % | 0.55 % | 0.69 % |
| Euro16 | – | 2.15 % | 1.43 % | 1.56 % | 1.25 % | 0.68 % | 0.92 % |
| *Average execution time in seconds* | | | | | | | |
| NSF15 | 256 | <0.001 | <0.001 | <0.001 | <0.001 | 7 | 75 |
| Euro16 | 34 | <0.001 | <0.001 | <0.001 | <0.001 | 12 | 43 |

**Table 3.9** Comparison of optimization algorithms for the RSA problem with anycast and unicast flows for UBN24 and Euro28 networks—average gap to results of the AFA algorithm

| No. of paths | No. of slices | FF (%) | MSF (%) | LSF (%) | TS (%) | SA (%) |
|---|---|---|---|---|---|---|
| *Network UBN24 with 40 Tbps traffic* | | | | | | |
| $k = 2$ | 407 | 51.0 | 5.7 | 6.9 | −4.3 | −4.9 |
| $k = 3$ | 336 | 56.6 | 5.1 | 5.2 | −4.3 | −5.3 |
| $k = 5$ | 287 | 60.1 | 3.0 | 3.5 | −1.4 | −2.9 |
| $k = 10$ | 277 | 62.7 | 1.9 | 3.0 | −2.7 | −2.1 |
| $k = 30$ | 274 | 63.5 | 2.0 | 3.8 | −1.6 | −2.6 |
| *Network Euro28 with 50 Tbps traffic* | | | | | | |
| $k = 2$ | 455 | 50.5 | 3.3 | 3.9 | −17.6 | −6.4 |
| $k = 3$ | 414 | 57.3 | 2.4 | 3.6 | −3.8 | −5.0 |
| $k = 5$ | 392 | 58.8 | 2.0 | 3.1 | −3.3 | −3.7 |
| $k = 10$ | 387 | 60.9 | 1.9 | 3.2 | −4.2 | −2.9 |
| $k = 30$ | 384 | 63.9 | 2.6 | 4.1 | −3.1 | −3.7 |

**Table 3.10** Average execution time in seconds of optimization algorithms for the RSA problem with anycast and unicast flows for UBN24 and Euro28 networks with $k = 30$

| | FF | MSF | LSF | AFA | TS | SA |
|---|---|---|---|---|---|---|
| UBN24 | 0.01 | 0.07 | 0.08 | 0.64 | 105 | 5690 |
| Euro28 | 0.01 | 0.13 | 0.15 | 1.33 | 59 | 1425 |

shows the average gap between each algorithm and the results of the AFA method as a function of the number of candidate paths. The second column of Table 3.9 includes the average results (number of slices) yielded by the AFA algorithm. Table 3.10 presents the average algorithm execution times.

The first clear conclusion is that the TS method provides the best results followed by the SA algorithm. Additionally, AFA outperforms all other greedy methods. However, metaheuristics need significantly more execution time than AFA and other simple heuristics. Another interesting observation is that the gap between AFA and metaheuristics decreases as the number of candidate paths increases. In our opinion this trend is due to the fact that the number of candidate paths strongly influences the size of the solution space. In particular, the solution space for networks with large traffic volumes and $k = 30$ is extremely large.

Moreover, Table 3.9 shows the influence of parameter $k$ (number of candidate paths) on spectrum usage (number of slices). The improvement between $k = 2$ and $k = 30$ is approx. 33 and 16 % for UBN24 and Euro28 networks, respectively. Recalling that the relatively short execution time of AFA and the acceptable execution time of TS reported in Table 3.10, we can conclude that it is better to use a large set of candidate paths since it has a major impact on spectrum usage while the running time of the heuristic remains satisfactory. More results and discussion showing comparisons of the heuristic algorithms can be found in [40, 43, 54, 57, 63].

## 3.3.4  Case Study

This section presents a case study run to examine the potential advantages of EONs for provisioning cloud computing traffic in comparison to Wavelength Switched Optical Networks (WSONs) implemented with the WDM technology [15]. All assumptions of the simulations were made according to close estimates of real requirements of national and international operators and in reference to data provided in the literature.

**Network Topologies**

The study used real-world networks: a pan-European network Euro28 (Fig. A.8, Table A.6) and a United States long-haul network US26 (Fig. A.10, Table A.7). Seven data center nodes were placed in each network, although we also tested scenarios with five and nine data centers (Table 3.11). In addition, each network was equipped with three interconnection points to other networks (e.g., locations of submarine cable landing stations) used to carry international traffic (Table 3.11). Decision on locations of data center nodes and interconnection points were made on the basis of data available at http://www.datacentermap.com/.

**Table 3.11** Location of data centers and interconnection points in Euro28 and US26 networks

| Case | Location |
| --- | --- |
| *Euro28 network* | |
| 5 DCs | Amsterdam, Frankfurt, London, Paris, Zurich |
| 7 DCs | Amsterdam, Frankfurt, London, Madrid, Paris, Warsaw, Zurich |
| 9 DCs | Amsterdam, Frankfurt, London, Madrid, Milan, Paris, Warsaw, Vienna, Zurich |
| Interconnection points | Dublin, Madrid, Athens |
| *US26 network* | |
| 5 DCs | Chicago, Houston, New York, San Francisco, Washington |
| 7 DCs | Atlanta, Chicago, Houston, Los Angeles, New York, San Francisco, Washington |
| 9 DCs | Atlanta, Chicago, Dallas, Denver, Houston, Los Angeles, New York, San Francisco, Washington |
| Interconnection points | San Francisco, Miami, New York |

**Service Demands**

Service traffic was modeled following the *Cisco Visual Networking Index* fore-casts for years 2012–2017 [69] and the *Cisco Global Cloud Index* for years 2011–2016 [70]. The traffic matrix in each network included A–Z service demands of four types (Table 3.12):

- City-City (CC) traffic representing all non-data center traffic calculated with 18 % CAGR.
- City-Data Center (CD) traffic representing all data center to user traffic calculated with 31 % CAGR.
- Data Center-Data Center (DD) traffic calculated with 32 % CAGR.
- International (IN) traffic leaving/entering the particular network calculated as a percentage of all network traffic.

The initial volume of traffic in 2012 was set to 20 Tb/s for the Euro28 network and 30 Tb/s for the US26 network. The proportions of traffic types in 2012 were based on the Cisco reports. The traffic values for subsequent years (i.e., 2014, 2016, 2018 and 2020) were calculated consistent with the CAGR of each traffic type according to the Cisco reports. The volume of international traffic was set as a percentage of all network traffic., i.e., for the Euro28 network the proportion of international traffic was 20 %, while the corresponding parameter for the US26 network was 10 %.

Similarly to [71], City-City traffic was created using a multivariable gravity model. More specifically, the total CC traffic (shown in Table 3.12) was allocated to each city pair $v$ and $w$ proportional to the product of their population given by formula $(P(v) * P(w))$ and inversely proportional to the distance between the cities given by $d_{vw}$. Note that the population does not reflect the population of the individual city but the population of the region covered by the city (e.g., country or state).

**Table 3.12** Summary of traffic volume for Euro28 and US26 networks in years 2012–2020

| Year | Traffic volume in Tb/s | | | | |
|---|---|---|---|---|---|
| | CC | CD | DD | IN | Total |
| *Euro28 network* | | | | | |
| 2012 | 2.1 | 10.0 | 4.0 | 4.0 | 20.0 |
| 2014 | 2.9 | 17.1 | 6.9 | 6.7 | 33.6 |
| 2016 | 4.1 | 29.3 | 12.0 | 11.3 | 56.7 |
| 2018 | 5.6 | 50.3 | 20.9 | 19.2 | 96.1 |
| 2020 | 7.9 | 86.3 | 36.5 | 32.7 | 163.3 |
| *US26 network* | | | | | |
| 2012 | 3.5 | 16.8 | 6.7 | 3.0 | 30.0 |
| 2014 | 4.9 | 28.8 | 11.6 | 5.0 | 50.4 |
| 2016 | 6.8 | 49.5 | 20.3 | 8.5 | 85.1 |
| 2018 | 9.5 | 84.9 | 35.3 | 14.4 | 144.1 |
| 2020 | 13.3 | 145.6 | 61.5 | 24.5 | 244.9 |

Since DC traffic is generated by businesses as well as individual users, its volume depends on both the population and the economy level. Consequently, the total DC traffic was distributed among all cities proportionally to the product of the city population and the GDP (Gross Domestic Product) value $(P(v) * G(v))$. The GDP parameter defines the economy level of the country (network Euro28) or state (network US26). Next, the CD traffic of each city was assigned to each data center proportionally to the distance between the city and the node with the data center. The distance was adjusted by the exponential factor $\varepsilon = 0.5$ to account for the fact that data center traffic is less locally-oriented. We assumed that the data centers are sufficiently provisioned with hardware (processing units, storage, etc.) and that they can serve all allocated workloads.

The distribution of the CD traffic gives the load of each data center denoted as $DD(v)$. The multivariable gravity model was used again to generate DD traffic. Traffic between data centers $v$ and $w$ was proportional to the product of their load $(DD(v) * DD(w))$ and inversely proportional to the distance between the nodes $d_{vw}$, once again adjusted by the exponential factor $\varepsilon = 0.5$.

Finally, international traffic was distributed to each city $v$ proportionally to the product of the population and the GDP $((P(v) * G(v)))$. The traffic of each city was divided equally to three interconnection points.

**Optical Scenarios**

The following four alternative optical transport network scenarios were examined in the case study:

- WSON-MLR—a wavelength switched optical network applying mixed-line-rate transmission with fixed 10, 40, and 100 Gb/s WDM transponders and the transmission distance limits equal to, respectively, 3200, 2300, and 2100 km (as in [72]).
- WSON-OFDM-MMF—a wavelength switched optical network with transponders implementing the adaptively between BPSK, QPSK, and m-QAM, where m belongs to 8, 16, 32, 64. Spectral efficiency is equal to $1, 2, \ldots, 6$ [b/s/Hz], respectively, for these modulation formats. We also used Polarization Division Multiplexing (PDM), which doubles the spectral efficiency
- EON-OFDM-SMF—an elastic optical network with BV-Ts implementing the PDM-OFDM technology (as in the WSON-OFDM-MMF scenario) and the QPSK (single) modulation format.
- EON-OFDM-MMF—an elastic optical network with BV-Ts implementing the PDM-OFDM technology and multiple modulation formats (as in WSON-OFDM-MMF).

WSON is based on the fixed 50 GHz ITU-T grid, while EON implements a flexible ITU-T grid of 6.25 GHz granularity. Furthermore, the WSON-OFDM-MMF and EON scenarios used three types of BV-Ts, each with a different capacity limit, respectively, 40, 100, and 400 Gb/s. The BV-Ts permits a for bit-rate adaptability with 10 Gb/s granularity. When applying the PDM-OFDM technology, the transmission model presented in [60] was used to estimate the transmission distance as a function of the modulation level selected and the transported bit-rate. A 12.5 GHz guard band between neighboring connections was introduced. In all scenarios, we assumed that the transmission reach is extended by regenerators, which were applied whenever necessary.

**Performance Metrics**

In our experiments, we analyzed the performance metrics cost, power consumption and spectrum usage performance metrics. The cost encompasses the CAPEX cost of equipment (transponders and regenerators) and one year OPEX costs of fiber leasing (as in [72]). All costs are presented in euros at current prices. Future cost predictions do not incorporate the inflation rate. Power consumption is determined according the energy requirements of all transponders and regenerators used in the network. Spectrum usage denotes the width of the spectrum in terms of the number of slices required in the network to establish all demands. Maximum spectrum usage is defined as the maximum required spectrum over all network links, and average spectrum usage is calculated as the average required spectrum in network links.

In the experiments, we was assumed that the regenerators did not convert the spectrum and modulation formats. Additionally, we relaxed the regenerator placement problem by using in-line signal regeneration. Therefore, the results in terms of cost and power consumption of regenerators are lower-bound estimates. Cost and power consumption assumptions were based on [72, 73]. More precisely, the costs were calculated relatively to the cost of a single WDM 10 Gb/s transponder estimated to be 2000 Euro. The fiber leasing cost was assumed to be 2000 Euro/km for a 20 year period. Therefore, the relative cost of a "dark" 50 GHz channel was set to 0.625 per km per year.

**Results**

The results were obtained using the AFA/AU/RSA method shown in Algorithm 3.5. The number of candidate paths was $k = 10$. We applied the anycast approach to provision the City-Data Center traffic. The main goal of the experiments was to compare the performance of various optical scenarios based on the WSON and EON concepts. Figures 3.5 and 3.6 show performance in terms of the network cost and power consumption for the Euro28 and US26 networks, respectively. It is clear that both networks provide similar results for the cost metric. More specifically, in 2012, the costs of using WSON and EON are comparable. However, in subsequent years, the EON approach provides a better performance; in 2020, WSON needs a significantly greater provision of costs compared to EON-OFDM-MMF, that is, in the WSON-MLR scenario, the cost is approx. greater by about 47 % and 54 % for Euro28 and US26 networks, respectively. For the WSON-OFDM-MMF scenario, the corresponding gaps are 27% and 38%, respectively. Additionally, in 2012 the EON-OFDM-SMF scenario is the most expensive, however, in following years it outperforms WSON scenarios and it reveals slightly higher cost overheads than EON-OFDM-MMF (EON-OFDM-MMF allows to use less costly modulation formats, while EON-OFDM-SMF does not). Differences in values of cost and power consumption spotted for both networks follows from different size of the networks and volume of traffic.

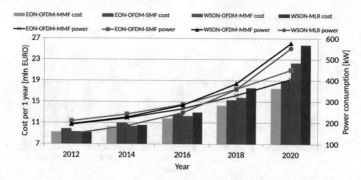

**Fig. 3.5** Comparison of EON-OFDM-MMF, EON-OFDM-SMF, WSON-OFDM-MMF, and WSON-MLR in the Euro28 network—cost and power consumption

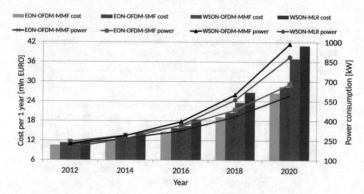

**Fig. 3.6** Comparison of EON-OFDM-MMF, EON-OFDM-SMF, WSON-OFDM-MMF, and WSON-MLR in the US26 network—cost and power consumption

For the power consumption metric, the trend is slightly different. In 2012, WSON-MLR consumes less power than other scenarios. An improvement of EON-OFDM-MMF is observed starting from 2018 and 2014 for Euro28 and US26 networks, respectively. Finally, in 2020 the EON-OFDM-MMF scenarios are significantly better than the WSON scenarios; in particular, WSON-MLR needs 36 and 49% more power, respectively, for Euro28 and US26 networks. The gap between EON-OFDM-SMF and EON-OFDM-MMF remains stable (approx. 10%) in the entire period.

Figures 3.7 and 3.8 report a comparison in terms of spectrum usage for Euro28 and US26 networks, respectively. EON-OFDM-MMF outperforms both WSON scenarios throughout period. In addition, the gap between these scenarios increases in subsequent years and in 2020 WSON-MLR requires more than 200% of the spectrum resources needed for EON-OFDM-MMF. Additionally, the gap between EON-OFDM-MMF and EON-OFDM-SMF increases in subsequent years. Recalling that traffic volume increases in time, the trends demonstrate that EON-OFDM-MMF is able to serve traffic with a higher spectral efficiency than WSON-MLR. The performance of WSON-OFDM-MMF falls between the EON-OFDM-MMF and WSON-MLR scenarios.

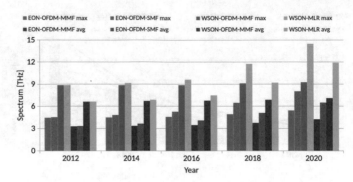

**Fig. 3.7** Comparison of EON-OFDM-MMF, EON-OFDM-SMF, WSON-OFDM-MMF, and WSON-MLR in the Euro28 network—spectrum usage

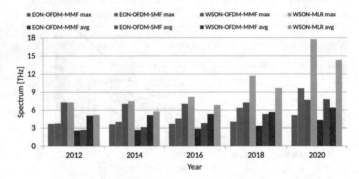

**Fig. 3.8** Comparison of EON-OFDM-MMF, EON-OFDM-SMF, WSON-OFDM-MMF, and WSON-MLR in the US26 network—spectrum usage

The next goal of experiments was to examine the potential advantages of using anycast instead of unicast transmission. We started by provisioning the City-Data Center traffic using anycasting, i.e., each CD demand may be assigned to any data center available in the network. Next, the same demands were served using the unicast approach, i.e., each demand was fixed to the closest DC node. To evaluate the potential gains of anycasting, we use an *anycast gain* parameter defined as the percentage difference between the results (cost, power consumption, spectrum) obtained for the unicast approach and the results (cost, energy, spectrum) obtained for the anycast approach. For instance, if the cost is five million euros for a particular network and traffic pattern for the unicast approach, and three million euros for the anycast approach, the corresponding value of anycast gain of the network cost is calculated as $(5 - 3)/3 = 66\%$. Figures 3.9 and 3.10 present the anycast gain for all metrics examined for the Euro28 and US26 networks, respectively. The results were obtained for the EON-OFDM-MMF case and traffic in 2020. Three different scenarios of data center locations with five, seven and nine DCs were considered. It is clear that in each case, anycasting provides an improvement of each performance metric. Furthermore,

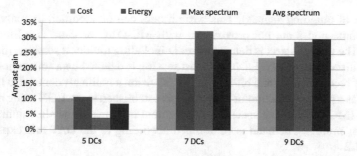

**Fig. 3.9** Anycast gain in 2020 for EON-OFDM-MMF in the Euro28 network

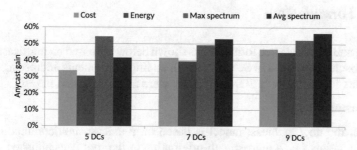

**Fig. 3.10** Anycast gain in year for EON-OFDM-MMF in the US26 network

if more data centers are located in the network, the anycast gain increases for all performance metrics.

The main conclusion of this case study is that using EON scenarios provides excellent performance in terms of the most important metrics in optical networks (i.e., cost, power consumption and spectrum usage) in comparison to the conventional WDM approach. Additionally, the advantages of the EON approach increase alongside traffic growth in subsequent years. More results and analysis can be found in [15].

## 3.4 Routing and Spectrum Allocation for Multicast Flows

Many popular online network services, (e.g., Content Delivery Networks, IP television, video streaming) can be provisioned in a scalable and cost-effective way using *all-optical multicasting*. As such, we address the subject of multicasting in EONs. More specifically, we propose two ILP formulations for the RSA problem with multicast flows, namely flow and candidate tree. We evaluate these models using numerical experiments and compare them with the optimization model proposed in [32].

We start by presenting the main assumptions of multicasting in EONs. To enable optical multicasting, network nodes are provisioned with *multicasting capable optical cross connects* (MC-OXCs) which are able to replicate the input data stream to multiple outputs [74, 75]. In the default EON multicasting model, we assume that all network nodes are *multicast capable* (MC), i.e., each network node is equipped with MC-OXC hardware. We also assume that MC-OXCs have no limit on the fanout (number of outgoing signals). This basic model can be modified to analyze cases where some nodes are not provisioned with MC-OXC or there is a limit on the fanout. The traffic matrix is defined as set $D$ including multicast sessions (demands) to be realized in the EON. Each multicast session is denoted by a source node (root) $s_d$, set of receivers $R(d)$ and bit-rate $h_d$. To provision a multicast demand, a light-tree defined as a point-to-multipoint connection is established in the EON using multicast capable (MC) nodes. The spectrum requirement in terms of the slice number for a particular multicast demand is determined according to the DAT rule explained in Sect. 3.1.2.

## 3.4.1   Formulations

The flow and candidate tree formulations shown below are based on multicast modeling approaches described in Sect. 1.2.3 with additional elements addressing specific constraints following from EONs and; they were first formulated in [36].

**Flow Formulation**

To recall, the flow multicast model is based on a unicast multicommodity node-link formulation, i.e., a unicast path originating at the root is established for each receiver in the multicast session, and all unicast paths form a multicast tree. Let binary variable $x_{edr}$ denote whether the unicast path from the root node to receiver $r \in R(d)$ uses link $e$ for demand (session) $d$. To define a multicast transmission for demand $d$, binary variable $x_{ed}$ denotes whether link $e$ is included in the multicast tree constructed to realize demand $d$.

The key innovation of this model over other papers on EON multicast modeling, e.g., [32] is that the spectrum requirement (number of required slices) is not known in advance, which follows directly from the DAT approach. A similar scenario was considered in model (3.2.1) formulated in the context of anycast flows. However, for anycast flows the transmission distance was defined as the length of a routing path used for a particular demand, and the modulation format was selected using the selected path length. This approach needs to be modified for multicast flows. More specifically, the multicast transmission can be viewed as a set of unicast transmissions from the root to each receiver included in a particular session. Using the DAT rule for multicasting, the modulation format is selected according to the most distant receiver in the tree. If the transmission distance to the furthest receiver of a particular multicast session exceeds the distance range of all modulation formats, regenerators are required. However, we do not address the regenerator placement problem in the context of multicasting in this section.

To model the spectrum usage, we apply the slice-based approach as in model (3.2.1). This decision is due to the fact that here we consider a case when the spectrum requirement of each demand is not given in advance, since the routing structure (tree)—and in consequence the transmission length—is determined during the optimization process. This makes using the channel-based approach somewhat problematic since channels of different sizes have to be considered, which may significantly increase the number of variables in the model.

The general notation used here to address the DAT rule in multicasting is similar to Sect. 3.2.1. More precisely, for each modulation format $m$ and demand (session) $d$, we use constant $b_{dm}$ denoting the maximum distance range supported for modulation format $m$ and bit-rate of demand $d$ and constant $n_{dm}$ denoting the number of slices required when modulation format $m$ is used for the bit-rate of demand $d$. Set $M$ includes all available modulation formats (i.e., 64-QAM, 32-QAM, 16-QAM, 8-QAM, QPSK and BPSK) sorted according to increasing values of the transmission range and the number of required slices. Furthermore, let $a_{dm}$ denote the lower bound of the distance range supported by modulation format $m$ and demand $d$. For $m = 1$, $a_{d1} = 0$, when $m > 1$, then $a_{dm} = b_{d(m-1)} + 1$.

Let us recall that when using the DAT rule, the spectrum requirement (number of slices) is not given as a constant and depends on the length of the selected routing path. For multicast demand $d$, we must take into account the length of the longest path from the root node to the receiver included in set $R(d)$. Therefore, variable $x_d$ denoting the length of demand $d$ is calculated as the maximum value of $\sum_{e \in E} l_e x_{edr}$ considering all receivers $r \in R(d)$. The path length given by variable $x_d$ selects a modulation format according to the DAT rule. Accordingly, $x_d$ is compared with the lower bound of the transmission range $a_{dm}$ of all modulation formats $m \in M$. Auxiliary binary variable $u_{dm}$ is used to check whether any modulation format $i \leq m$ can be applied to demand $d$ when using the DAT rule. Keeping consistent with the values of $n_{dm}$, constant $h_{dm} = n_{dm} - n_{d(m-1)}$ denotes the number of additional slices required for demand $d$ if modulation format $m$ is applied instead of modulation format $m - 1$. In conclusion, the number of slices required for demand $d$ on the selected routing path is given by formula $u_d = \sum_{m \in M} u_{dm} h_{dm}$.

---

### EON/M/RSA/Spectrum/Flow/Slice-based

**sets**

| | |
|---|---|
| $V$ | nodes |
| $E$ | links |
| $\delta^+(v)$ | links leaving node $v$ |
| $\delta^-(v)$ | links entering node $v$ |
| $D$ | multicast demands |
| $R(d)$ | receivers in demand (session) $d \in D$ |
| $M$ | modulation formats |

**constants**

| | |
|---|---|
| $h_d$ | volume (requested bit-rate) of demand $d$ |
| $b_{dm}$ | maximum distance range supported for modulation format $m$ and demand $d$ (km) |
| $a_{dm}$ | lower bound of the distance range supported for modulation format $m$ and demand $d$. If $m = 1$, then $a_{d1} = 0$. If $m > 1$, then $a_{dm} = b_{d(m-1)} + 1$ (km) |

$n_{dm}$    number of slices required for demand $d$ (with bit-rate $h_d$) using modulation
        format $m$

$h_{dm}$    number of additional slices required for demand $d$ if modulation format $m$ is
        applied instead of modulation format $m - 1$, $h_{dm} = n_{dm} - n_{d(m-1)}$

$l_e$      length of link $e$ (km)

$l^{max}$   maximum distance of a path in the network (km)

$n$       number of slices available on each fiber link

$s_d$      source (root) node of demand $d$

## variables

$x_{edr}$   $=1$, if in demand $d$ path to receiver $r \in R(d)$ uses link $e$; 0, otherwise (binary)

$x_{ed}$    $=1$, if demand $d$ uses link $e$; 0, otherwise (binary)

$x_d$      length of a tree created for demand $d$ (continuous)

$u_{dm}$    $=1$, if any modulation format $i \leq m$ can be applied to demand $d$ according to
        DAT; 0, otherwise (binary)

$u_d$      number of slices required for demand $d$ (integer)

$o_{di}$    $=1$, if the starting slice of demand $d$ is smaller than that of demand $i$; 0,
        otherwise (binary)

$c_{di}$    $=1$, if demands $d$ and $i$ use common link(s); 0, otherwise (binary)

$w_d$      indicates the starting slice used for demand $d$ (integer)

$y_d$      indicates the ending slice used for demand $d$ (integer)

$y$       indicates the maximum slice used in the network (integer)

## objective

$$\text{minimize} \quad F = y \tag{3.4.1a}$$

## constraints

$$\sum_{e \in \delta^+(v)} x_{edr} - \sum_{e \in \delta^-(v)} x_{edr} = \begin{cases} +1 & \text{if } v = s_d \\ -1 & \text{if } v = r, \quad v \in V, d \in D, r \in R(d) \\ 0 & \text{otherwise} \end{cases} \tag{3.4.1b}$$

$$x_{edr} \leq x_{ed}, \quad d \in D, e \in E, r \in R(d) \tag{3.4.1c}$$

$$x_d \geq \sum_{e \in E} l_e x_{edr}, \quad d \in D, r \in R(d) \tag{3.4.1d}$$

$$l^{max} u_{dm} \geq (x_d - a_{dm}), \quad d \in D, m \in M \tag{3.4.1e}$$

$$u_d \geq \sum_{m \in M} u_{dm} h_{dm}, \quad d \in D \tag{3.4.1f}$$

$$c_{di} \geq x_{ed} + x_{ei} - 1, \quad e \in E, d, i \in D : d \neq i \tag{3.4.1g}$$

$$o_{di} + o_{id} = 1, \quad d, i \in D : d \neq i \tag{3.4.1h}$$

$$y_i - w_d + 1 \leq n(1 + o_{di} - c_{di}), \quad d, i \in D : d \neq i \tag{3.4.1i}$$

$$y_d - w_i + 1 \leq n(2 - o_{di} - c_{di}), \quad d, i \in D : d \neq i \tag{3.4.1j}$$

$$y_d - w_d + 1 \geq u_d, \quad d \in D \tag{3.4.1k}$$

$$y \geq y_d, \quad d \in D. \tag{3.4.1l}$$

Objective function (3.4.1a) denotes the spectrum usage and is defined similarly to the previous RSA models, i.e., the aim is to minimize the maximum slice index used in the network to realize all multicast demands. Constraint (3.4.1b) defines unicast paths connecting root node $s_d$ and each receiver of demand (session) $d$ included in set $R(d)$. More precisely, by using variable $x_{edr}$ a node-link formulation of multicommodity flows is applied ensuring that a unicast path is established between the root node and each receiver. Condition (3.4.1c) is in the model to construct a multicast tree defined by variables $x_{ed}$. In other words, constraint (3.4.1c) guarantees that each link included in at least one unicast path established from the root node to a receiver must be included in the tree. Constraints (3.4.1d)–(3.4.1f) follow the DAT rule—to define variables $x_d$ denoting the number of slices required for demand $d$. More precisely, inequality (3.4.1d) is used to calculate variable $x_d$ denoting the length of the tree created for demand $d$ (i.e., length of the longest unicast path in the tree). Next, in (3.4.1e) and (3.4.1f) the appropriate modulation format is selected (variable $u_{dm}$), and consequently the number of required slices is obtained (variable $u_d$). Constraints (3.4.1g)–(3.4.1l) are used to control the spectrum usage according to the slice-based approach and these constraints are the same as in model EON/A/RSA/Spectrum/Node-link defined in (3.2.1).

**Candidate Tree Formulation**

The second formulation of multicasting in EONs is based on the candidate tree formulation. To recall, this approach assumes that for each multicast demand (session), there is a set of pre-calculated candidate tree structures which originate at the root node and include all receivers as tree nodes. As each routing structure (candidate tree) is given in advance, we can easily find the length of the tree (longest path from root to receiver) and calculate the number of slices required for demand $d$ realized on tree $p$ (defined by constant $n_{dp}$) using the DAT rule. We use the slice-based approach to model the spectrum. In general, the following model is similar to the anycast model EON/A/RSA/Spectrum/Slice-based formulated in (3.2.2).

**EON/M/RSA/Spectrum/Candidate Tree/Slice-based**

**sets**

| | |
|---|---|
| $E$ | links |
| $S$ | slices |
| $D$ | multicast demands (sessions) |
| $P(d)$ | candidate trees for realizing demand $d$ |

**constants**

$\delta_{edp}$  =1, if link $e$ belongs to tree $p$ realizing demand $d$; 0, otherwise
$n_{dp}$     requested number of slices for demand $d$ on tree $p$

**variables**

$x_{dp}$  =1, if candidate tree $p$ is used to realize demand $d$; 0, otherwise (binary)
$y_{ed}$  =1, if demand $d$ uses link $e$; 0, otherwise (binary)
$u_d$     number of slices required for demand $d$ (integer)
$o_{di}$  =1, if the starting slice of demand $d$ is smaller than that of demand $i$; 0, otherwise (binary)
$c_{di}$  =1, if demands $d$ and $i$ use common link(s); 0, otherwise (binary)
$w_d$     indicates the starting slice used for demand $d$ (integer)
$y_d$     indicates the ending slice used for demand $d$ (integer)
$y$       indicates the maximum slice used in the network (integer)

**objective**

$$\text{minimize} \quad F = y \tag{3.4.2a}$$

**constraints**

$$\sum_{p \in P(d)} x_{dp} = 1, \quad d \in D \tag{3.4.2b}$$

$$\sum_{p \in P(d)} \delta_{edp} x_{dp} \leq y_{ed}, \quad d \in D, e \in E \tag{3.4.2c}$$

$$\sum_{p \in P(d)} n_{dp} x_{dp} \leq u_d, \quad d \in D \tag{3.4.2d}$$

$$c_{di} \geq y_{ed} + y_{ei} - 1, \quad e \in E, d, i \in D : d \neq i \tag{3.4.2e}$$

$$o_{di} + o_{id} = 1, \quad d, i \in D : d \neq i \tag{3.4.2f}$$

$$y_i - w_d + 1 \leq |S|(1 + o_{di} - c_{di}), \quad d, i \in D : d \neq i \tag{3.4.2g}$$

$$y_d - w_i + 1 \leq |S|(2 - o_{di} - c_{di}), \quad d, i \in D : d \neq i \tag{3.4.2h}$$

$$y_d - w_d + 1 \geq u_d, \quad d \in D \tag{3.4.2i}$$

$$y \geq y_d, \quad d \in D. \tag{3.4.2j}$$

The objective function (3.4.2a) is the same as in the previous model (3.4.1). Equation (3.4.2b) ensures that exactly one candidate tree is selected to realize a multicast demand. Constraint (3.4.2c) defines variable $y_{ed}$ which denotes whether the multicast demand $d$ uses link $e$. Constraint (3.4.2d) is used to calculate the number of slices required for demand $d$. To control the spectrum usage, we use the same constraints as in the previous model (3.4.1). The model can be easily adapted to other objective functions as shown in Sect. 3.3.

Candidate tree modeling provides a number of interesting advantages when compared to the flow model or the EON multicast model presented in [32], known as *path* for ease of reference. Firstly, we briefly analyze the applicability of the ILP models in the context of various constraints following from multicasting in optical networks. We already considered that all network nodes are multicast capable. Nevertheless, in *sparse splitting networks* some nodes may be multicast incapable (MI), i.e., not able to replicate and multicast optical signals [74, 76]. In such scenarios, in order to account for MI nodes, the flow and path models require additional constraints, which may increase the size and solution time of these models. On the other hand, the CT model does not need any changes, since we can address the requirement of using MI nodes by generating proper candidate trees. An additional constraint in optical multicasting is that MC nodes have a limited fanout, i.e., the number of outgoing signals [74]. Once again, the CT model does not need any modifications as the fanout constraint can be included directly in the tree generation process. In contrast, flow and path models require new constraints to account for this limitation. Lastly, the CT approach (both the ILP model and any heuristic based on the CT approach) can be tuned easily to select the best trade-off between solution quality in terms of spectrum usage and the running time according to the size of the problem instance by choosing the number of candidate trees for each session. In turn, in the flow formulation it is impossible to scale the size of the model; in the case of the path formulation, the scalability is limited compared to the CT model, since we can only select the number of candidate paths for each receiver of the multicast session.

### *3.4.2 Algorithms*

**Minimum Spanning Tree (MST) Algorithm**

The MST/M/RSA method presented in Algorithm 3.11 is a simple greedy algorithm proposed for the RSA problem with multicast flows. According to [32, 76], the general idea of the MST algorithm is analogous to the First Fit method shown in Algorithm 3.2. Instead of allocating the unicast/anycast demand to the shortest routing path, the MST/M/RSA method allocates the multicast demand (session) to the minimum spanning tree. The input data for the MST algorithm is the network topology represented as set of links $E$, set of multicast demands $D$, minimum spanning tree for each demand included in set $P(d)$ and set of candidate channels $C(d, p)$ for each demand $d \in D$ and tree $p \in P(d)$. The size of each channel $c \in C(d, p)$ is equal to the spectrum requirement $n_{dp}$ (i.e., number of slices included in the channel) calculated for demand $d$ using tree $p$ according to a selected physical EON model.

As in the case of the FF/AU/RSA algorithm, algorithm MST/M/RSA processes all demands in a single run without any special ordering of the demands. The main loop of the MST/M/RSA algorithm (lines 2–7) is defined to process all demands.

The routing tree for demand $d$ is selected as the minimum spanning tree included in set $P(d)$ on position with index 1 (line 4). To find the spectrum allocation on the lowest possible spectrum range, function FF_SA yields the best available spectrum channel (line 5). Note that the only modification required in the FF_SA method (Algorithm 3.1) is to assume that set $E(p)$ contains all links included in tree $p$. Function *Allocate_Demand()* ensures that a particular demand is allocated on the selected path and channel (line 6). Finally, the value of the objective function is calculated (line 8). The complexity of the MST/M/RSA algorithm is $O(|D|\,|C|\,|E|)$.

---

**Algorithm 3.11** MST/M/RSA (Minimum Spanning Tree for M/RSA problem)

---

**Require:** set of edges $E$, set of multicast demands (sessions) $D$, sets $P(d)$ with a minimum spanning
   tree for each demand $d \in D$, candidate channels $C(d, p)$ for each demand $d \in D$ and tree $p \in$
   $P(d)$
**Ensure:** routing and spectrum allocation for each demand $d \in D$ included in vectors *tree* and
   *channel*, value of objective function
1: **procedure** $MST/M/RSA(D, P(d), C(d, p))$
2:   **for** $i := 1$ **to** $|D|$ **do**
3:     $d := Member(D, i)$
4:     $tree[d] := Member(P(d), 1)$
5:     $channel[d] := FF\_SA(C(d, tree[d]), tree[d])$
6:     $Allocate\_Demand(d, tree[d], channel[d])$
7:   **end for**
8:   $Find\_Objective\_Function(tree[], channel[])$
9: **end procedure**

---

## Adaptive Frequency Assignment (AFA) Algorithm

The AFA/M/RSA algorithm [36] is based on the CT approach and is an extension of the AFA/AU/RSA method proposed in Algorithm 3.5 for the RSA problem with anycast and unicast flows. To recall, the main idea behind the AFA method is to adaptively choose ordering of demands and next process all demands one-by-one using this sequence. The main difference between AFA/M/RSA and AFA/AU/RSA is that the latter uses new ordering metrics designed especially for multicast demands. In spite of each demand being assigned a metric equal to the minimum value of the requested number of slices required for a particular demand as in AFA/AU/RSA (i.e., $n_d = n_{dp}$), four new metrics are used for ordering multicast demands. The metrics are demand bit-rate ($n_d = h_d$), demand bit-rate multiplied by the number of receivers ($n_d = h_d \cdot |R(d)|$), demand bit-rate multiplied by the number of slices required for the first candidate tree ($n_d = h_d \cdot n_{d1}$), and bit-rate multiplied by the number of slices required for the first candidate tree and multiplied by the number of receivers ($n_d = h_d \cdot |R(d)| \cdot n_{d1}$). The allocation of demands in the AFA algorithm is repeated for each metric and the best result is the saved as the closing solution of AFA. We do not present the AFA/M/RSA pseudocode, since the modifications of AFA/AU/RSA (Algorithm 3.5) required to obtain AFA/M/RSA are straightforward.

Furthermore, as a result of using CT modeling of multicast flows, other heuristic and metaheuristic algorithms formulated for the RSA problem with joint optimization

of anycast and unicast demands and reported in Sect. 3.3.2 can be modified to solve the
RSA problem with multicast flows. The only required adjustment to the algorithms
is analysis of candidate trees for multicast demands instead of candidate paths used
in the context of anycast or unicast demands.

### 3.4.3  Numerical Results

**Comparison of Algorithms**

The main goal of numerical experiments reported in this section is to compare the
performance of various optimization approaches for the RSA problem with multicast flows defined in (3.4.2). We compare three formulations: the flow model defined
in (3.4.1), the path model proposed in Sect. II.A of [32], and the candidate tree
(CT) model defined in (3.4.2). All models were implemented in CPLEX [62]. We
also analyze two heuristics: the AFA/M/RSA algorithm with 1000 candidate trees
for each multicast demand and the MST/M/RSA algorithm. Three representative
network topologies were examined: the German national network DT14 (Fig. A.2,
Table A.2), the US network NSF15 (Fig. A.4, Table A.3) and the pan-European
Euro16 (Fig. A.5, Table A.4). Ten sets of multicast demands with number of ses-
sion $|D| = 7, 9, 11, 15, 20$ were generated at random for each network topology.
The bit-rate of each session was randomly selected in the range 10–200 Gb/s. The
average number of receivers in each session was set to five. An important parameter of the path model is $K$ denoting the number of candidate paths used for each
source-destination pair. The path model was executed with $K = 2, 3, 4$. We applied
10, 20 and 50 candidate trees for each demand in the CT model. The candidate trees
were generated using the algorithm proposed in [77]. A time limit of 1 h was set for
solving each problem instance using the CPLEX solver, with all other settings left at
default. The experiments were run on a PC with IntelCore i7-2620M CPU and 4GB
RAM.

The physical model of EON presented in [60] was used to estimate the transmission range of modulation formats. It should be noted that additional physical
impairments resulting from using optical splitting to provide multicasting were not
included in the model, since to the best of our knowledge there is no reliable research
on this issue in the context of EONs. However, the models formulated in the context
of multicast flows and heuristic algorithms are generic and can be adapted to any
new physical multicasting models which may be developed in the future.

Tables 3.13, 3.14 and 3.15 show the performance of all optimization approaches,
for the DT14, NSF15 and Euro16 networks, respectively. Each table presents two
performance metrics for each optimization approach, namely, the average number
of spectrum slices (value of the objective function) and the average execution time
given in seconds.

It is clear that for low numbers of multicast demands (sessions), the flow
model outperforms all other approaches. This is due to the fact that the flow model

**Table 3.13** Comparison of various optimization approaches for the RSA problem with multicast flows for the DT14 network

| $|D|$ | Flow | Path(2) | Path(3) | Path(4) | CT(10) | CT(20) | CT(50) | AFA(1000) | MST |
|---|---|---|---|---|---|---|---|---|---|
| *Average number of slices* | | | | | | | | | |
| 7 | 11.8 | 16.8 | 14.0 | 14.0 | 13.0 | 12.0 | 11.8 | 12.8 | 23.8 |
| 9 | 15.2 | 20.0 | 16.8 | 16.6 | 15.4 | 14.6 | 14.0 | 15.4 | 27.0 |
| 11 | – | 24.4 | 19.6 | 19.4 | 18.6 | 17.4 | 16.2 | 18.8 | 31.4 |
| 15 | – | 30.8 | 26.6 | 26.4 | 26.0 | 25.6 | 25.8 | 26.2 | 42.8 |
| 20 | – | 39.6 | 35.8 | 36.6 | 35.0 | 36.4 | 37.6 | 34.4 | 51.8 |
| *Average execution time in seconds* | | | | | | | | | |
| 7 | 2540.5 | 0.1 | 1.5 | 10.1 | 5.8 | 10.2 | 153.3 | 0.2 | 0.0 |
| 9 | 3600.0 | 1.0 | 19.4 | 555.2 | 39.4 | 457.1 | 2440.1 | 0.3 | 0.0 |
| 11 | – | 8.0 | 739.1 | 1192.8 | 1159.0 | 3122.5 | 3600.0 | 0.2 | 0.0 |
| 15 | – | 831.9 | 2239.5 | 2464.6 | 3037.1 | 3600.5 | 3600.0 | 1.3 | 0.0 |
| 20 | – | 2235.1 | 3241.5 | 3600.0 | 2688.3 | 3600.0 | 3600.0 | 1.0 | 0.0 |

**Table 3.14** Comparison of various optimization approaches for the RSA problem with multicast flows for the NSF15 network

| $|D|$ | Flow | Path(2) | Path(3) | Path(4) | CT(10) | CT(20) | CT(50) | AFA(1000) | MST |
|---|---|---|---|---|---|---|---|---|---|
| *Average number of slices* | | | | | | | | | |
| 7 | 25.4 | 39.4 | 34.2 | 33.0 | 30.8 | 27.6 | 27.4 | 27.6 | 51.4 |
| 9 | 28.2 | 44.6 | 39.0 | 37.6 | 35.6 | 32.6 | 30.0 | 37.4 | 60.4 |
| 11 | 41.6 | 52.8 | 47.4 | 46.2 | 47.6 | 41.8 | 38.8 | 41.4 | 75.6 |
| 15 | – | 65.2 | 61.0 | 60.0 | 64.0 | 54.8 | 53.6 | 54.8 | 100.0 |
| 20 | – | 88.4 | 81.4 | 79.8 | 78.0 | 72.8 | 72.8 | 69.2 | 114.2 |
| *Average execution time in seconds* | | | | | | | | | |
| 7 | 2907.1 | 0.2 | 9.3 | 5.9 | 1.2 | 4.1 | 94.7 | 0.2 | 0.0 |
| 9 | 3153.9 | 2.8 | 133.0 | 509.9 | 18.8 | 752.0 | 1703.0 | 0.3 | 0.0 |
| 11 | 3600.0 | 12.2 | 1474.8 | 2002.7 | 787.5 | 2302.1 | 2975.0 | 1.0 | 0.0 |
| 15 | – | 667.9 | 2168.8 | 1843.8 | 1467.2 | 3349.8 | 3600.0 | 1.5 | 0.0 |
| 20 | – | 1810.8 | 2300.8 | 2636.6 | 1198.9 | 3600.0 | 3600.0 | 3.1 | 0.0 |

incorporates all possible routing trees, while the other ILP models and heuristics are based on limited subsets of candidate trees. However, as the number of demands increases, the flow model lacks scalability and it cannot find feasible results within the 1 h run-time limit, while the best results in terms of the spectrum usage are provided by the CT(50) approach. Finally, for the highest number of sessions ($|D| = 20$), the AFA algorithm outperforms all ILP models in terms of spectrum usage. Moreover, AFA only needs a few seconds of execution time, while ILP models run for tens of minutes. The main conclusion of these results is that ILP modeling for multicasting

**Table 3.15** Comparison of various optimization approaches for the RSA problem with multicast flows for the Euro16 network

| $|D|$ | Flow | Path(2) | Path(3) | Path(4) | CT(10) | CT(20) | CT(50) | AFA(1000) | MST |
|---|---|---|---|---|---|---|---|---|---|
| *Average number of slices* | | | | | | | | | |
| 7 | 16.8 | 21.8 | 18.6 | 18.0 | 20.0 | 19.0 | 16.8 | 17.4 | 28.4 |
| 9 | 23.0 | 26.0 | 22.2 | 21.2 | 23.2 | 22.2 | 20.4 | 20.6 | 34.2 |
| 11 | – | 33.4 | 28.2 | 27.2 | 29.6 | 27.6 | 26.0 | 27.2 | 43.8 |
| 15 | – | 37.4 | 31.8 | 32.4 | 34.6 | 34.4 | 30.6 | 32.0 | 51.0 |
| 20 | – | 42.6 | 41.0 | 46.2 | 40.6 | 42.4 | 44.2 | 39.2 | 60.4 |
| *Average execution time in seconds* | | | | | | | | | |
| 7 | 2903.2 | 0.2 | 1.4 | 21.0 | 1.2 | 9.2 | 6.9 | 0.2 | 0.1 |
| 9 | 2986.5 | 1.0 | 65.7 | 871.1 | 7.9 | 550.7 | 1216.5 | 0.2 | 0.1 |
| 11 | – | 2.9 | 448.4 | 807.9 | 8.5 | 375.7 | 1163.8 | 0.7 | 0.1 |
| 15 | – | 13.2 | 1782.7 | 2450.7 | 379.7 | 405.8 | 3286.5 | 1.3 | 0.1 |
| 20 | – | 1279.2 | 3571.5 | 3600.1 | 3374.5 | 3600.0 | 3600.0 | 1.2 | 0.1 |

in EONs faces scalability problems for larger problem instances, and in such scenarios the heuristic AFA algorithm is the best choice. Finally, it should be noted that the path model proposed in [32] is outperformed by our ILP models for all of cases.

More detailed analysis of the results shows that the performance of our optimization approaches depends on network topology. In particular, the greatest differences between the path and CT models are reported for the NSF15 network, while for other networks the gaps between these two models are minor. Additionally, in the context of NSF15 and $|D| = 20$, AFA outperforms other methods to a greater extent compared to other networks. This is mainly due to the fact that on average link, lengths in the NSF15 network are greater than in the Euro15 and DT14 networks. Therefore, following the DAT rule, NSF15 needs less effective modulation formats, therefore higher numbers of slices are required for demands. As such, a more effective allocation of even one multicast demand can provide a major saving of the amount of overall spectrum required in the network.

## 3.5 Routing and Spectrum Allocation for Multicast and Unicast Flows

The previous section examined the RSA problem with multicast traffic only. This section considers a corresponding RSA problem with joint multicast and unicast flows [56]. The same optical networks are currently able to carry different types of flows, therefore joint optimization of various types of flows can significantly improve consumption of network resources. The idea of formulating the RSA problem with joint multicast and unicast flows is analogous to the optimization of joint anycast and unicast flows introduced in Sect. 3.3. To simplify the notation, the *structure* term is introduced to denote the routing path used to realize unicast demands and the tree used to realize multicast demands. In other words, routing structure $p$ included in set $P(d)$ is either a path (if $d$ is unicast) or a tree (if $d$ is multicast). This approach uses the same formulation of constraints for both multicast and unicast demands.

### 3.5.1   Formulations

This section presents two ILP formulations for the RSA problem with joint multicast and unicast flows. The models differ in terms how optical spectrum usage is defined in the model, or how the slice-based model and channel-based model are formulated. Both models are based on the candidate tree concept, since—as shown above in Sect. 3.4—this formulation outperforms other approaches in the context of multicast modeling in EONs.

#### Slice-Based Formulation

The slice-based model is analogous to the EON/M/RSA/Spectrum/Candidate Tree model defined in (3.4.2). The only modification is that two types of demands (multicast and unicast) are included in set $D$ and the structure is used instead of a tree.

---

**EON/MU/RSA/Spectrum/Candidate Tree/Slice-based**

**sets**

$E$      links
$S$      slices
$D$      multicast and unicast demands
$P(d)$   candidate structures (trees or paths) for realizing demand $d$

**constants**

$\delta_{edp}$   $=1$, if link $e$ belongs to structure $p$ realizing demand $d$; 0, otherwise
$n_{dp}$   requested number of slices for demand $d$ on structure $p$
$n$     number of slices available on each fiber link

**variables**

$x_{dp}$   $=1$, if candidate structure $p$ is used to realize demand $d$; 0, otherwise (binary)
$y_{ed}$   $=1$, if demand $d$ uses link $e$; 0, otherwise (binary)
$u_d$    number of slices required for demand $d$ (integer)
$o_{di}$   $=1$, if the starting slice of demand $d$ is smaller than that of demand $i$; 0, otherwise (binary)
$c_{di}$   $=1$, if demands $d$ and $i$ use common link(s); 0, otherwise (binary)

$w_d$   indicates the starting slice used for demand $d$ (integer)
$y_d$   indicates the ending slice used for demand $d$ (integer)
$y$    indicates the maximum slice used in the network (integer)

**objective**

$$\text{minimize} \quad F = y \tag{3.5.1a}$$

**constraints**

$$\sum_{p \in P(d)} x_{dp} = 1, \quad d \in D \tag{3.5.1b}$$

$$\sum_{p \in P(d)} \delta_{edp} x_{dp} \leq y_{ed}, \quad d \in D, e \in E \tag{3.5.1c}$$

$$\sum_{p \in P(d)} n_{dp} x_{dp} \leq u_d, \quad d \in D \tag{3.5.1d}$$

$$c_{di} \geq y_{ed} + y_{ei} - 1, \quad e \in E, d, i \in D : d \neq i \tag{3.5.1e}$$

$$o_{di} + o_{id} = 1, \quad d, i \in D : d \neq i \tag{3.5.1f}$$

$$y_i - w_d + 1 \leq n(1 + o_{di} - c_{di}), \quad d, i \in D : d \neq i \tag{3.5.1g}$$

$$y_d - w_i + 1 \leq n(2 - o_{di} - c_{di}), \quad d, i \in D : d \neq i \tag{3.5.1h}$$

$$y_d - w_d + 1 \geq u_d, \quad d \in D \tag{3.5.1i}$$

$$y \geq y_d, \quad d \in D. \tag{3.5.1j}$$

The objective function and all constraints of the model described above are equivalent to model (3.4.2), so for a more detailed discussion on the model see Sect. 3.4.1.

**Channel-Based Formulation**

The second model formulated for the RSA problem with joint multicast and unicast flows uses the channel concept for spectrum modeling. In general, the formulation is based on model (3.3.1) and more details on the formulation can be found in Sect. 3.3.1.

--------

**EON/MU/RSA/Spectrum/Link-path/Channel-based**

**sets**

$E$      links
$S$      slices
$D$      demands (multicast and unicast)
$P(d)$   candidate structures (trees or paths) for realizing demand $d$
$C(d, p)$ candidate channels for demand $d$ allocated on structure $p$

**constants**

$\delta_{edp}$    $=1$, if link $e$ belongs to structure $p$ realizing demand $d$; 0, otherwise
$n_{dp}$         requested number of slices for demand $d$ on structure $p$
$\gamma_{dpcs}$  $=1$, if channel $c$ associated with demand $d$ on structure $p$ uses slice $s$; 0, otherwise

**variables**

$x_{dpc}$  $=1$, if channel $c$ on candidate structure $p$ is used to realize demand $d$; 0, otherwise (binary)
$y_{es}$   $=1$, if slice $s$ is occupied on link $e$; 0, otherwise (binary)
$y_s$      $=1$, if slice $s$ is occupied on any network link; 0, otherwise (binary)

**objective**

$$\text{minimize} \quad F = \sum_{s \in S} y_s \tag{3.5.2a}$$

**constraints**

$$\sum_{p \in P(d)} \sum_{c \in C(d,p)} x_{dpc} = 1, \quad d \in D \tag{3.5.2b}$$

$$\sum_{d \in D} \sum_{p \in P(d)} \sum_{c \in C(d,p)} \gamma_{dpcs}\delta_{edp}x_{dpc} \leq y_{es}, \quad e \in E, s \in S \tag{3.5.2c}$$

$$\sum_{e \in E} y_{es} \leq |E| \, y_s, \quad s \in S \tag{3.5.2d}$$

Both models proposed for the RSA problem with joint multicast and unicast flows can be formulated with other objective functions as shown in Sect. 3.3 in the context of the RSA problem with joint anycast and unicast traffic.

## 3.5.2  Algorithms

As well as the AFA method, we also present results of a simple heuristic algorithm combining the minimal spanning tree method (Algorithm 3.11) with the First Fit (FF) method (Algorithm 3.2). In the MST/FF method, the multicast demands are allocated using the MST algorithm first, followed by running the FF to allocate unicast demands.

### 3.5.3    Numerical Results

The first goal of the numerical experiments was to compare the slice-based and channel-based formulations presented above. Secondly, we investigated the effectiveness of heuristic algorithms against results yielded by ILP modeling using the CPLEX solver [62]. The third goal of the experiments was to study issues related to the use of multicasting in EONs [55, 56].

Since ILP models have a low scalability, our experiments with the CPLEX solver used three smaller networks: the German national network (Fig. A.2, Table A.2), the US network NSF15 (Fig. A.4, Table A.3) and the pan-European network Euro16 (Fig. A.5, Table A.4). For additional experiments with the AFA heuristic, larger topologies were applied, i.e., the US backbone network US26 (Fig. A.9, Table A.7) and a pan-European network Euro28 (Fig. A.7, Table A.6). Traffic patterns including both unicast and multicast demands were generated at random, with the unicast demand bit-rate selected from the range 10–400 Gb/s, and the multicast demand bit-rate selected from the range 10–200 Gb/s. If not stated otherwise, we assumed that each traffic matrix is split equally between unicast and multicast traffic. The volume of multicast traffic was calculated as the traffic received by all participants of the session (receivers). For instance, if the multicast bit-rate is 200 Gb/s and the session includes five receivers, the total received traffic is 1 Tb/s. The volume of an unicast demand was simply equal to its bit-rate. For smaller networks, each multicast session included five receivers, while for larger networks each session had between five and 15 receivers. For smaller networks, we created 25 different sets of traffic patterns for each value of the overall traffic (4, 5 and 6 Tb/s). We used the k-shortest path algorithm to create sets of candidate paths for unicast demands. In the case of candidate tree generation, we applied the algorithm proposed in [77].

The main physical assumptions of the EONs model were the same as in Sect. 3.4.3. A time limit of 1 h was set for CPLEX to solve each problem instance, while all other solver settings were left as default. Experiments were run on a PC with IntelCore i7-2620M CPU and 4GB RAM.

**Comparison of ILP Models**

Following initial experiments showing the low scalability of the ILP models run in CPLEX, the settings for further experiments were determined to be the number of candidate paths for unicast demands $k = 2$ and the number of candidate trees for multicast demands $t = 10$. We use the running time of the CPLEX solver as the main performance metric in the comparison between slice-based and channel-based models. Similarly to the comparison of slice-based and channel-based models in the context of anycast flows reported in Sect. 3.2.2, we noted that the performance of the channel-based model strongly depends on the number of slices available in the network denoted by parameter $|S|$. To illustrate this issue, we conducted the following experiment. First, we calculated the optimal result (i.e., minimum number of required slices) for two randomly selected traffic patterns of the Euro16 network with overall traffic equal to 6 Tb/s. Next, we ran both models starting with $|S|$ equal

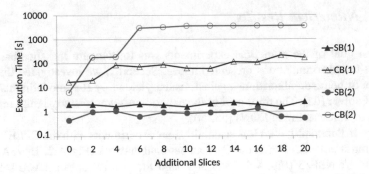

**Fig. 3.11** Execution time of slice-based and channel-based model as a function of the number of additional slices for the Euro16 network

to the optimal value and then increasing $|S|$ by two slices up to 20. Results presented in Fig. 3.11 clearly show that the execution time of the slice-based (SB) model does not depend on the value of $|S|$. In turn, the running time of the channel-based (CB) model increases almost exponentially with the number of additional slices (the y-axis is in the logarithmic scale). Note that curve CB(2) curve is constant starting from six slices, due to reaching the 1 h limit of the CPLEX solver. In consequence, the AFA algorithm was used to provide a reasonable upper bound on the objective function, and this value was used as $|S|$ in further experiments with CPLEX.

Table 3.16 presents a comparison of both models. The results shown are the average value of the objective function and the average optimality gap for each model, as well as presenting a comparison of execution times in three categories: SB < CB

**Table 3.16** Comparison of ILP models for the RSA problem with multicast and unicast flows

| Traffic | Obj. function | | Optimality gap | | Execution time | | |
|---|---|---|---|---|---|---|---|
| | SB | CB | SB (%) | CB (%) | SB < CB | SB > CB | SB = CB = 1 h |
| *DT14 network* | | | | | | | |
| 4 Tb/s | 16.96 | 16.96 | 0.44 | 0.41 | 21 | 3 | 1 |
| 5 Tb/s | 22.88 | 22.88 | 0.00 | 0.32 | 25 | 0 | 0 |
| 6 Tb/s | 22.48 | 22.32 | 6.87 | 1.80 | 12 | 8 | 5 |
| *NSF15 network* | | | | | | | |
| 4 Tb/s | 51.76 | 51.76 | 0.00 | 0.25 | 25 | 0 | 0 |
| 5 Tb/s | 56.16 | 56.08 | 0.15 | 3.75 | 24 | 0 | 1 |
| 6 Tb/s | 59.60 | 60.36 | 2.34 | 6.57 | 15 | 1 | 9 |
| *Euro16 network* | | | | | | | |
| 4 Tb/s | 37.76 | 37.76 | 0.27 | 0.00 | 23 | 2 | 0 |
| 5 Tb/s | 47.52 | 47.52 | 0.24 | 1.46 | 24 | 0 | 1 |
| 6 Tb/s | 57.12 | 57.12 | 0.00 | 1.43 | 25 | 0 | 0 |

(the slice-based model provides the solution faster than the channel-based model), SB > CB (the situation is reversed) and SB = CB = 1 h (number of cases in which both models reach the 1 h limit execution time without finding the optimal solution). Note that the slice-based model significantly outperforms the channel-based model in terms of execution time; in 194 out of 225 cases the slice-based model finds the solution faster, while the channel-based model only succeeds in 14 cases. In general, both models yield similar values for the objective function, although in some cases the 1 h running time stops the CPLEX solver prior to obtaining the optimal solution. Additionally, as network traffic increases, the execution time and the number of 1 h limit cases also increases.

**Performance of Heuristic Algorithms**

The next goal of the experiments was to evaluate two heuristic algorithms, AFA/MU/RSA and MST-FF/MU/RSA, in comparison with results yielded by CPLEX. We used the same networks and demands patterns as in the comparison of ILP models. Note that some of the results returned by CPLEX did not have the optimality guarantee; however, the average optimality gap of all CPLEX results was lower than 1 %. For the AFA algorithm, we report results obtained with a range of sets of candidate paths and trees. Firstly, AFA is run with the same number of candidate paths and trees as CPLEX, i.e., $k = 2$ and $t = 10$, denoted as AFA(2, 10). However, since the AFA algorithm provides good scalability in contrast to CPLEX, we also use AFA(10, 200), AFA(20, 600) and AFA(30, 1000).

Table 3.17 shows the average gaps to CPLEX results obtained for different traffic volume and network topology. The worst performance is returned by the simplest MST-FF method. The AFA(2,10) algorithm using the same candidate structures as

**Table 3.17** Comparison of heuristic algorithms for the RSA problem with multicast and unicast flows—average gap to CPLEX(2, 10)

| Traffic | MST-FF (%) | AFA(2, 10) (%) | AFA(10, 200) (%) | AFA(20, 600) (%) | AFA(30, 1000) (%) |
|---|---|---|---|---|---|
| *DT14 network* | | | | | |
| 4 Tb/s | 26.2 | 13.5 | −2.8 | −8.2 | −9.0 |
| 5 Tb/s | 28.8 | 11.7 | 1.2 | −3.6 | −3.6 |
| 6 Tb/s | 23.0 | 11.4 | −13.4 | −19.8 | −21.0 |
| *NSF15 network* | | | | | |
| 4 Tb/s | 14.7 | 8.7 | −3.6 | −15.5 | −16.5 |
| 5 Tb/s | 21.1 | 11.6 | −12.7 | −15.8 | −16.6 |
| 6 Tb/s | 24.6 | 14.5 | 5.9 | 2.0 | 0.7 |
| *Euro16 network* | | | | | |
| 4 Tb/s | 24.9 | 5.7 | −20.5 | −21.6 | −22.9 |
| 5 Tb/s | 24.4 | 10.7 | −10.4 | −13.3 | −13.3 |
| 6 Tb/s | 19.8 | 7.1 | −21.0 | −28.2 | −30.0 |

the CPLEX solver falls between 5.7 and 14.5 % from CPLEX results. However, the effectiveness of AFA improves as the number of candidate path and trees increases, and AFA(30,1000) significantly outperforms CPLEX, in particular in Euro16 (on average, the results show a 22 % improvement). In terms of execution time, the MST/FF and AFA(2, 10) heuristics return results in under 1 s. The average running times of CPLEX, when using the more time-effective SB model, are 548, 549 and 114 s for DT14, NSF15 and Euro16 networks, respectively (running times of the CB method are significantly higher). The main conclusion of results shown in Table 3.17 is that the AFA method offers a significantly higher scalability than ILP models, therefore AFA provides better results in terms of spectrum usage in a reasonable time than CPLEX modeling [56].

**Multicasting in EONs**

This section presents results obtained for the larger Euro28 and US26 networks using the AFA/MU/RSA heuristic. We start by examining the performance of AFA as a function of the number of candidate structures. Ten random traffic patterns with an overall volume of 40 Tb/s and including 50 % of multicast and 50 % of unicast traffic were generated at random for each network. Figure 3.12 shows the results of the average number of slices obtained with AFA using seven different combinationss of candidate structures sets from $k = 2$ and $t = 10$ (denoted as (2, 10)) to $k = 30$ and $t = 1000$ (denoted as (30, 1000). We can see that increasing the sets of candidate structures reduces the spectrum consumption. The gap between (2, 10) and (30, 1000) is 24 and 32 % for Euro28 and US26 networks, respectively.

The next experiment evaluated the potential gains of using multicast transmission in place of unicast. To start with, we used multicasting to provision the same traffic pattern (i.e., data is sent from the root to each receiver via a multicast transmission). Unicast transmission was then applied for each root-receiver pair (i.e., a single unicast demand was used to serve each transmission from the root to the receiver). Figure 3.13 presents results obtained for the US26 and Euro28 networks with the overall traffic volume equal to 40 Tb/s as a function of multicast traffic proportion. For instance, a multicast proportion of 30 % means that 12 Tb/s of network traffic is provisioned

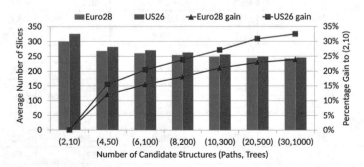

**Fig. 3.12** Average number of slices as a function of candidate structures number for US26 and Euro28 networks with overall traffic of 40 Tb/s

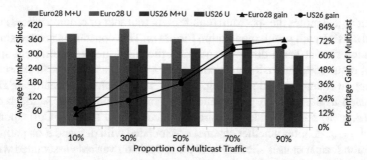

**Fig. 3.13** Multicast versus unicast—average number of slices as a function of the proportion of multicast traffic in US26 and Euro28 networks with overall traffic set to 40 Tb/s

by multicasting and the remaining 70 % of network traffic is unicast. Figure 3.13 includes four series of results presented as columns, showing two results for each network: using multicast transmission (M + U) and using pure unicast transmission (U). Moreover, two additional curves refer to the gain in spectrum usage resulting from using multicast transmission instead of pure unicast transmission. It is clear that as the proportion of multicast traffic increases, the gain of using multicasting increases almost linearly; for the 90 % case, the multicast approach needs approx. 70 % less spectrum resources than the unicast approach.

## 3.6 Routing, Modulation and Spectrum Allocation for Anycast and Unicast Flows

This section focuses on the Routing, Modulation and Spectrum Allocation (RMSA) optimization problem with joint anycast and unicast flows [44]. The RMSA problem is an extended version of the classical RSA problem. More specifically, as well as routing and spectrum allocation, RMSA also involves selecting a modulation format for each demand to be established in the EON. To this end, it is assumed that set $M$ includes several modulation formats which can be used in the network. We present a formulation of the AU/RMSA problem with heuristic algorithms and illustrative results.

### 3.6.1 Formulations

We use the channel-based approach to formulate the AU/RMSA problem. Using the A/RSA model defined in (3.2.2) as a basis, we can construct the corresponding model using the slice-based approach. The model is analogous to the AU/RSA problem (3.3.1). The main difference is that various modulation formats are available in set $M$.

Constant $n_{dpm}$ denotes the spectrum requirement (number of required slices) for each demand $d$ using path $p$ with modulation format $m$. This constant is needed since each modulation format has a different spectral efficiency, therefore different spectrum width is required to provision a demand with a particular bit-rate. In consequence, according to channel-based modeling, let $C(d, p, m)$ include all candidate channels for demand $d$ allocated on path $p$ and using modulation format $m$. Each channel $c \in C(d, p, m)$ has the width (number of slices) equal to $n_{dpm}$. Correspondingly, constant $\gamma_{dpcsm}$ denotes whether channel $c$ associated with demand $d$ on path $p$ under modulation format $m$ uses slice $s$. Finally, the decision variable associated with each demand includes the option of selecting a modulation format $m \in M$, i.e., $x_{dpmc}$ is 1 if channel $c$ on candidate path $p$ under modulation format $m$ is used to realize demand $d$ and 0 otherwise.

### EON/AU/RMSA/Spectrum/Link-path/Channel-based

**sets**

| | |
|---|---|
| $E$ | links |
| $S$ | slices |
| $D$ | demands (anycast and unicast) |
| $D^{DS}$ | anycast downstream demands |
| $M$ | modulation formats |
| $P(d)$ | candidate paths for flows realizing demand $d$. If $d$ is a unicast demand, the candidate path connects end nodes of the demand. If $d$ is an anycast upstream demand, the candidate path connects the client node and the DC node. If $d$ is a downstream demand, the candidate path connects the DC node and the client node |
| $C(d, p, m)$ | candidate channels for demand $d$ allocated on path $p$ and using modulation format $m$ |

**constants**

| | |
|---|---|
| $\delta_{edp}$ | $=1$, if link $e$ belongs to path $p$ realizing demand $d$; 0, otherwise |
| $n_{dpm}$ | requested number of slices for demand $d$ on path $p$ using modulation format $m$ |
| $\gamma_{dpcsm}$ | $=1$, if channel $c$ associated with demand $d$ on path $p$ under modulation format $m$ uses slice $s$; 0, otherwise |
| $\tau(d)$ | index of a demand associated with demand $d$. If $d$ is a downstream demand, then $\tau(d)$ must be an upstream demand and vice versa |
| $o(p)$ | origin node of path $p$ |
| $t(p)$ | destination node of path $p$ |

**variables**

| | |
|---|---|
| $x_{dpmc}$ | $=1$, if channel $c$ on candidate path $p$ under modulation format $m$ is used to realize demand $d$; 0, otherwise (binary) |

$y_{es}$     $=1$, if slice $s$ is occupied on link $e$; 0, otherwise (binary)

$y_s$     $=1$, if slice $s$ is occupied on any network link; 0, otherwise (binary)

**objective**

$$\text{minimize} \quad F = \sum_{s \in S} y_s \tag{3.6.1a}$$

**constraints**

$$\sum_{p \in P(d)} \sum_{m \in M} \sum_{c \in C(d,p,m)} x_{dpmc} = 1, \quad d \in D \tag{3.6.1b}$$

$$\sum_{d \in D} \sum_{m \in M} \sum_{p \in P(d)} \sum_{c \in C(d,p,m)} \gamma_{dpcsm} \delta_{edp} x_{dpmc} \leq y_{es}, \quad e \in E, s \in S \tag{3.6.1c}$$

$$\sum_{e \in E} y_{es} \leq \mid E \mid y_s, \quad s \in S \tag{3.6.1d}$$

$$\sum_{p \in P(d)} \sum_{m \in M} \sum_{c \in C(d,p,m)} o(p) x_{dpmc}$$

$$= \sum_{p \in P(\tau(d))} \sum_{m \in M} \sum_{c \in C(\tau(d),p,m)} t(p) x_{\tau(d)pmc}, \quad d \in D^{DS}. \tag{3.6.1e}$$

The objective function (3.6.1a) minimizes spectrum usage as in previous models. Equation (3.6.1b) ensures that exactly one candidate path, one modulation format and one candidate channel are selected for each demand $d$. To ensure that a slice on a particular link can be allocated to at most one lightpath, Eq. (3.6.1c) is added to the model. Constraint (3.6.1d) defines variable $y_s$ which indicates whether there is at least one link where slice $s$ is occupied. Lastly, constraint (3.6.1e) guarantees that both associated anycast demands use candidate paths connected to the same DC node. Note that [44, 47] formulate and discuss the AU/RMSA problems with objective functions of average spectrum, cost and power consumption.

## 3.6.2 Algorithms

Since the AU/RMSA problem defined in (3.6.1) is similar to the AU/RSA problem (3.3.1), the algorithms described in Sect. 3.3.2 can be adapted easily to address additional constraints resulting from the modulation selection. This section includes two algorithms: the AFA/AU/RMSA algorithm that is a modification of the AFA/AU/RSA method shown in Algorithm 3.5, and TS/AU/RMSA which is an extended version of the TS/AU/RSA method presented in Algorithm 3.9.

**Adaptive Frequency Assignment (AFA) Algorithm**

The main idea behind the AFA/AU/RMSA method shown in Algorithm 3.12 is to adaptively select a sequence of processed demands in order to minimize the objective function. However, due to modulation allocation considerations, the solution space of the RMSA problem is increased compared to the pure RSA problem. Therefore, the demand ordering process is modified (lines 3–9). In particular, for each demand $d \in D$ the following metrics are calculated. Firstly, let $n_d := \min\limits_{p \in P(d), m \in M} \{n_{dpm}\}$ denote the minimum value of $n_{dpm}$ taking into account all possible paths included in set $P(d)$ and available modulation formats included in set $M$. Secondly, let $l_d := \min\limits_{p \in P(d)} \{l_{dp}\}$ denote the minimum length of candidate paths calculated as the hop number constant $l_{dp}$ represents the hop number of path $p \in P(d)$. Finally, a new metric $a_d := \lceil n_d \cdot l_d / DIV \rceil$ is introduced, where $DIV$ is a tuning parameter of the algorithm. Notice that metric $a_d$ is a function of two elements: path length and spectrum usage. Metric $a_d$ is used to divide all demands into subsets $D(a_d)$ with the same value of $a_d$.

The main loop of the algorithm (lines 10–31) processes sets $D(a_d)$ in decreasing order of $a_d$, similarly as in Algorithm 3.5. Since modulation format needs to be selected alongside routing and spectrum allocation, a new function *FPMC Spectrum*() is used to find the best possible allocation for current demand $d$. Function *FPMCSpectrum*() can be obtained by a small modification of the *FPCSpectrum*() defined in Algorithm 3.3. More specifically, as well as considering all possible routing paths $p \in P(d)$ and spectrum channels $c \in C(d, p, m)$, function *FPMCSpectrum*() must also check all available modulation formats $m \in M$. Moreover, functions *Best_Allocation*() and *Allocate_Demand*() need to be modified slightly to enable processing of various modulation formats, although the modifications are straightforward. The complexity of algorithm AFA/AU/RMSA is given by $O(|D|^2 |P| |M| |C| |E|)$, where here $|M|$ denotes the number of modulation formats.

Note that using the general concept of the AFA/AU/RMSA algorithm makes it easy to modify other greedy methods such as FF, LPF and MSF to address new constraints following from the additional dimension of modulation format selection incorporated in the RMSA problems.

**Tabu Search Algorithm**

Similarly, metaheuristic methods developed for the RSA problem can be adapted to solve corresponding RMSA problems. This section presents the TS/AU/RMSA algorithm based on the TS/AU/RSA method formulated as Algorithm 3.9. To address new requirements related to the selection of a modulation format for each demand, the RSA version of the TS algorithm needs to be modified slightly. Firstly, solution encoding is updated with additional information on the selected modulation format for each demand, i.e., it is assumed that for each demand $d$ the solution is coded by tuple $(p_d, m_d, seq_d)$, where $p_d$ denotes the path selected to serve demand $d$, $m_d$ stands for the modulation format chosen for demand $d$, and $seq_d$ denotes the sequence number of demand $d$ (demands are allocated one-by-one according to this order). Thus, the solution is described as the following vector:

---

**Algorithm 3.12** AFA/AU/RMSA (Adaptive Frequency Assignment for AU/RMSA problem)

---

**Require:** set of edges $E$, set of anycast and unicast demands $D$, sets $P(d)$ with candidates paths for each demand $d \in D$, set $M$ with available modulation formats, candidate channels $C(d, p, m)$ for each demand $d \in D$, path $p \in P(d)$ and modulation format $m$, tuning parameter $DIV$

**Ensure:** routing, modulation and spectrum allocation for each demand $d \in D$ included in vectors *path*, *mod* and *channel*, value of objective function

1: **procedure** $AFA/AU/RMSA(D, P(d), C(d, p))$
2:    $a_{max} := 0$
3:    **for** $i := 1$ **to** $|D|$ **do**
4:      $d := Member(D, i)$
5:      $n_d := \min\limits_{p \in P(d), m \in M} \{n_{dpm}\}, l_d := \min\limits_{p \in P(d)} \{l_{dp}\}$
6:      $a_d := \lceil n_d \cdot l_d / DIV \rceil$
7:      $D(a_d) := D(a_d) \cup \{d\}$
8:      **if** $a_d > a_{max}$ **then** $a_{max} := a_d$
9:    **end for**
10:   **for** $a := a_{max}$ **to** 1 **do**
11:     **while** $D(a) \neq \emptyset$ **do**
12:       **for** $i := 1$ **to** $|D(a)|$ **do**
13:         $d := Member(D, i)$
14:         **if** $Established(d) = FALSE$ **then**
15:           **if** $Type(d) = ANYCAST$ **then**
16:             **if** $Established(\tau(d)) = FALSE$ **then**
17:               $\{path[d], mod[d], channel[d]\} := FPMCSpectrum(P(d), C(d, p, m))$
18:             **else**
19:               $r := server[\tau(d)]$
20:               $\{path[d], mod[d], channel[d]\} := FPMCSpectrum(P(d, r), M, C(d, p, m))$
21:             **end if**
22:           **end if**
23:           **if** $Type(d) = UNICAST$ **then**
24:             $\{path[d], mod[d], channel[d]\} := FPMCSpectrum(P(d), M, C(d, p, m))$
25:           **end if**
26:         **end if**
27:       **end for**
28:       $d^* := Best\_Allocation(D(a), path, mod, channel)$
29:       $Allocate\_Demand(d^*, path[d^*], mod[d^*], channel[d^*])$
30:     **end while**
31:   **end for**
32:   $Find\_Objective\_Function(path[], mod[], channel[])$
33: **end procedure**

---

$$X = [(p_1, m_1, seq_1), (p_2, m_2, seq_2), \ldots, (p_{|D|}, m_{|D|}, seq_{|D|})]. \qquad (3.6.2)$$

Moreover, in the RMSA/AU problem the set of TS moves compared to algorithm TS/AU/RSA is extended with a new move operation which involves swapping the modulation format. This operation simply changes the modulation format for a particular demand using the set of all available modulation formats. Other swap operations defined for the TS/AU/RSA method presented in Algorithm 3.9, namely, demand order swap, DC node swap, and path swap, remain unchanged. A more detailed description of the TS/AU/RMSA method can be found in [47].

### 3.6.3   Numerical Results

This section compares different optimization approaches in the context of AU/RMSA problems with objective functions related to spectrum, cost and power consumption. The following algorithms were examined: branch and bound method implemented in the CPLEX solver, FF, MSF, LPF, AFA and TS. Due to the complexity of the RMSA problem, we were only able to find the optimal solution using the CPLEX solver for relatively small problem instances. Experiments were performed using a PC with IntelCore i7-2620M CPU and 4GB RAM.

#### Simulation Setup

To compare heuristic methods against optimal results, we used three representative network topologies: the German national network DT14 (Fig. A.2, Table A.2), the US network NSF15 (Fig. A.4, Table A.3) and the pan-European network Euro16 (Fig. A.5, Table A.4). We chose networks with different physical diameters in order to analyze the impact of distance-adaptive modulation formats on algorithm performance. The rationale is that transmission distance is a key parameter which strongly influences the selection of modulation formats. We considered scenarios with two, three and four DC nodes, and we investigated two location scenarios for each number of nodes. Therefore, six ($3 \times 2$) different server location scenarios were analyzed for each topology. Network traffic, i.e., sets with unicast and anycast demands, were generated at random. The volume of a unicast demand was chosen from the range 10–200 Gb/s, the range was 10–400 Gb/s for anycast demands. We investigated six different traffic scenarios in terms of the anycast ratio parameter (see Sect. 3.3.3), i.e., 0, 20, 40, 60, 80 and 100 %. The assumptions of the EON regarding the transmission model and the definition of network cost and power consumption were the same as Sect. 3.3.4.

#### Comparison of Algorithms

We tuned the AFA/AU/RMSA and TS/AU/RMSA algorithms to find the best combinations of input parameters. The only tuning parameter of the AFA/AU/RMSA method is $DIV$. To recall, the main aim of the $DIV$ parameter is adjusting the metric of each demand which is used to order the allocation demands. The initial experiments run for selected traffic patterns examined $DIV$ values: 3, 4, 5, 6, 7, 8, 9, and 10. Since $DIV = 5$ returned the best results on average, we used this value in further experiments. Therefore, all results of the AFA/AU/RMSA method reported below are acquired with $DIV = 5$. For information on tuning the TS/AU/RMSA algorithm refer to [47].

The main part of experiments compared different algorithms applied to the AU/RMSA problems with objective functions related to cost, power consumption, maximum spectrum usage and average spectrum usage. Table 3.18 shows the performance of the FF, MSF, LPF, AFA and TS methods presented as the average optimality gap compared to optimal results provided by the CPLEX solver. It is clear that the TS algorithm significantly outperforms other heuristics and provides results close to optimal results. It should be mentioned that the FF method was unable to obtain feasible results for problems with cost and power consumption objective functions. This is because in these functions, the number of available frequency slices is limited and thus the FF algorithm defaulting to the shortest path was unable to allocate all demands [47].

The average processing time of all analyzed greedy heuristic methods was less than 1 ms. The TS algorithm required longer time, but in general the average processing time of TS was shorter than the execution time of the the CPLEX solver, especially for the spectrum oriented objective functions. More details can be found in [44, 47].

### 3.6.4 Case Study

We used the TS/AU/RMSA algorithm described above to run a case study to examine the impact of distance-adaptive modulation formats on EON performance. We also investigated how the optimization of one particular performance metric impacts other performance metrics [44]. All main assumptions of the simulations were similar to the case study presented in Sect. 3.3.4; we recall key elements below.

**Table 3.18** Comparison of heuristic algorithms for the RMSA problem with anycast and unicast flows—average optimality gap

| Function | FF (%) | MSF (%) | LPF (%) | AFA (%) | TS (%) |
|---|---|---|---|---|---|
| *DT14 network* | | | | | |
| Cost | – | 14.4 | 14.5 | 0.0 | 0.0 |
| Power consumption | – | 17.6 | 18.0 | 0.3 | 0.2 |
| Maximum spectrum | 28.3 | 12.7 | 12.1 | 11.1 | 0.1 |
| Average spectrum | 23.0 | 17.3 | 17.5 | 4.6 | 0.2 |
| *NSF15 network* | | | | | |
| Cost | – | 60.1 | 59.1 | 8.8 | 1.0 |
| Power consumption | – | 53.3 | 52.2 | 2.0 | 1.1 |
| Maximum spectrum | 43.0 | 10.4 | 18.8 | 13.3 | 1.8 |
| Average spectrum | 33.3 | 19.3 | 20.2 | 6.0 | 0.9 |
| *Euro16 network* | | | | | |
| Cost | – | 57.3 | 56.2 | 13.2 | 1.6 |
| Power consumption | – | 50.5 | 49.3 | 7.4 | 1.5 |
| Maximum spectrum | 39.2 | 12.3 | 10.0 | 8.8 | 1.0 |
| Average spectrum | 34.6 | 24.4 | 19.0 | 6.4 | 0.4 |

**Assumptions**

We assumed that EON uses BV-Ts implementing the PDM-OFDM technology with multiple modulation formats selected from BPSK, QPSK, and x-QAM, where x belongs to 8, 16, 32, 64 and spectral efficiency is equal to $1, 2, \ldots, 6$ [b/s/Hz], respectively. Three types of BV-Ts were available, each characterized by a different bit-rate limit (40, 100, and 400 Gb/s, respectively). The EON transmission model was based on [60]. We applied a 12.5 GHz guard band between neighboring connections. The transmission reach was extended using regenerators applied as necessary and if the transmission reach of the lowest-level modulation format was shorter than the path length. In-line signal regeneration was allowed; however, the regenerators did not convert the spectrum and modulation formats. Thus, the results of cost and power consumption related to regenerators were lower-bound estimates.

Two real-world topologies were used in this study: the German national network DT14 with four data centers (Fig. A.2, Table A.2) and the US long-haul network US26 with seven data centers (Fig. A.10, Table A.7). The traffic model was exactly the same as in Sect. 3.3.4 and it was calculated according to the *Cisco Visual Networking Index* [69] and *Cisco Global Cloud Index* [70] forecasts. The traffic was computed for 2014–2018, with the initial value in year set to 5 and 20 Tb/s for DT14 and US26 networks, respectively.

The network cost and power consumption performance metrics were estimated according to data from papers [72, 78–81]. Network cost includes the CAPEX cost of equipment (transponders, regenerators) and one year OPEX cost of fiber leasing. The power consumption was calculated according to the sum of all transponder and regenerator energy requirements.

**Results**

The major goal of the numerical experiments was to compare the performance of the BPSK, QPSK and x-QAM modulation formats, where x belongs to 8, 16, 32, 64. The experiment methodology was as follows. For each demand set (unique in terms of network topology and year), the RMSA problem was solved using one of six modulation formats. The same problem was then solved using all six modulation formats. Figures 3.14, 3.15 and 3.16 present the results obtained for cost, power consumption and spectrum performance metrics, respectively.

Analysis of the results obtained for the cost function (Fig. 3.14) indicates that for the US26 network, using a higher modulation format increases cost; for instance, for

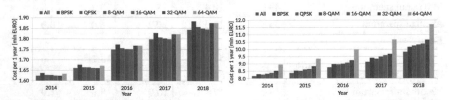

**Fig. 3.14** Network cost as a function of modulation formats for networks DT14 (*left*) and US26 (*right*)

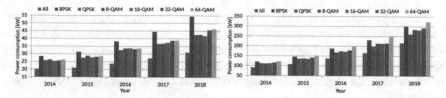

**Fig. 3.15**  Power consumption as a function of modulation formats for networks DT14 (*left*) and US26 (*right*)

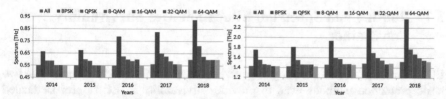

**Fig. 3.16**  Spectrum usage as a function of modulation formats for networks DT14 (*left*) and US26 (*right*)

64-QAM, the network cost is up to 16 % higher than when all modulation formats can be used. In turn, for the DT14 network, the corresponding gap is much smaller (up to 2 %). This is mainly due to network size, i.e., US26 is significantly larger with greater distances between communicating nodes compared to DT14. Therefore, using higher modulation formats such as 64-QAM for more distant node pairs requires more regenerators, increasing network cost. On the other hand, in the DT14 network the distances are significantly shorter, and using 64-QAM does not usually require using additional costly regenerators.

The results reported for power consumption (Fig. 3.15) show different trends. For the US26 network, using a single modulation format requires between 16 and 40 % more energy than all modulation scenarios. The worst results are observed for BPSK and 64-QAM, while the best performance is offered by QPSK. For DT14, the highest power consumption is seen in BPSK (up to 43 % gap for all modulation scenarios for 2018), while the remaining modulations show a similar performance with a 20–35 % gap compared to all modulation scenarios. This is due to selecting the power consumption model which assumes that power consumption of a transponder depends on the number of subscribers and the modulation format used.

For performance related to spectrum usage (Fig. 3.16), the differences between results obtained for different modulation formats are the greatest. This is because of relatively significant differences in spectral efficiency between individual modulation formats. Accordingly, the less effective BPSK modulation format needs up to 36 % more spectrum compared to the most flexible scenario allowing all modulation formats. Nevertheless, the most spectrally-efficient modulation formats (e.g., 64-QAM) support significantly smaller transmission ranges, and additional regenerators are needed for greater distances, increasing network cost and power consumption.

For all performance metrics, as traffic increases in subsequent years, the gaps between modulation format-related scenarios also increase; this means that as network traffic increases, using all available modulation formats becomes increasingly beneficial. More results, discussion and analysis regarding the problem of using different modulation formats in EONs with anycast and unicast flows can be found in [44, 47].

## 3.7  Routing and Spectrum Allocation with Survivability Constraints

EONs can be protected similarly to general connection-oriented networks. The two basic protection methods used to provide survivability in EONs are: Dedicated Path Protection (DDP) and Shared Backup Path Protection (SBBP), introduced in Sect. 2.5. To recall, the key idea behind both approaches is to provide two failure disjoint routing paths (working and backup). The main difference between DPP and SBPP is that the SBPP method makes it possible to share optical spectrum resources between backup paths of demands which do not fail at the same time due to a network failure. Various issues related to the optimization of EONs protected with the DPP approach including ILP formulations, algorithms and performance evaluation are presented in [41, 42, 45, 48–50, 59, 82–86]. In turn, the SBPP method applied in EONs is shown in [31, 33, 34, 39, 41, 42, 58, 82–84, 86–88]. Moreover, the p-cycle technique for protecting EONs is examined in [37, 38, 83, 89, 90].

A new protection approach of *bandwidth squeezing* is developed in the context of EONs. The working path provisions the full requested bit-rate, while the backup path activated in the event of working path failure is used to protect (or recover) a part of the requested bit-rate. This reduction in the bit-rate served after the failure is coordinated with a class-based network control function which squeezes the traffic outside the committed service profile without degrading the in-profile traffic. As a consequence, the number of frequency slices which are actually recovered from a failure on the backup path can be reduced compared to the spectrum consumed on the working path. The main advantage of bandwidth squeezing in EONs is that there are more opportunities to sustain the connectivity in the event of network failure compared to fixed grid networks, where the granularity of bandwidth allocation is fixed and thus both working and backup paths must use the same amount of spectrum resources. Note that the bandwidth squeezing approach can be applied for various protection methods including DPP, SBPP and restoration [20, 34, 41, 42, 53, 91–93].

This section includes formulations of the RSA problem for both DPP and SBPP scenarios, descriptions of heuristic algorithms and results of numerical experiments. EON protection is only addressed in the context of unicast flows; however modifications required to tackle anycast and multicast flows are relatively straightforward. Specific issues related to survivability of anycast flows in EONs can be found in [45, 85, 94].

### 3.7.1 Formulations

The formulation is based on the link-path approach with channel-based spectrum modeling; it is analogous to models (3.2.3) and (3.3.1). However, the notation is modified slightly to address the DPP protection used in the model. More specifically, for each demand $d \in D$, set $Q(d)$ includes $k$ pairs $(p, q)$ of disjoint routing paths, where $p$ is a working path and $q$ denotes the backup path. Accordingly, a new constant $\beta_{edq}$ denotes whether link $e$ belongs to backup path $q$ realizing demand $d$ in a similar way to using $\delta_{edp}$ to define working paths. We use symbols $C(d, p)$ and $C(d, q)$ to denote sets including candidate channels for working or backup paths, respectively. The width of each candidate channel is calculated according to the requested number of slices for demand $d$ on path $p$ (constant $n_{dp}$). However, to account for the bandwidth squeezing method, the number of slices required for the backup path $n_{dq}$ can be lower than $n_{dp}$ and thus the channels included in set $C(d, q)$ are generated accordingly. Constants $\gamma_{dpcs}$ and $\alpha_{dqcs}$ define channels for working path $p$ or backup path $q$, respectively.

Since working and backup paths selected for demand $d$ can allocate different spectrum channels, two decision variables are required to model the routing and spectrum allocation. As well as variable $x_{dpc}$ denoting the selection of working path $p$ and channel $c$ for demand $d$, a new variable $z_{dqc}$ indicates which channel is allocated for the backup path $q$. Note that for a particular demand, the selected working path and backup path must belong to a single pair $(p, q)$ included in set $Q(d)$ to ensure that working and backup paths are disjoint. To facilitate modeling of spectrum usage, variables $w_{es}$ and $b_{es}$ denote whether slice $s$ on link $e$ is allocated to working or backup path, respectively.

---

#### EON/U/RSA/DPP/Spectrum/Link-path/Channel-based

**sets**

| | |
|---|---|
| $E$ | links |
| $S$ | slices |
| $D$ | demands (anycast and unicast) |
| $Q(d)$ | candidate pairs of link disjoint paths $(p, q)$ for flows realizing demand $d$ |
| $C(d, p)$ | candidate channels for demand $d$ allocated on working path $p$ |
| $C(d, q)$ | candidate channels for demand $d$ allocated on backup path $q$ |

**constants**

| | |
|---|---|
| $\delta_{edp}$ | $=1$, if link $e$ belongs to working path $p$ realizing demand $d$; 0, otherwise |
| $\beta_{edq}$ | $=1$, if link $e$ belongs to backup path $q$ realizing demand $d$; 0, otherwise |
| $n_{dp}$ | requested number of slices for demand $d$ on path $p$ |
| $\gamma_{dpcs}$ | $=1$, if channel $c$ associated with demand $d$ on working path $p$ uses slice $s$; 0, otherwise |

$\alpha_{dqcs}$   =1, if channel $c$ associated with demand $d$ on backup path $q$ uses slice $s$; 0, otherwise

**variables**

$x_{dpc}$   =1, if channel $c$ on working path $p$ is used to realize demand $d$; 0, otherwise (binary)

$z_{dqc}$   =1, if channel $c$ on backup path $q$ is used to realize demand $d$; 0, otherwise (binary)

$w_{es}$   =1, if slice $s$ is occupied on link $e$ for working flows; 0, otherwise (binary)

$b_{es}$   =1, if slice $s$ is occupied on link $e$ for backup flows; 0, otherwise (binary)

$y_{es}$   =1, if slice $s$ is occupied on link $e$; 0, otherwise (binary)

$y_s$   =1, if slice $s$ is occupied on any network link; 0, otherwise (binary)

**objective**

$$\text{minimize} \quad F = \sum_{s \in S} y_s \tag{3.7.1a}$$

**constraints**

$$\sum_{p \in Q(d)} \sum_{c \in C(d,p)} x_{dpc} = 1, \quad d \in D \tag{3.7.1b}$$

$$\sum_{d \in D} \sum_{p \in Q(d)} \sum_{c \in C(d,p)} \gamma_{dpcs} \delta_{edp} x_{dpc} \leq w_{es}, \quad e \in E, s \in S \tag{3.7.1c}$$

$$\sum_{d \in D} \sum_{q \in Q(d)} \sum_{c \in C(d,p)} \alpha_{dqcs} \beta_{edq} z_{dqc} \leq b_{es}, \quad e \in E, s \in S \tag{3.7.1d}$$

$$w_{es} + b_{es} \leq y_{es}, \quad e \in E, s \in S \tag{3.7.1e}$$

$$\sum_{e \in E} y_{es} \leq |E| \, y_s, \quad s \in S \tag{3.7.1f}$$

$$\sum_{c \in C(d,p)} x_{dpc} = \sum_{c \in C(d,q)} y_{dqc}, \quad d \in D, (p, q) \in Q(d). \tag{3.7.1g}$$

The objective of optimization (3.7.1a) is to minimize the number of allocated slices. Equation (3.7.1b) ensures that for each demand $d$ exactly one working path and exactly one candidate channel are selected. To find the allocation of slices to working paths and to meet the guarantee that a slice on a particular link can be allocated to at most one lightpath, constraint (3.7.1c) is added to the model. Inequality (3.7.1d) controls the allocation of slices to backup paths. Constraint (3.7.1e) ensures that a slice on a particular link can be allocated to at most one lightpath considering both working and backup paths. Inequality (3.7.1f) defines variable $y_s$. The last constraint (3.7.1g) ensures that to realize a particular demand with DPP protection both working and backup paths belong to the same pair of paths $(p, q)$ from set $Q(d)$. Moreover, condition (3.7.1g) assigns different channels to working and backup paths of a particular demand (different channel (DCh) approach). Note that if the same channel (SCh) approach is used (i.e., the same spectrum channel is allocated on both working and backup paths), constraint (3.7.1g) should be substituted with the following constraint:

$$x_{dpc} = y_{dqc}, \quad d \in D, (p, q) \in Q(d), c \in C(d, p). \tag{3.7.1h}$$

The DPP method can be used in the context of single link failures and single node failures. To guarantee single link failure protection, working and backup paths $(p, q)$ included in sets $Q(d)$ must be link disjoint. In turn, to tackle single node failure protection working and backup paths $(p, q)$ must be calculated to be node disjoint. Other types of failure can be addressed in a similar way.

The authors of [58] present an RSA problem with SBPP protection. The ILP model is similar to the DPP model defined in (3.7.1). To address shared protection, instead of $b_{es}$ a new variable $b_{egs}$ is used to indicate whether slice $s$ is occupied on link $e$ for restoration after link $g$ failure. Next, constraints (3.7.1d) and (3.7.1e) are replaced with the two following constraints:

$$\sum_{d \in D} \sum_{q \in Q(d)} \sum_{c \in C(d,p)} \delta_{gdp} \alpha_{dqcs} \beta_{edq} z_{dqc} \leq b_{egs}, \quad e \in E, g \in E, e \neq g, s \in S \tag{3.7.2a}$$

$$w_{es} + b_{egs} \leq y_{es}, \quad e \in E, g \in E, e \neq g, s \in S. \tag{3.7.2b}$$

These modifications refer to the case of SBPP without stub release, assuming that spectrum resources on the failure-affected paths are not used after a failure. The opposite case known as stub release uses spare resources available after the failure for backup connections. For a formulation of a model with the stub release scenario see [58].

Below we present a short discussion on modifying the RSA/DPP model defined in (3.7.1) to enable optimization of anycast flows in EONs protected by DPP methods. The unicast formulation shown above requires a single important change to include anycast traffic. In the case of anycast upstream demands, set $Q(d)$ should contain pairs of disjoint paths from the client node to one of the DC nodes. In turn, for anycast downstream demands, the pair of disjoint paths included in $Q(d)$ should start at one of the DC nodes and lead to the client node. Since each DC node provides the same content or service, the working and backup paths of the same pair $(p, q)$ of disjoint paths can use different DC nodes, as discussed in Sect. 2.6. For more details on issues related to survivability aspects of RSA with anycast flows refer to [45].

## 3.7.2  Algorithms

In general, the RSA/DPP problem is similar to the classical RSA problem. Therefore, the heuristic algorithms formulated in the context of RSA problems (see Sect. 3.3.2) can be modified easily to address additional constraints following from the DPP protection method. In this section, we present several heuristics formulated for the RSA/DPP model defined in (3.7.1). The algorithms are designed to optimize unicast flows; however, to obtain the versions with anycast flows, the modifications are analogous to the construct of algorithms presented in Sect. 3.3.2 for the RSA problem with joint anycast and unicast flows.

### First Fit (FF) Algorithm

The FF/U/RSA/DPP method (Algorithm 3.13) is based on classical First Fit (Algorithm 3.2). The key difference is that in order to address the DPP protection, algorithm FF/U/RSA/DPP processes pairs of disjoint paths included in set $Q(d)$ in place of single candidate paths. Moreover, it is assumed that the *paths* and *channels* vectors include two fields, refering to working path and backup paths respectively. For instance, *paths.w*[d] denotes the working path selected for demand $d$ and *paths.b*[d] denotes the backup path selected for demand $d$. The FF/U/RSA/DPP algorithm allocates the working path (lines 4–6) first, and next it finds the backup path (lines 7–9). The complexity of the FF/U/RSA/DPP algorithm is the same as in the basic version of the FF method defined in Algorithm 3.2, i.e., $O(|D| |C| |E|)$. Using the FF/U/RSA/DPP algorithm, it is easy to modify other greedy heuristics such as the LPF and MSF methods for the RSA problem with DPP protection or SBPP protection [50, 58, 59].

---

**Algorithm 3.13** FF/U/RSA/DPP (First Fit for U/RSA/DPP problem)

---

**Require:** set of edges $E$, set unicast demands $D$, sets $Q(d)$ with pairs $(p, q)$ of disjoint paths for
    each demand $d \in D$, candidate channels $C(d, p)$ and $C(d, q)$ for each demand $d \in D$ and paths
    $(p, q) \in Q(d)$

**Ensure:** routing and spectrum allocation for each demand $d \in D$ included in vectors *paths* and
    *channels*, value of objective function

1: **procedure** $FF/U/RSA/DPP(D, Q(d), C(d, p), C(d, q))$
2:  **for** $i := 1$ **to** $|D|$ **do**
3:    $d := Member(D, i)$
4:    $paths.w[d] := Member(Q(d), 1)$
5:    $channels.w[d] := FF\_SA(C(d, paths.w[d]), paths.w[d])$
6:    $Allocate\_Demand(d, paths.w[d], channels.w[d])$
7:    $paths.b[d] := Member(Q(d), 1)$
8:    $channels.b[d] := FF\_SA(C(d, paths.b[d]), paths.b[d])$
9:    $Allocate\_Demand(d, paths.b[d], channels.b[d])$
10: **end for**
11: $Find\_Objective\_Function(paths[], channels[])$
12: **end procedure**

---

**Adaptive Frequency Assignment (AFA) Algorithm**

The modification of the AFA/AU/RSA method shown in Algorithm 3.5 for RSA/DPP needs additional insight. The AFA/U/RSA/DPP method described in Algorithm 3.14 processes demands in several separate loops according to the value $n_d$ (line 5). Next, all demands included in sets $D(n)$ (demands with the same value of $n_d$) are processed in one loop (lines 10–20) in decreasing order of $n_d$. To select the next demand for allocation, the algorithm analyzes all not established demands still included in set $D(n)$. Function *FPCSpectrumDPP*() called in line 14 is a modified version of function *FPCSpectrum*() presented in Algorithm 3.3. The modification is required to process pairs of paths $(p, q)$ instead of single candidate paths $p$. In fact, function *FPCSpectrumDPP*() finds the best pair of paths $(p, q)$ for demand $d$ in terms of the spectrum usage defined as the maximum slice number required in the network to allocate pair of paths $(p, q)$ for demand $d$. Next, function *Best_AllocationDPP*($D(n)$, *paths*, *channels*) is called to determine the best demand $d^\star$ for the next allocation of both working and backup path from all not established demands still included in set $D(n)$ (line 17). As in the previous version of AFA, a special collision metric is applied in the *Best_AllocationDPP* function if demands return the same value of the spectrum usage. In particular, for each link $e \in E$ we define $c_e = \sum_{d \in D}(\sum_{p \in Q(d)} \delta_{edp} n_d + \sum_{q \in Q(d)} \beta_{edq} n_d)$. Metric $c_e$ is used to estimate the number of slices which may be allocated to link $e$ taking into account all candidate pairs of working and backup paths for each demand. Next, we define $l_p = \sum_{e \in p} c_e$ and $l_q = \sum_{e \in q} c_e$ as lengths of paths $p$ and $q$ calculated according to metric $c_e$. Finally, metric $l_d = \frac{1}{|Q(d)|} \sum_{(p,q) \in Q(d)} (l_p + l_q)$ denotes the collision metric of demand $d$. Note that $l_d$ is defined as the average length of candidate pairs of paths for demand $d$ in terms of metric $c_e$. The complexity of algorithm AFA/U/RSA/DPP is the same as that reported for the AFA method described in Algorithm 3.5, namely, $O(|D|^2 |P| |C| |E|)$.

For further information on the DPP version of AFA, see [50, 59]. In turn, the AFA algorithm in the context of the SBPP protection is presented and evaluated in [58].

**Tabu Search Algorithm**

A Tabu Search (TS) algorithm for the U/RSA/DPP problem with DPP protection (referred to as TS/U/RSA/DPP) is similar to the TS/AU/RSA method proposed for the AU/RSA problem and described in Algorithm 3.9 [59]. The TS/U/RSA/DPP method uses a solution representation similar to that defined in (3.3.5), i.e., the solution is encoded by a demand allocation order (demands are allocated one-by-one according to this order) and a pair of routing paths (working and backup) for each demand. To evaluate a particular solution, a method similar to Algorithm 3.7 is used with a minor adjustment considering pairs of paths in place of single routing paths. Just as TS/AU/RSA, the TS/U/RSA/DPP method needs a starting solution and any heuristic method discussed above can be used for this purpose.

Consistent with the solution representation, TS/U/RSA/DPP uses two types of move operations: demand order swap and pair of paths swap. The former is the same as in the TS/AU/RSA method and involves swapping two randomly chosen

---

**Algorithm 3.14** AFA/U/RSA/DPP (Adaptive Frequency Assignment for U/RSA/DPP problem)

---

**Require:** set of edges $E$, set unicast demands $D$, sets $Q(d)$ with pairs $(p, q)$ of disjoint paths for each demand $d \in D$, candidate channels $C(d, p)$ and $C(d, q)$ for each demand $d \in D$ and paths $(p, q) \in Q(d)$

**Ensure:** routing and spectrum allocation for each demand $d \in D$ included in vectors *paths* and *channels*, value of objective function

**Ensure:** routing and spectrum allocation for each demand $d \in D$ included in vectors *path* and *channel*, value of objective function

1: **procedure** $AFA/U/RSA/DPP(D, Q(d), C(d, p), C(d, q))$
2:   $n_{max} := 0$
3:   **for** $i := 1$ **to** $|D|$ **do**
4:     $d := Member(D, i)$
5:     $n_d := \min_{p \in Q(d)} \{n_{dp}\}$
6:     $D(n_d) := D(n_d) \cup \{d\}$
7:     **if** $n_d > n_{max}$ **then** $n_{max} := n_d$
8:   **end for**
9:   **for** $n := n_{max}$ **to** 1 **do**
10:     **while** $D(n) \neq \emptyset$ **do**
11:       **for** $i := 1$ **to** $|D(n)|$ **do**
12:         $d := Member(D, i)$
13:         **if** $Established(d) = FALSE$ **then**
14:           $\{paths.w[d], channels.w[d]\} := FPCSpectrumDPP(Q(d), C(d, p))$
15:         **end if**
16:       **end for**
17:       $d^\star := Best\_AllocationDPP(D(n), paths, channels)$
18:       $Allocate\_Demand(d^\star, paths.w[d^\star], channels.w[d^\star])$
19:       $Allocate\_Demand(d^\star, paths.b[d^\star], channels.b[d^\star])$
20:     **end while**
21:   **end for**
22:   $Find\_Objective\_Function(paths[], channels[])$
23: **end procedure**

---

demands, changing the demand ordering used in the allocation process. The latter is analogous to the path swap operation of TS/AU/RSA; however, here a pair of paths is changed for a particular demand. Using these modifications, the pseudocode shown in Algorithm 3.9 can be adapted to obtain the TS/U/RSA/DPP algorithm. A more comprehensive treatment of the TS algorithm for the RSA problem with DPP protection is given in [59].

## 3.7.3   Numerical Results

This section presents numerical results performed to evaluate algorithms proposed for the U/RSA/DPP problem formulated in (3.7.1), and examines the performance of the DPP and SBPP protection methods in EONs.

**Table 3.19** Comparison of optimization algorithms for the RSA problem with DPP protection for the INT9 network—average optimality gap, lengths of 95 % confidence intervals and average execution time

| Scenario | CPLEX | FF | MSF | LSF | AFA | TS |
|---|---|---|---|---|---|---|
| *Average optimality gap* | | | | | | |
| SC | – | 11 % | 3.98 % | 7.67 % | 2.83 % | 0.32 % |
| DC | – | 10.1 % | 4.90 % | 7.94 % | 3.69 % | 1.08 % |
| *Lengths of 95 % confidence intervals* | | | | | | |
| SC | – | 2.06 % | 1.27 % | 1.83 % | 1.09 % | 0.26 % |
| DC | – | 1.81 % | 1.38 % | 1.96 % | 1.29 % | 0.69 % |
| *Average execution time in seconds* | | | | | | |
| SC | 22.6 | <0.002 | <0.002 | <0.002 | <0.002 | 2.1 |
| DC | 9.4 | <0.002 | <0.002 | <0.002 | <0.002 | 2.1 |

**Simulation Setup**

We examined three topologies: INT9 (Fig. A.1, Table A.1), NSF15 (Fig. A.4, Table A.3) and UBN24 (Fig. A.6, Table A.5). Optimal results were obtained with the ILP model implemented in the CPLEX solver [62]. The evaluation was performed on an Intel $i5$ 3.3 GHz 16 GB computer. For the INT9 network, the number of slices available in the network was $|S| = 48$, while for larger networks $|S| = 1500$, which is sufficiently large to allocate all demands. Candidate pairs of working and backup paths were generated as link disjoint to protect the network against a single link failure. Pairs of paths were calculated as shortest paths, taking into account the overall length of both paths assuming $|Q(d)| = 2, 3, 5, 10, 30$. The spectrum requested for each demand was generated at random using a uniform distribution. For the INT9 topology, set $D$ included 15 demands, while for larger networks $|D| = 210$ for NSF15 and $|D| = 552$ for UBN24. If not stated otherwise, the results are averaged over 100 randomly generated demand sets [50, 59].

**Comparison of Algorithms**

We compared the FF, LPF, MSF, AFA and TS algorithms. For a detailed information on tuning of TS method see [59]. Table 3.19 presents performance results of heuristic algorithms in comparison to optimal results yielded by the CPLEX solver for the INT9 network. In the experiments, the number of pairs of paths is $k = 2$. Since the CPLEX solver reached optimality in a $2h$ period for just 88 out of 150 cases, the results were averaged over these 88 demand sets. We can see that the TS method outperforms other heuristics and it achieves near-optimal solutions in the examined cases. The computation time of FF, LSF, MSF and AFA is below 2 ms on average. The TS method requires a considerably shorter execution time compared to CPLEX [50, 59].

Table 3.20 shows the performance of heuristic methods for larger networks NSF15 and UBN24 with the number of candidate pairs of disjoint paths equal to

**Table 3.20** Comparison of optimization algorithms for the RSA problem with DPP protection for the NSF15 and UBN24 networks—average gap to minimum obtained result

| k | FF (%) | MSF (%) | LSF (%) | AFA (%) | TS/SRT (%) | TS/AFA (%) |
|---|--------|---------|---------|---------|------------|------------|
| *NSF15 network* | | | | | | |
| 2 | 19.1 | 12.4 | 6.3 | 3.0 | 1.2 | **0.0** |
| 3 | 21.9 | 11.8 | 6.2 | 2.7 | 1.3 | **0.0** |
| 5 | 22.5 | 12.5 | 6.9 | 3.4 | 1.6 | **0.0** |
| 10 | 28.8 | 8.6 | 4.1 | 0.6 | 2.6 | **0.1** |
| 30 | 31.8 | 11.0 | 3.7 | 0.1 | 4.1 | **0.0** |
| *UBN24 network* | | | | | | |
| 2 | 17.3 | 5.5 | 3.8 | 0.6 | 1.7 | **0.0** |
| 3 | 24.7 | 5.8 | 5.2 | 0.4 | 5.4 | **0.0** |
| 5 | 32.1 | 7.1 | 7.9 | 0.7 | 10.2 | **0.0** |
| 10 | 38.2 | 8.2 | 9.2 | 1.0 | 10.2 | **0.0** |
| 30 | 39.1 | 4.5 | 2.8 | 0.6 | 6.6 | **0.0** |

$k = 2, 3, 5, 10, 30$. We report results of two versions of the TS method. The TS algorithms differ in terms of the method applied to obtain the initial solution, i.e., TS/SRT and TS/AFA denote TS methods which use the output of SRT and AFA as the initial solution, respectively. The SRT method is a simple greedy method described in [59]. The reason for using two versions of the TS method was to examine the influence of the initial solution on the performance of TS. Note that the considered problem instances were too large to obtain optimal results using the CPLEX solver. To compare the algorithms, we used the following procedure. First, every heuristic was executed for every unique test scenario. Next, we found the best algorithm result. Finally, we calculated the percentage gap to the minimum (best) result for each algorithm. 100 unique demand sets were tested for each case (topology, number of candidate pairs of paths).

We can see that the TS/AFA algorithm outperforms all other methods since the average gap to the best obtained result is almost always 0 %. More detailed analysis of the results shows that the TS/AFA yields the best result in 985 of 1000 tested cases reported in Table 3.20. In contrast, the TS/SRT performance is significantly worse compared to TS/AFA, and the gap between the methods increases as the candidate pairs of paths increase. This confirms that the quality of the initial solution is significant for the performance of the TS method. The second best method is AFA, which means that TS/AFA can improve the solution provided by AFA for all tested cases. It should be noted that the gap between AFA and TS/AFA decreases as the number of candidate pairs of paths increases. These results indicate that the TS method experiences scalability problems when the solution space (i.e., number of candidate pairs of paths) increases. In terms of execution time, simple greedy methods FF, MSF and LPF need less than 1 s. In turn, AFA requires up to 15 s, while the TS method consumes up to 500 s. In general, the execution time of each tested method increases with the network size and the number of candidate pairs of paths.

More discussion on the results related to DPP protection can be found in [50, 59], while [58] includes a comparison of various heuristic methods applied to the RSA problem with SBPP protection.

**Performance of DPP and SBPP**

The next goal was to compare two protection scenarios DPP and SBPP against results obtained for the no protection case, i.e., only working paths were established in the network and there were no backup paths. Figures 3.17 and 3.18 present the average number of slices required for each scenarios with different numbers of candidate pairs of paths. The results for the DPP method are yielded by the TS/AFA method, while the results for SBPP and no protection are provided by the AFA method. The results are averaged over 100 different demand sets.

For the most interesting case with $k = 30$, the additional spectrum (slices) needed in the network to provide the SBPP protection is 53 and 52 % for the NSF15 and UBN24 networks, respectively. In the case of the DPP protection, the corresponding numbers are 120 and 108 %. Differences observed between the networks are mainly

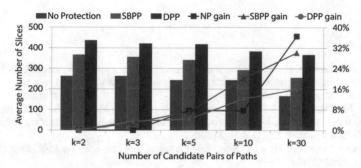

**Fig. 3.17** DPP and SBPP versus the no protection scenario for the NSF15 network and different number of candidate pairs of paths

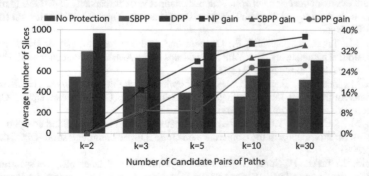

**Fig. 3.18** DPP and SBPP versus the no protection scenario for the UBN24 network and different number of candidate pairs of paths

due to the fact that the NSF15 topology is more sparse (average node degree is 3.07) compared to the UBN24 topology (average node degree is 3.58). Consequently, in the case of NSF15, the backup paths are on average longer compared to working paths and consume more spectrum resources in the network. For the SBPP scenario, the gap between the networks is smaller, since the ability to share the backup capacity considerably reduces the influence of longer backup paths.

Another interesting observation concerning Figs. 3.17 and 3.18 refers to the influence of the number of candidate pairs of paths. As the number of candidate pairs of paths increases, the required number of slices decreases for both networks all analyzed scenarios. The greatest improvement between $k = 2$ and $k = 30$ is seen for the no protection case (approx. 37 % for both networks). In the case of the SBPP approach, the gap between $k = 2$ and $k = 30$ is 30 and 34 % for the NSF15 and UBN24 networks, respectively. Finally, for the DPP approach the improvement between $k = 2$ and $k = 30$ is 16 and 26 % for NSF15 and UBN24, respectively. Again, variations in results obtained for both networks follow from their topological properties. The main conclusion of the results is that when solving the RSA problem in EONS, it is worth having a large set of candidate paths, since it has a great impact on spectrum usage. Additional results are presented in [50, 58, 59].

# References

1. Jinno, M., Kozicki, B., Takara, H., Watanabe, A., Sone, Y., Tanaka, T., Hirano, A.: Distance-adaptive spectrum resource allocation in spectrum-sliced elastic optical path network. IEEE Commun. Mag. **48**(8), 138–145 (2010)
2. Jinno, M., Takara, H., Kozicki, B., Tsukishima, Y., Sone, Y., Matsuoka, S.: Spectrum-efficient and scalable elastic optical path network: architecture, benefits, and enabling technologies. IEEE Commun. Mag. **47**(11), 66–73 (2009)
3. Simmons, J.: Optical Network Design and Planning, 2nd edn. Springer Publishing Company, Incorporated (2014)
4. Jinno, M., Takara, H., Kozicki, B., Tsukishima, Y., Yoshimatsu, T., Kobayashi, T., Miyamoto, Y., Yonenaga, K., Takada, A., Ishida, O., Matsuoka, S.: Demonstration of novel spectrum-efficient elastic optical path network with per-channel variable capacity of 40 Gb/s to over 400 Gb/s. In: Proceedings of the 34th European Conference on Optical Communication (ECOC), pp. 1–2 (2008)
5. Jinno, M., Takara, H., Kozicki, B.: Concept and enabling technologies of spectrum-sliced elastic optical path network (slice). In: Proceedings of the Asia Communications and Photonics Conference (ACP 2009), pp. 1–2 (2009)
6. Jinno, M., Takara, H., Kozicki, B.: Dynamic optical mesh networks: drivers, challenges and solutions for the future. In: Proceedings of the 35th European Conference on Optical Communication (ECOC 2009), pp. 1–4 (2009)
7. Jinno, M., Tsukishima, Y.: Virtualized optical network (von) for agile cloud computing environment. In: Proceedings of the Conference on Optical Fiber Communication (OFC 2009), pp. 1–3 (2009)
8. Kozicki, B., Takara, H., Sone, Y., Watanabe, A., Jinno, M.: Distance-adaptive spectrum allocation in elastic optical path network (slice) with bit per symbol adjustment. In: Proceedings of the Conference on Optical Fiber Communication (OFC 2010), pp. 1–3 (2010)

9. Kozicki, B., Takara, H., Tsukishima, Y., Yoshimatsu, T., Kobayashi, T., Yonenaga, K., Jinno, M.: Optical path aggregation for 1-tb/s transmission in spectrum-sliced elastic optical path network. IEEE Photonics Technol. Lett. **22**(17), 1315–1317 (2010)

10. Kozicki, B., Takara, H., Yoshimatsu, T., Yonenaga, K., Jinno, M.: Filtering characteristics of highly-spectrum efficient spectrum-sliced elastic optical path (slice) network. In: Proceedings of the Conference on Optical Fiber Communication (OFC 2009), pp. 1–3 (2009)

11. ITU-T Recommendation G.694.1 (ed. 2.0). Spectral grids for WDM applications: DWDM frequency grid (2012)

12. Farrel, A., King, D., Li, Y., Zhang, F.: Generalized labels for the flexi-grid in lambda switch capable (lSC) label switching routers. Technical Report draft-ietf-ccamp-flexigrid-lambda-label-03.txt, IETF (2015)

13. Chandrasekhar, S., Liu, X., Zhu, B., Peckham, D.W.: Transmission of a 1.2-tb/s 24-carrier no-guard-interval coherent ofdm superchannel over 7200-km of ultra-large-area fiber. In: Proceedings of the 35th European Conference on Optical Communication (ECOC 2009), vol. 2009, Supplement, pp. 1–2 (2009)

14. Hugues-Salas, E., Zervas, G., Simeonidou, D., Kosmatos, E., Orphanoudakis, T., Stavdas, A., Bohn, M., Napoli, A., Rahman, T., Cugini, F., Sambo, N., Frigerio, S., D'Errico, A., Pagano, A., Riccardi, E., Lopez, V., Fernandez-Palacios Gimenez, J.P.: Next generation optical nodes: the vision of the european research project idealist. IEEE Commun. Mag. **53**(2), 172–181 (2015)

15. Klinkowski, M., Walkowiak, K.: On the advantages of elastic optical networks for provisioning of cloud computing traffic. IEEE Netw. **27**(6), 44–51 (2013)

16. Lord, A., Wright, P., Mitra, A.: Core networks in the flexgrid era. J. Lightw. Technol. **33**(5), 1126–1135 (2015)

17. Ruiz, M., Velasco, L., Lord, A., Fonseca, D., Pioro, M., Wessaly, R., Fernandez-Palacios, J.P.: Planning fixed to flexgrid gradual migration: drivers and open issues. IEEE Commun. Mag. **52**(1), 70–76 (2014)

18. Tomkos, I., Azodolmolky, S., Sole-Pareta, J., Careglio, D., Palkopoulou, E.: A tutorial on the flexible optical networking paradigm: state of the art, trends, and research challenges. Proc. IEEE **102**(9), 1317–1337 (2014)

19. Yu, X., Tornatore, M., Xia, M., Wang, J., Zhang, J., Zhao, Y., Zhang, J., Mukherjee, B.: Migration from fixed grid to flexible grid in optical networks. IEEE Commun. Mag. **53**(2), 34–43 (2015)

20. Zhang, G., De Leenheer, M., Morea, A., Mukherjee, B.: A survey on ofdm-based elastic core optical networking. IEEE Commun. Surv. Tutor. **15**(1), 65–87 (2013)

21. B. Chatterjee, N. Sarma, and E. Oki. Routing and spectrum allocation in elastic optical networks: a tutorial. IEEE Commun. Surv. Tutor. **PP**(99), 1–1 (2015)

22. Christodoulopoulos, K., Tomkos, I., Varvarigos, E.: Spectrally/bitrate flexible optical network planning. In: Proceedings of the 36th European Conference and Exhibition on Optical Communication (ECOC 2010), pp. 1–3 (2010)

23. Christodoulopoulos, K., Tomkos, I., Varvarigos, E.: Elastic bandwidth allocation in flexible OFDM based optical networks. IEEE J. Lightw. Technol. **29**(9), 1354–1366 (2011)

24. Klinkowski, M., Walkowiak, K.: Routing and spectrum assignment in spectrum sliced elastic optical path network. IEEE Commun. Lett. **15**(8), 884–886 (2011)

25. Klinkowski, M., Walkowiak, K., Jaworski, M.: Off-line algorithms for routing, modulation level, and spectrum assignment in elastic optical networks. In: Proceedings of the 13th International Conference on Transparent Optical Networks (ICTON 011), pp. 1–6 (2011)

26. Velasco, L., Klinkowski, M., Ruiz, M., Comellas, J.: Modeling the routing and spectrum allocation problem for flexgrid optical networks. Photonic Netw. Commun. **24**(3), 177–186 (2012)

27. Wang, Y., Cao, X., Hu, Q.: Routing and spectrum allocation in spectrum-sliced elastic optical path networks. In: Proceedings of the 2011 IEEE International Conference on Communications (ICC 2011), pp. 1–5 (2011)

28. Wang, Y., Cao, X., Hu, Q., Pan, Y.: Towards elastic and fine-granular bandwidth allocation in spectrum-sliced optical networks. IEEE/OSA J. Opt. Commun. Netw. **4**(11), 906–917 (2012)

29. Wang, Y., Cao, X., Pan, Y.: A study of the routing and spectrum allocation in spectrum-sliced elastic optical path networks. In: Proceedings of the IEEE INFOCOM 2011, Shanghai, China (2011)
30. Capucho, J.H.L., Resendo, L.C.: Ilp model and effective genetic algorithm for routing and spectrum allocation in elastic optical networks. In: Proceddings of the 2013 SBMO/IEEE MTT-S International Microwave Optoelectronics Conference (IMOC 2013), pp. 1–5 (2013)
31. Eira, A., Pedro, J., Pires, J.: Optimized design of shared restoration in flexible-grid transparent optical networks. In: Proceedings of the Conference on Optical Fiber Communication (OFC 2012), pp. 1–3 (2012)
32. Gong, L., Zhou, X., Liu, X., Zhao, W., Lu, W., Zhu, Z.: Efficient resource allocation for all-optical multicasting over spectrum-sliced elastic optical networks. IEEE/OSA J. Opt. Commun. Netw. 5(8), 836–847 (2013)
33. Shen, G., Wei, Y., Yang, Q.: Shared backup path protection (sbpp) in elastic optical transport networks. In: Proceedings of the Asia Communications and Photonics Conference (ACP 2012), pp. PAF4C.6. Optical Society of America (2012)
34. Shen, G., Wei, Y., Bose, S.: Optimal design for shared backup path protected elastic optical networks under single-link failure. IEEE/OSA J. Opt. Commun. Netw. 6(7), 649–659 (2014)
35. Tornatore, M., Rottondi, C., Goscien, R., Walkowiak, K., Rizzelli, G., Morea, A.: On the complexity of routing and spectrum assignment in flexible-grid ring networks [invited]. IEEE/OSA J. Opt. Commun. Netw. 7(2), A256–A267 (2015)
36. Walkowiak, K., Goscien, R., Klinkowski, M., Wozniak, M.: Optimization of multicast traffic in elastic optical networks with distance-adaptive transmission. IEEE Commun. Lett. 18(12), 2117–2120 (2014)
37. Wei, Y., Xu, K., Jiang, Y., Zhao, H., Shen, G.: Optimal design for p-cycle-protected elastic optical networks. Photonic Netw. Commun. pp. 1–12 (2015)
38. Wei, Y., Xu, K., Zhao, H., Shen, G.: Applying p-cycle technique to elastic optical networks. In: Proceedings of the 2014 International Conference on Optical Network Design and Modeling (ONDM 2014), pp. 1–6 (2014)
39. Yang, C., Hua, N., Zheng, X.: Shared path protection based on spectrum reserved matrix model in bandwidth-variable optical networks. In: Proceedings of the 7th International ICST Conference on Communications and Networking in China (CHINACOM 2012), pp. 256–261 (2012)
40. Aibin, M., Walkowiak, K.: Simulated annealing algorithm for optimization of elastic optical networks with unicast and anycast traffic. In: Proceedings of the 16th International Conference on Transparent Optical Networks (ICTON 2014), pp. 1–4 (2014)
41. Castro, A., Velasco, L., Comellas, J., Junyent, G.: On the benefits of multi-path recovery in flexgrid optical networks. Photonic Netw. Commun. 28(3), 251–263 (2014)
42. Castro, A., Velasco, L., Ruiz, M., Comellas, J.: Single-path provisioning with multi-path recovery in flexgrid optical networks. In: 2012 4th International Congress on Ultra Modern Telecommunications and Control Systems and Workshops (ICUMT), pp. 745–751 (2012)
43. Goscien, R., Klinkowski, M., Walkowiak, K.: A tabu search algorithm for routing and spectrum allocation in elastic optical networks. In: Proceedings of the 16th International Conference on Transparent Optical Networks (ICTON 2014), pp. 1–4 (2014)
44. Goscien, R., Walkowiak, K., Klinkowski, M.: Distance-adaptive transmission in cloud-ready elastic optical networks. IEEE/OSA J. Opt. Commun. Netw. 6(10), 816–828 (2014)
45. Goscien, R., Walkowiak, K., Klinkowski, M.: Joint anycast and unicast routing and spectrum allocation with dedicated path protection in elastic optical networks. In: Proceedings of the 10th International Conference on the Design of Reliable Communication Networks (DRCN 2014), pp. 1–8 (2014)
46. Goscien, R., Walkowiak, K., Klinkowski, M.: On the regenerators usage in cloud-ready elastic optical networks with distance-adaptive modulation formats. In: Proceedings of the 2014 European Conference on Optical Communication (ECOC 2014), pp. 1–3 (2014)
47. Goscien, R., Walkowiak, K., Klinkowski, M.: Tabu search algorithm for routing, modulation and spectrum allocation in elastic optical network with anycast and unicast traffic. Comput. Netw. 79, 148–165 (2015)

48. Klinkowski, M.: An evolutionary algorithm approach for dedicated path protection problem in elastic optical networks. Cybern. Syst. **44**(6–7), 589–605 (2013)
49. Klinkowski, M.: A genetic algorithm for solving RSA problem in elastic optical networks with dedicated path protection. In: Herrero, A., Snel, V., Abraham, A., Zelinka, I., Baruque, B., Quintin, H., Luis Calvo, J., Sedano, J., Corchado, E. (eds.) International Joint Conference CISIS12-ICEUTE12-SOCO12 Special Sessions. Advances in Intelligent Systems and Computing, vol. 189, pp. 167–176. Springer, Berlin (2013)
50. Klinkowski, M., Walkowiak, K.: Offline rsa algorithms for elastic optical networks with dedicated path protection consideration. In: Proceedings of the 4th International Congress on Ultra Modern Telecommunications and Control Systems and Workshops (ICUMT 2012), pp. 670–676 (2012)
51. Klinkowski, M., Walkowiak, K., Goscien, R.: Optimization algorithms for data center location problem in elastic optical networks. In: Proceedings of the 15th International Conference on Transparent Optical Networks (ICTON 2013), pp. 1–5 (2013)
52. Kmiecik, W., Goscien, R., Walkowiak, K., Klinkowski, M.: Two-layer optimization of survivable overlay multicasting in elastic optical networks. Opt. Switch. Netw. **14** Part 2(0):164–178 (2014). Special Issue on RNDM 2013
53. Paolucci, F., Castro, A., Cugini, F., Velasco, L., Castoldi, P.: Multipath restoration and bitrate squeezing in sdn-based elastic optical networks [invited]. Photonic Netw. Commun. **28**(1), 45–57 (2014)
54. Walkowiak, K., Goscien, R., Klinkowski, M.: On minimization of the spectrum usage in elastic optical networks with joint unicast and anycast traffic. In: Proceedings of the 2013 Asia Communications and Photonics Conference (ACP 2013), pp. AF4G.1. Optical Society of America (2013)
55. Walkowiak, K., Goscien, R., Tornatore, M., Klinkowski, M.: Impact of fanout and transmission reach on performance of multicasting in elastic optical networks. In: Proceedings of the 2015 Optical Fiber Communications Conference and Exhibition (OFC 2015), pp. 1–3 (2015)
56. Walkowiak, K., Goscien, R., Wozniak, M., Klinkowski, M.: Joint optimization of multicast and unicast flows in elastic optical networks. In: Proceedings of the 2015 IEEE International Conference on Communications (ICC 2015) (2015)
57. Walkowiak, K., Klinkowski, M.: Joint anycast and unicast routing for elastic optical networks: modeling and optimization. In: Proceedings of the 2013 IEEE International Conference on Communications (ICC 2013), pp. 3909–3914 (2013)
58. Walkowiak, K., Klinkowski, M.: Shared backup path protection in elastic optical networks: modeling and optimization. In: Proceedings of the 9th International Conference on the Design of Reliable Communication Networks (DRCN 2013), pp. 187–194 (2013)
59. Walkowiak, K., Klinkowski, M., Rabiega, B., Goscien, R.: Routing and spectrum allocation algorithms for elastic optical networks with dedicated path protection. Opt. Switch. Netw. **13**, 63–75 (2014)
60. Politi, C., Anagnostopoulos, V., Matrakidis, C., Stavdas, A., Lord, A., Lopez, V., Fernandez-Palacios, J.P.: Dynamic operation of flexi-grid OFDM-based networks. In: Proceedings of the Conference on Optical Fiber Communication (OFC 2012), Los Angeles, USA (2012)
61. Bocoi, A., Schuster, M., Rambach, F., Kiese, M., Bunge, C.-A., Spinnler, B.: Reach-dependent capacity in optical networks enabled by ofdm. In: Proceedings of the Conference on Optical Fiber Communication (OFC 2009), pp. 1–3 (2009)
62. IBM. ILOG CPLEX optimizer. http://www.ibm.com
63. Goscien, R.: Joint optimization of anycast and unicast flows in elastic optical networks using tabu search algorithm (in polish). Master's thesis, Wroclaw University of Technology, Department of Systems and Computer Networks (2013). Supervisor: K. Walkowiak
64. Glover, F.: Tabu search fundamentals and uses. Technical report, University of Colorado (1995)
65. Pioro, M., Medhi, D.: Routing, Flow, and Capacity Design in Communication and Computer Networks. Morgan Kaufmann, San Francisco (2004)
66. X. Yang. Nature-Inspired Optimization Algorithms. Elsevier (2014)

67. Kirkpatrick Jr., S., Gelatt, C.D., Vecchi, M.P.: Optimization by simulated annealing. Science **220**, 671–680 (1983)
68. El-Ghazali, T.: Metaheuristics-From Design to Implementation. Wiley, Hoboken (2009)
69. Cisco. Cisco visual networking index: Forecast and methodology, 2012–2017 (2013). White paper. http://www.cisco.com/c/en/us/solutions/service-provider/visual-networking-index-vni/index.html
70. Cisco. Cisco global cloud index: Forecast and methodology, 2011–2016 (2012). White paper. http://www.cisco.com/c/en/us/solutions/service-provider/global-cloud-index-gci/index.html
71. Deore, A., Turkcu, O., Ahuja, S., Hand, S.J., Melle, S.: Total cost of ownership of wdm and switching architectures for next-generation 100Gb/s networks. IEEE Commun. Mag. **50**(11), 179–187 (2012)
72. Palkopoulou, E., Angelou, M., Klonidis, D., Christodoulopoulos, K., Klekamp, A., Buchali, F., Varvarigos, E., Tomkos, I.: Quantifying spectrum, cost, and energy efficiency in fixed-grid and flex-grid networks. IEEE/OSA J. Opt. Commun. Netw. **4**(11), B42–B51 (2012)
73. Klekamp, A., Gebhard, U., Ilchmann, F.: Energy and cost efficiency of adaptive and mixed-line-rate ip over dwdm networks. J. Lightw. Technol. **30**(2), 215–221 (2012)
74. Peng, Y., Hu, W., Sun, W., Wang, X., Jin, Y.: Impairment constraint multicasting in translucent wdm networks: architecture, network design and multicasting routing. Photonic Netw. Commun. **13**(1), 93–102 (2007)
75. Sambo, N., Meloni, G., Berrettini, G., Paolucci, F., Malacarne, A., Bogoni, A., Cugini, F., Poti, L., Castoldi, P.: Demonstration of data and control plane for optical multicast at 100 and 200 Gb/s with and without frequency conversion. IEEE/OSA J. Opt. Commun. Netw. **5**(7), 667–676 (2013)
76. Ding, A., Poo, G.: A survey of optical multicast over WDM networks. Comput. Commun. **26**(2), 193–200 (2003)
77. Walkowiak, K., Kasprzak, A., Wozniak, M.: Algorithms for calculation of candidate trees for efficient multicasting in elastic optical networks. In: Proceedings of the 17th International Conference onTransparent Optical Networks (ICTON 2015), pp. 1–4 (2015)
78. Klekamp, A., Gebhard, U., Ilchmann, F.: Efficiency of adaptive and mixed-line-rate ip over dwdm networks regarding capex and power consumption [invited]. IEEE/OSA J. Opt. Commun. Netw. **4**(11), B11–B16 (2012)
79. Rambach, F., Konrad, B., Dembeck, L., Gebhard, U., Gunkel, M., Quagliotti, M., Serra, L., Lopez, V.: A multilayer cost model for metro/core networks. IEEE/OSA J. Opt. Commun. Netw. **5**(3), 210–225 (2013)
80. Rizzelli, G., Morea, A., Tornatore, M., Rival, O.: Energy efficient traffic-aware design of on-off multi-layer translucent optical networks. Comput. Netw. **56**(10), 2443–2455 (2012). Green communication networks
81. Vizcaino, J., Ye, Y., Monroy, I.: Energy efficiency analysis for flexible-grid ofdm-based optical networks. Comput. Netw. **56**(10), 2400–2419 (2012). Green communication networks
82. Ellinas, G., Papadimitriou, D., Rak, J., Staessens, D., Sterbenz, J., Walkowiak, K.: Practical issues for the implementation of survivability and recovery techniques in optical networks. Opt. Switch. Netw. **14**, Part 2(0), 179–193 (2014). Special Issue on RNDM 2013
83. Shen, G., Guo, H., Bose, S.: Survivable elastic optical networks: survey and perspective (invited). Photonic Netw. Commun. pp. 1–17 (2015)
84. Takagi, T., Hasegawa, H., Sato, K., Tanaka, T., Kozicki, B., Sone, Y., Jinno, M.: Algorithms for maximizing spectrum efficiency in elastic optical path networks that adopt distance adaptive modulation. In: Proceedings of the 36th European Conference and Exhibition on Optical Communication (ECOC 2010), pp. 1–3 (2010)
85. Walkowiak, K., Goscien, R., Kmiecik, W., Klinkowski, M.: Content distribution in elastic optical networks with dedicated path protection. In: Proceedings of the 6th International Workshop on Reliable Networks Design and Modeling (RNDM 2014), pp. 116–122 (2014)
86. Zhang, J., Lv, Ch., Zhao, Y., Chen, B., Li, X., Huang, Sh, Gu, W.: A novel shared-path protection algorithm with correlated risk against multiple failures in flexible bandwidth optical networks. Opt. Fiber Technol. **18**(6), 532–540 (2012)

87. Shao, X., Yeo, Y., Xu, Z., Cheng, X., Zhou, L.: Shared-path protection in ofdm-based optical networks with elastic bandwidth allocation. In: Proceedings of the Conference on Optical Fiber Communication (OFC 2012), pp. 1–3 (2012)
88. Wang, C., Shen, G., Bose, S.K.: Distance adaptive dynamic routing and spectrum allocation in elastic optical networks with shared backup path protection. J. Lightw. Technol. **PP**(99), 1–1 (2015)
89. Chen, X., Ji, F., Zhu, Z.: Service availability oriented p-cycle protection design in elastic optical networks. IEEE/OSA J. Opt. Commun. Netw. **6**(10), 901–910 (2014)
90. Wu, J., Liu, Y., Yu, C., Wu, Y.: Survivable routing and spectrum allocation algorithm based on p-cycle protection in elastic optical networks. Optik—Int. J. Light Electron Opt. **125**(16), 4446–4451 (2014)
91. Castro, A., Ruiz, M., Velasco, L., Junyent, G., Comellas, J.: Path-based recovery in flexgrid optical networks. In: Proceedings of the 14th International Conference on Transparent Optical Networks (ICTON 2012), pp. 1–4 (2012)
92. Sone, Y., Watanabe, A., Imajuku, W., Tsukishima, Y., Kozicki, B., Takara, H., Jinno, M.: Highly survivable restoration scheme employing optical bandwidth squeezing in spectrum-sliced elastic optical path (slice) network. In: Proceedings of the Conference on Optical Fiber Communication (OFC 2009), pp. 1–3 (2009)
93. Talebi, S., Alam, F., Katib, I., Khamis, M., Salama, R., Rouskas, G.: Spectrum management techniques for elastic optical networks: a survey. Opt. Switch. Netw. **13**, 34–48 (2014)
94. Goscien, R., Walkowiak, K., Klinkowski, M.: Gains of anycast demand relocation in survivable elastic optical networks. In: Proceedings of the 6th International Workshop on Reliable Networks Design and Modeling (RNDM 2014), pp. 109–115 (2014)

# Chapter 4
# Overlay Networks

In this chapter, we concentrate on the optimization of overlay networks. Firstly, we define the concept of overlay networks. Next, we formulate several optimization problems related to overlay networks and fundamental in the context of cloud computing and content-oriented services as ILP models. Additionally, we explain various solutions for selected optimization problems, as well as presenting results of numerical experiments.

## 4.1 Introduction

Computer and communication networks—including the Internet and wireless 3G and 4G networks—have been rapidly developing in recent years. In consequence, many new challenges and difficulties have appeared in the technical and business domains; they include the vast diversity of new network services, the need for cooperation between many domains using various physical technologies, growing numbers of Internet users, increasing traffic volumes, high competition between ICT companies, demand for very low times to market for new applications, security and reliability requirements, etc. In many cases, classical networking solutions based on TCP/IP protocols are insufficient for the fast and cost-effective deployment of new services demanded by the market. As such, the concept of overlay networks has been gaining increasing attention since the turn of the century [1–7].

An *overlay network* is built on top of an existing underlying network, which is responsible for supplying basic networking functions such as routing and forwarding to provide connectivity between overlay nodes. The majority of overlay networks are built in the application layer on top of the TCP/IP layers. Note that the overlay nodes are connected to each other via logical (overlay) links spanning many physical links of the underlying networks using various technologies and protocols. The key benefit of the overlay approach is that limitations of the underlay can be overcome directly in the overlay layer, since overlay systems are flexible and able to implement a range of networking approaches. It should be stressed that since overlay services are

© Springer International Publishing Switzerland 2016
K. Walkowiak, *Modeling and Optimization of Cloud-Ready and Content-Oriented Networks*, Studies in Systems, Decision and Control 56, DOI 10.1007/978-3-319-30309-3_4

used based on the default, point-to-point connectivity provided by underlying layers, there is no need to cooperate with parties responsible for the underlying networks when a new overlay service is issued or an existing overlay service is modernized. In fact, overlays do not require or cause any changes to the underlying network. For instance, any failure identified in the overlay system can be fixed by the overlay network itself, i.e., the overlay routing is changed in order to omit the broken elements of the network [7].

The concept of overlay networks can be classified as a *network virtualization* technique which decouples the roles of the traditional Internet Service Providers (ISPs) into two independent entities: infrastructure providers (InPs) responsible for the physical infrastructure, and service providers (SPs) creating virtual networks (VNs) by aggregating resources from multiple InPs and offering end-to-end services. As well as overlay networks, examples of network virtualization solutions are Virtual Private Networks (VPNs) and Virtual Local Area Networks [1, 2, 7, 8].

An interesting example of overlay networks are Peer-to-Peer (P2P) systems. According to statistics, in the mid-2000s, BitTorrent and other P2P systems generated more than 50 % of consumer Internet traffic [2, 5–7, 9]. P2P systems assume that each node (peer) represents both a server (producer providing data to other nodes) and a client (consumer receiving data from other nodes). Therefore, a node of the P2P system can be referred to as a *servent*, combining of the first part of the word *server* and the second part of the word *client*. P2P systems can be classified as unstructured and structured. In unstructured systems, data is stored without any specific structure, while structured P2P systems use the concept of a Distributed Hash Tables (DHTs) which provide a special structure of data distributed among many peers. Unstructured P2P systems can be divided into centralized P2P, pure P2P and hybrid P2P. Centralized P2P systems (e.g., Napster) use a central server storing information (e.g., IP addresses) about the location of content at particular peers. Pure P2P systems do not use centrals server and they depend on flooding information on the desired content over the network (e.g., Gnutella 0.4 and Freenet). Finally, hybrid P2P systems employ a hierarchy of superpeers, i.e., servers which keep information on content location (e.g., Gnutella 0.6) [2, 3, 5–7].

A popular example of a P2P-based file distribution system is the BitTorrent protocol [2, 5, 7, 10]. Files distributed in the BitTorrent system are divided into pieces of a fixed size (blocks), typically between 64 KB and 4 MB each. BitTorrent is based on a centralized service known as *tracker* which stores information on which nodes have particular blocks. A peer $v$ wishing to download a file receives a random list of nodes which have the requested file. Next, node $v$ requests pieces of the file contacting the peers included in the list. When peer $v$ downloads some of the pieces, it can upload them to other peers. Since the goal of the BitTorrent system is to provide effective file sharing, peers are encouraged to not only download files, but also to upload, which is achieved by a tit-for-tat strategy.

Another example using the idea of overlay networks are Content Delivery Networks (CDNs) which cache content and enable its distribution on a massive scale. The highest volume of content provided in CDNs is video [11]. Video streaming in lower networking layers using multicast transmission poses many challenges due to technical and business limitations. For instance, to implement multicast transmission in the optical or network layer, the corresponding hardware (switches, routers, etc.) must be equipped with additional capabilities. In some cases, the absence of this functionality limits multicasting. Secondly, multicast addressing schemes in lower layers constrain scalability, and multicast use is usually reduced to a single service provider. Moreover, there questions remain regarding security management, flow control, congestion control and different system configurations depending on the ISPs. Finally, the absence of a business model supporting inter-ISPs multicasting usually limits IP multicast solutions to a single ISP. A cost-effective, flexible and scalable alternative to video streaming in lower networking layers is the basis of *overlay multicasting*. Overlay multicasting, also known as application layer multicasting or P2P multicasting, uses a multicast tree consisting of overlay nodes (end hosts). The links of the tree are overlay links between overlay nodes. In contrast to multicasting in lower layers, the uploading (non-leaf) node in the tree is not a router or switch but an overlay node which is also a normal end host (receiver). Overlay multicasting can be used to deliver a wide range of data including elastic content (e.g., data files) and streaming content with specific bit rate requirements (e.g., media streaming) [2, 5, 7, 12–20].

In addition, overlay networks are used in the context of distributed computing systems. These types of systems are widely applied in academic and business domains to compute tasks requiring vast processing power not available on a single machine. Distributed computing systems can be classified as grid computing systems and P2P computing systems. Note that a grid is established by organizations and institutions, and contains a number of specialized machines (servers) connected by a high-capacity computer network. A dedicated network can be used, or the elements of the grid systems are simply connected by an overlay network to provide the connectivity. Building the grid system is a sophisticated task regarding in technical and financial aspects [21–23]. In contrast, P2P computing systems, also known as public-resource computing systems or global computing systems, are built using many private machines which are most frequently home computers (PCs or Macs) or even gaming consoles. Users wishing to participate in a P2P computing system simply install special software and register with a selected computing project. The users then receive data chunks to be processed and send back the results. P2P computing systems are based on overlay networks such as the Internet, i.e., participating nodes are connected to the system with regular access links. A popular P2P computing project is SETI@home, which searches for extra-terrestrial intelligence [5, 24–27].

For a comprehensive survey on various aspects related to overlay networks refer to [2, 3, 5–7].

## 4.2 Network Design for Overlay Multicasting

This section presents a network design problem related to overlay multicasting [28, 29]. The goal is to optimize multicast flows and select access link capacity of overlay nodes in order to minimize network cost and satisfy the requirement that each network node receives the information requested.

### 4.2.1  Formulations

Our overlay network is modeled on an analysis of real overlay systems and previous research on the subject. The network is defined by a set of nodes $V$ connected to the overlay system by access links. The overlay system is based on a single substrate network (e.g., the Internet) and overlay nodes communicate directly with each other within a full mesh topology. The models can be modified to address the construction of an overlay network over multiple substrate networks. It is assumed that capacity constraints are checked for access links only, since the node capacity constraint is typically sufficient in overlay networks. More precisely, each node is connected to the overlay network with an access link with a limited capacity, while the underlying core network is considered as overprovisioned and thus the only bottlenecks are access links [30–36]. For each node $v \in V$, set $K(v)$ includes candidate access links to be selected. Each link $k \in K(v)$ is described by download capacity $d_{vk}$, upload capacity $u_{vk}$ and cost $\xi_{vk}$. Note that cost $\xi_{vk}$ can be interpreted in various ways, e.g., leasing cost per month, deployment cost, and power consumption cost. In turn, link capacity is defined separately for downstream and upstream directions, since in overlay systems users (nodes) can use asymmetric access links. As well as participating in overlay multicasting, overlay nodes usually also participate in other network services and resources. Constants $a_v$ and $b_v$ denote the download and upload background traffic for each node $v \in V$, respectively. Background traffic must be taken into account when the capacity constraints are checked. The optimization goal is to select every node type of an access link (binary variable $y_{vk}$) in order to provision overlay multicasting with background traffic and to minimize the network cost defined as the cost of all access links used.

The overlay network is used to deliver streaming data from a root node to receivers using overlay multicasting. To improve performance of the system, data is divided into several streams and each stream is sent using a separate multicast tree. Consequently, each node downloads portions of the multicast stream via different routes (trees). This approach provides several benefits. Firstly, the upload capacity of nodes can be used more efficiently, since multiple substreams have a lower granularity of bit-rate than single streams carrying all the data. Secondly, when only one streaming tree is used many nodes act as downloaders only and do not contribute to increasing the upload capacity. The system's fairness can be improved, i.e., each participant in the system uploads a similar volume of data as it receives. Finally, using multiple streaming trees improves system resilience [37–39] (for more details see Sect. 4.3). It should be noted that special coding solutions have been developed for multiple streaming tree transmissions [5, 30, 40–42]. Let $T$ denote a set including all trees (streams), and for each tree constant $h_t$ denotes the bit-rate assigned to a particular tree. All trees use the same root (streaming node). Various modeling approaches can be used to construct the trees. Below we present two models which use multicast flow formulation and multicast level formulation (for more information on modeling of multicast flows refer to Sect. 1.2.3).

We also consider hop-constrained multicasting, i.e., there is an upper limit on the number of hops on a path between the root node and any receiver [43]. The main reason for this is to improve the QoS parameters of overlay multicasting including network reliability and transmission delay. Note that the hop limit indicates an upper limit on the number of levels of the multicast tree. For more details on levels in multicasting refer to Sect. 1.2.3.

**Flow Formulation**

The first model uses the flow formulation of multicast flows. Since we consider an overlay network, the basic formulation presented in Sect. 1.2.3, must be modified slightly. Since overall links are defined as pairs of nodes $(v, w)$, the variable denoting whether a path from the root node to receiver $r$ in tree $t$ using link $(v, w)$ is defined as $x_{vwtr}$. Similarly, let $x_{vwt}$ denote whether overlay link $(v, w)$ is included in tree $t$.

**OVR/M/ND/Cost/Flow**

**sets**

| | |
|---|---|
| $V$ | nodes (peers) |
| $K(v)$ | access link types for node $v$ |
| $L$ | levels |
| $T$ | trees |
| $R(t)$ | receivers of tree $t$ |

**constants**

| | |
|---|---|
| $a_v$ | download background traffic of node $v$ (kb/s) |
| $b_v$ | upload background traffic of node $v$ (kb/s) |
| $d_{vk}$ | download capacity of access link type $k$ for node $v$ (kb/s) |
| $u_{vk}$ | upload capacity of access link type $k$ for node $v$ (kb/s) |
| $\xi_{vk}$ | cost of access link type $k$ for node $v$ |
| $s$ | streaming node (tree root) |
| $h_t$ | streaming rate of tree $t$ (kb/s) |

**variables**

| | |
|---|---|
| $x_{vwtr}$ | =1, if in tree $t$ the streaming path from the root to receiver $r$ includes an overlay link from node $v$ to node $w$ (no other overlay nodes in between); 0, otherwise (binary) |
| $x_{vwt}$ | =1, if overlay link from node $v$ to node $w$ (no other overlay nodes in between) is included in tree $t$; 0, otherwise (binary) |
| $y_{vk}$ | =1, if node $v$ is connected to the overlay network by an access link type $k$; 0, otherwise (binary) |

**objective**

$$\text{minimize} \quad F = \sum_{v \in V} \sum_{k \in K(v)} \xi_{vk} y_{vk} \qquad (4.2.1a)$$

**constraints**

$$\sum_{v \in V} x_{vwtr} - \sum_{v \in V} x_{wvtr} = \begin{cases} +1 & \text{if} \quad v = r \\ -1 & \text{if} \quad v = s, \quad w \in V, t \in T, r \in R(t) \\ 0 & \text{otherwise} \end{cases} \tag{4.2.1b}$$

$$x_{vwtr} \le x_{vwt}, \quad v \in V, w \in V, t \in T, r \in R(t) \tag{4.2.1c}$$

$$\sum_{k \in K(v)} y_{vk} = 1, \quad v \in V \tag{4.2.1d}$$

$$a_v + \sum_{t \in T} h_t \le \sum_{k \in K(v)} y_{vk} d_{vk}, \quad v \in V \setminus \{s\} \tag{4.2.1e}$$

$$b_v + \sum_{w \in V} \sum_{t \in T} x_{vwt} h_t \le \sum_{k \in K(v)} y_{vk} u_{vk}, \quad v \in V \tag{4.2.1f}$$

$$\sum_{v \in V} \sum_{w \in V} x_{vwtr} \le |L|, \quad t \in T, r \in R(t). \tag{4.2.1g}$$

The objective function (4.2.1a) is the cost of access links of the overlay multicasting network. Equation (4.2.1b) using the node-link formulation ensures that in every tree $t \in T$, a unicast path is established from the root node $s$ to each receiver $r \in R(t)$. Inequality (4.2.1c) defines variables $x_{vwt}$ which create multicast trees. Equality (4.2.1d) is in the model to guarantee that exactly one access link is selected for every overlay node. Conditions (4.2.1e) and (4.2.1f) are download and upload capacity constraints, respectively. The left-hand side of both inequalities represents the overall flow including background traffic and multicasting traffic downloaded and uploaded by an overlay node, respectively. In the case of the download capacity, the flow delivered to the node does not depend on the configuration of multicast trees, since every node $v \in V \setminus \{s\}$ downloads bit-rates of all trees and the background traffic. The right-hand side of both constraints denotes the download and upload capacity assigned to a particular node, respectively. The model includes capacity constraints formulated for access links only, since the underlying network is assumed to be overprovisioned. However, the formulation can be modified easily to account for capacity bottlenecks between overlay nodes as shown in [20]. Finally, constraint (4.2.1g) defines the upper limit on the number of levels (hops) in the path from the root to every receiving node.

**Level Formulation**

The second model uses the level formulation to model multicast flows. Since, multiple trees are established in the network, the flow variable is $x_{vwtl}$ denotes whether there is a link from node $v$ to node $w$ (no other overlay nodes in between) in multicast tree $t$ and node $v$ is located on level $l$.

**OVR/M/ND/Cost/Level**
**variables (additional)**

$x_{vwtl}$ $=1$, if there is a link from node $v$ to node $w$ (no other overlay nodes in between)
in multicast tree $t$ and node $v$ is located on level $l$; 0, otherwise (binary)

**objective**

$$\text{minimize} \quad F = \sum_{v \in V} \sum_{k \in K(v)} \xi_{vk} y_{vk} \tag{4.2.2a}$$

**constraints**

$$\sum_{v \in V} \sum_{l \in L} x_{vstl} = 0, \quad t \in T \tag{4.2.2b}$$

$$\sum_{v \in V} \sum_{l \in L} x_{vrtl} = 1, \quad t \in T, r \in R(t) \tag{4.2.2c}$$

$$\sum_{w \in V} x_{vwt1} = 0, \quad v \in W \setminus \{s\}, t \in T \tag{4.2.2d}$$

$$x_{vwtl} \leq \sum_{u \in V} x_{uv(l-1)t}, \quad v \in V, w \in V, t \in T, l \in L \setminus \{1\} \tag{4.2.2e}$$

$$\sum_{k \in K(v)} y_{vk} = 1, \quad v \in V \tag{4.2.2f}$$

$$a_v + \sum_{t \in T} h_t \leq \sum_{k \in K(v)} y_{vk} d_{vk}, \quad v \in V \setminus \{s\} \tag{4.2.2g}$$

$$b_v + \sum_{w \in V} \sum_{t \in T} \sum_{l \in L} x_{vwtl} h_t \leq \sum_{k \in K(v)} y_{vk} u_{vk}, \quad v \in V. \tag{4.2.2h}$$

The objective function is formulated in the same way as in the flow model. New constraints follow directly from the level modeling. To recall, equality (4.2.2b) ensures that root node $s$ cannot download multicast flows. Next equality (4.2.2c) guarantees that for every tree $t \in T$ every receiving node $r \in R(t)$ is connected to the tree. Condition (4.2.2d) is in the model to ensure that only root node $s$ can be a parent node on level $l=1$. Inequality (4.2.2e) expresses the requirement for every tree $t \in T$ that node $v$ cannot upload multicast flows to any other node $w$ on level $l$ if node $v$ is not located on level $l - 1$ of the multicast tree. Equality (4.2.2g) ensures that only one access link type is selected for every node. Conditions (4.2.2g) and (4.2.2h) define download and upload access link capacity constraints, respectively.

### 4.2.2   Algorithms

The OVR/M/ND problem formulated in previous section is $\mathcal{N}\mathcal{P}$-complete, since it can be reduced to the Hop-Constrained Minimum Spanning Tree Problem shown to be $\mathcal{N}\mathcal{P}$-complete in [44]. Heuristic algorithms are needed to enable solving large problem instances. The three algorithms described below are the greedy method [29], Lagrangian relaxation [28], and the evolutionary algorithm [45]. Note that the algorithms are created in the context of the level formulation given by (4.2.2).

**Greedy Algorithm**

Algorithm 4.1 shows a greedy method of solving the OVR/M/ND problem [29]. The algorithm constructs multicast trees by adding new nodes to the trees one-by-one and increasing network link capacity if required. It is assumed that sets $X$ and $Y$ include variables $x_{vwtl}$ and $y_{vk}$ equal to 1. In other words, sets $X$ and $Y$ are used to represent the solution of the problem. Let $h = \sum_{t \in T} h_t$ denote the overall streaming rate of all trees included in set $T$. We can assume that for every node $v \in V$, set $K(v)$ includes access link types sorted according to increasing values of the cost.

The algorithm starts by initializing all variables $x_{vwtl}$ and $y_{vk}$ (line 2–8). More specifically, set $X$ is left empty while access link types included in set $Y$ are selected as follows. Firstly, for the root node $s$ the link capacity is selected as the cheapest option ensuring the upload capacity is not lower than the background traffic $b_s$ plus the streaming rate of all trees $h$ (lines 3–4). Secondly, for every remaining node $v \in V \setminus \{s\}$, the access link is chosen as the most cost effective option downloading the required traffic including background traffic $a_v$ and the overall streaming rate $h$.

The next operation ensures that for every tree $t \in T$ there is at least one connection from the root node $s$ to another node (lines 9–16). Two sets are created for this purpose; set $A$ includes all trees sorted in decreasing order of streaming rate $h_t$, and set $B$ contains all nodes excluding the root node sorted in decreasing order of the residual upload capacity (lines 9–10). The first node from set $B$ is assigned one-by-one to the first tree included in set $A$, and so on until at least one node is connected to the root for every tree $t \in T$.

Next, the main loop of the algorithm is executed (lines 18–36) in order to construct all multicast trees. Function $Find\_Best\_Allocation(X, Y, l)$ returns the best configuration (i.e., tree $t$, parent node $v$ and children node $w$) for the next allocation on level $l$. Using the current state of the problem encoded in sets $X$ and $Y$, the following procedure is executed. Tree $t$ enabling at least one new transfer and the lowest number of nodes connected to the tree is selected first. Next, parent node $v$ is chosen with enough residual upload capacity, located on level $l$ in tree $t$. If more than one feasible parent node exists, the algorithm selects the node with the highest value of residual upload capacity. Finally, child node $w$ which is not connected to tree $t$ is calculated. If there is more than one feasible child node, the node residual upload capacity is again used as the criterion. If the transfer on level $l$ is impossible due to limited resources of link capacity, the level is incremented (line 23). Two options arise if all levels are already analyzed ($l > |L|$). Firstly, the transfer is completed, i.e.,

**Algorithm 4.1** GA/OVR/M/ND (Greedy Algorithm for Overlay Network Design with Multicast Flows)

---

**Require:** set of overlay nodes $V$, sets of candidate access links $K(v)$, set of streaming trees $T$
**Ensure:** selection of access links and overlay multicast routing denoted by solutions $X$ and $Y$,
    value of objective function

1: **procedure** $GA/OVR/M/ND(V, K(v), T)$
2:  $X := \emptyset, Y := \emptyset$
3:  $k := \min\{i \in K(s) : u_{si} \geq b_s + h\}$
4:  $Y := Y \cup \{y_{sk}\}$
5:  **for** $v \in V \backslash \{s\}$ **do**
6:    $k := \min\{i \in K(v) : d_{vi} \geq a_v + h\}$
7:    $Y := Y \cup \{y_{vk}\}$
8: **end for**
9:  $A := Sort\_Trees\_Dec\_Rate(T)$
10:  $B := Sort\_Nodes\_Dec\_Residual(V \backslash \{s\})$
11:  $i := 1$
12: **for** $t \in A$ **do**
13:  $w := Member(B, i)$
14:  $X := X \cup \{x_{swt1}\}$
15:  $i := i + 1$
16: **end for**
17:  $l := 1, test := 0$
18: **repeat**
19:  **repeat**
20:    $(v, w, t) := Find\_Best\_Allocation(X, Y, l)$
21:    $X := X \cup \{x_{vwtl}\}$
22:  **until** $Is\_Possible\_Transfer(X, Y, l) = TRUE$
23:  $l := l + 1$
24:  **if** $l > |L|$ **then**
25:    **if** $Is\_Completed(X, Y) = TRUE$ **then**
26:      $test := 1$
27:    **else**
28:      **if** $Is\_Possible\_Update(X, Y) = TRUE$ **then**
29:        $Y := Update\_Capacity(X, Y)$
30:        $l := 1$
31:      **else**
32:        **return** $(FALSE, NULL, NULL)$
33:      **end if**
34:    **end if**
35:  **end if**
36: **until** $test < 1$
37: **return** $(TRUE, X, Y)$
38: **end procedure**

---

all nodes are connected to every tree. Then, the algorithm stops and returns a feasible solution (line 26). Otherwise, the algorithm tries to update the link capacities and continue the construction of trees starting from the first level (lines 29–30). However, if the update of access link capacity is impossible, the algorithm returns information that it cannot find a feasible solution (line 32). Function $Update\_Capacity(X, Y)$ updates the capacity of a single node. As such, we examine all nodes for which the capacity can be increased. If more than one such node exist, an additional criterion is applied. Accordingly, different combinations of the following two metrics are used: node average level considering all trees, and relative cost of upload capacity increase given by the formula $(u_{v(k+1)} - u_{vk})/(\xi_{v(k+1)} - \xi_{vk})$.

The maximum complexity of algorithm GA/OVR/M/ND is $O(|V|^3 |T| |L| |K|)$, where $|V|$ denotes the number of nodes, $|T|$ is the number of trees, $|L|$ denotes the number of levels and $|K|$ is the number of candidate access links per node. Note that the maximum complexity of function $Update\_Capacity(X, Y)$ is $O(|V| |K|)$. It is clear that in the worst case, this function can be called for every combination of parent node, child node, level and tree.

**Lagrangian Relaxation**

The next algorithm is based on the Lagrangian relaxation (LR) approach combined with the subgradient optimization approach [28]. More general information on this method can be found in Sect. 2.2.2. In order to formulate a dual problem to model (4.2.2a)–(4.2.2h), constraint (4.2.2h) is relaxed using vector $\lambda = (\lambda_1, \lambda_2, \ldots, \lambda_{|V|})$ of positive Lagrangian multipliers $\lambda_v$ for each node $V \in V$. Consequently, the following Lagrangian relaxation of problem OVR/M/ND/Cost/Level is formulated.

**OVR/M/ND/Cost/Level/LR**
**objective**

$$\text{minimize} \quad \varphi(\lambda) = \sum_{v \in V} \sum_{k \in K(v)} y_{vk}(\xi_{vk} - \lambda_v u_{vk}) + \sum_{v \in V} \lambda_v b_v$$
$$+ \sum_{v \in V} \sum_{w \in W} \sum_{l \in L} \sum_{t \in T} \lambda_v x_{vwtl} h_t \qquad (4.2.3a)$$

**constraints** (4.2.2b)–(4.2.2g)

Problem (4.2.3) can be broken down into two subproblems. The first subproblem includes variables $y_{vk}$ only and is stated as follows.

**OVR/M/ND/Cost/Level/LR1**
**objective**

$$\text{minimize} \quad \varphi_1(\lambda) = \sum_{v \in V} \sum_{k \in K(v)} y_{vk}(\xi_{vk} - \lambda_v u_{vk}) \qquad (4.2.4a)$$

**constraints**

$$\sum_{k \in K(v)} y_{vk} = 1, \quad v \in V \tag{4.2.4b}$$

$$a_v + \sum_{t \in T} h_t \leq \sum_{k \in K(v)} y_{vk} d_{vk}, \quad v \in V \setminus \{s\}. \tag{4.2.4c}$$

Model (4.2.4) can be broken down into $|V|$ subproblems, one for each node $v \in V$. Since the number of access link types $k \in K(v)$ is relatively low, problem (4.2.4) can be solved for each separate node $v$ by an inspecting all possible values of $k$. The procedure is very simple: assuming that access link types are sorted according to increasing values of cost, the inspection starts with $k = 1$ and it is continued for subsequent values of $k$ (link types) until constraint (4.2.4c) is satisfied.

The second subproblem of (4.2.3) contains variables $x_{vwtl}$ only and is formulated as follows.

**OVR/M/ND/Cost/Level/LR2**
**objective**

$$\text{minimize} \quad \varphi_2(\lambda) = \sum_{v \in V} \sum_{w \in W} \sum_{l \in L} \sum_{t \in T} \lambda_v x_{vwtl} h_t \tag{4.2.5a}$$

**constraints**

$$\sum_{v \in V} \sum_{l \in L} x_{vstl} = 0, \quad t \in T \tag{4.2.5b}$$

$$\sum_{v \in V} \sum_{l \in L} x_{vrtl} = 1, \quad t \in T, r \in R(t) \tag{4.2.5c}$$

$$\sum_{w \in V} x_{vwt1} = 0, \quad v \in W \setminus \{s\}, t \in T \tag{4.2.5d}$$

$$x_{vwtl} \leq \sum_{u \in V} x_{uv(l-1)t}, \quad v \in V, w \in V, t \in T, l \in L \setminus \{1\}. \tag{4.2.5e}$$

Problem (4.2.5) can be broken down with respect to $t$ into $|T|$ subproblems, i.e., one problem for every $t \in T$. The solution of this problem is relatively straightforward. The key observation is that according to the formulation of the objective function (4.2.5a), $\lambda_v$ represents the cost of using node $v$ as the parent node in the multicast tree. Therefore, the solution procedure finds node $v$ with the lowest value of $\lambda_v$ and assigns all other nodes directly to this node. If the cheapest node is the root node $s$, then all other nodes are connected to the root. Otherwise, the cheapest node $v$ is first connected to the root node, followed by assigning all other nodes to $v$. The solution leads to a situation where the cheapest node $v$ is the parent node of all other peers except the root node and itself. Because the upload capacity of every node in the network is limited by the maximum capacity of access link types, in most cases the solution does not uphold the upload capacity constraint (4.2.2h) of the primal problem. Therefore, an additional upload capacity constraint on upload flow is added to problem (4.2.5). Let $u_v^{max}$ denote the maximum upload capacity which can be selected for node $v$ according to access link types included in set $K(v)$. The additional constraint is formulated as follows:

$$b_v + \sum_{w \in V} \sum_{t \in T} \sum_{l \in L} x_{vwtl} h_t \leq u_v^{max}, \quad v \in V. \tag{4.2.5f}$$

Since problem (4.2.5a)–(4.2.5f) is $\mathcal{NP}$-complete (like the Hop-Constrained Minimum Spanning Tree Problem problem [44]), Algorithm 4.2 presents a heuristic algorithm to calculate the lower bound of this problem. Algorithm GA/OVR/M/ND/LR2 saturates subsequent nodes sorted according to increasing values of $\lambda_v$ up to the maximal upload capacity limit of each node. The upload flow of node $v$ denoted as $f_v$ is initialized with the background traffic (line 3). Since root node $s$ must be connected to every tree $t \in T$ with at least one overlay link, the upload flow of $s$ and objective function $F$ is updated accordingly (lines 4–7). Set $A$ includes nodes sorted according to increasing values of $\lambda_v$ (line 8). The allocation process (lines 10–22) starts with the cheapest node (line 9). More precisely, for each tree $t \in T$ subsequent nodes from set $A$ are saturated and the objective function $F$ is updated, accordingly.

Methods solving the Lagrangian relaxation problem (4.2.3) mean that a subgradient algorithm as proposed in [46] can be applied. A detailed pseudocode of the subgradient method is included in Sect. 2.2.2 of this book. Since some of the constraints are relaxed, the solution may not be infeasible. To build a feasible solution, heuristic GA/OVR/M/ND presented in Algorithm 4.1 is applied with a small modification: instead of initializing access link capacities as described in lines 4–8, values of variables $y_{vk}$ yielded by the subgradient search are used.

**Evolutionary Algorithm**

Another heuristic proposed for problem OVR/M/ND/Cost (4.2.2) is the evolutionary algorithm proposed in [45]. A key issue in developing the EA is devising an encoding scheme enabling the efficient performance of the algorithm [46–51]. A popular encoding approach is to include all problem variables in the chromosome. In the case of model (4.2.2), the number of variables is high, and they are bound by numerous complex constraints. In consequence, using all variables encoded in the chromosome means that a predominant part of the solution space is not feasible in terms of the problem constraints. Therefore, the EA may encounter major problems finding feasible solutions, even using methods of fixing the solution in order to satisfy all constraints such as penalty functions or repair functions. As such, we propose the following encoding scheme. The chromosome contains information related to multicast routing only (denoted by variable $x_{vwtl}$). Information related to access link capacity (denoted by variable $y_{vk}$) is determined later according to the multicast configuration. The chromosome numbers $|V||T|$ genes, i.e., for every tree $t \in T$, there is one gene for each node $v \in V$. The value of the gene denoted as $c_{vt}$ represents the number of child nodes (outgoing links) of node $v$ in tree $t$:

**Algorithm 4.2** GA/OVR/M/ND/LR2 (Greedy Algorithm for Solving Lagrangian Relaxation Subproblem 2)

---

**Require:** set of overlay nodes $V$, streaming tree $t \in T$, vector of Lagrangian multipliers $\lambda$
**Ensure:** value of objective function (4.2.5a)
1: **procedure** $GA/OVR/M/ND/LR2(V, t, \lambda)$
2:   $F = 0$
3:   **for** $v \in V$ **do** $f_v := b_v$
4:   **for** $t \in T$ **do**
5:     $f_s := f_s + h_t$
6:     $F := F + \lambda_s h_t$
7:   **end for**
8:   $A := Sort\_Nodes\_Inc\_Lambda(V \backslash \{s\})$
9:   $i := 1, v := Member(A, i)$
10:  **for** $t \in T$ **do**
11:    $n := 1$
12:    **repeat**
13:     **if** $f_v + h_t \leq u_v^{max}$ **then**
14:      $f_v = f_v + h_t$
15:      $F := F + \lambda_v h_t$
16:      $n := n + 1$
17:     **else**
18:      $i := i + 1$
19:      $v := Member(A, i)$
20:     **end if**
21:    **until** $n > |V| - 1$
22:  **end for**
23:  **return** $F$
24: **end procedure**

---

$$X = [c_{11}, c_{21}, \ldots, c_{|V|1}, c_{12}, c_{22}, \ldots, c_{|V|2}, \ldots, c_{1|T|}, c_{2|T|}, \ldots, c_{|V||T|}]. \quad (4.2.6)$$

For instance, considering a network consisting of ($|V| = 6$) nodes and one tree ($|T| = 1$), chromosome $X = [2, 1, 1, 1, 0, 0]$ denotes a tree in which node $v = 1$ has 2 children, and nodes $v = 2$, $v = 3$ and $v = 4$ have one child, while nodes $v = 5$ and $v = 6$ are leaf nodes and do not have outgoing links. Note that the sum of all gene values related to one tree must equal ($|V| - 1$), since all nodes excluding the root node must be connected to the tree.

The main advantage of this encoding scheme is that the solution space is significantly reduced. For instance, in the case of ten nodes, one tree and five levels, the chromosome encoded by (4.2.6) includes only ten genes and the solution space of the proposed encoding includes 51,770 possible solutions. To calculate this value, all integer partitions of number 9 must be considered with all possible permutations of partitions on ten positions. In contrast, if each gene of the chromosome represents a single variable $x_{vwtl}$, the number of genes in the chromosome is 500 and the size of the solution space is approx. $10^{24}$.

Another important issue to be addressed is the memory required to store chromosomes of a single population. Considering number of nodes $|V| = 200$, number of trees $|T| = 10$ and number of levels $|L| = 10$, a chromosome using encoding

(4.2.6) requires approx. 2 KB only (one byte per gene). The encoding with each gene representing variable $x_{vwtl}$ needs 4 MB (one byte per gene). Assuming a population containing 500 chromosomes, the corresponding memory requirements are 1 MB and 2 GB for the encoding schemes, respectively.

Since the encoding scheme does not define directly the precise configuration of the multicast tree or the link capacity, the following procedure is used to find the objective value of a particular chromosome $X$ denoting one unique solution. First, root node $s$ is placed on level $l = 1$. Next, all genes excluding the root node are analyzed to find the node with the highest value of $c_{vt}$ (number of children of node $v$ in tree $t$). This node is connected directly to the root node of the current tree and is located on level $l = 2$. The procedure is repeated for subsequent nodes sorted in decreasing order of $c_{vt}$. When the root node is saturated, i.e., no more nodes can be connected to the root due to the value of $c_{st}$, the next level of the tree is started, and the procedure assigns subsequent nodes to nodes already connected to the tree in order to satisfy the limit of outgoing connections given by $c_{vt}$. If the level limit is violated, all nodes located on levels $l > |L|$ are reconnected randomly to other nodes in order to satisfy the level constraint. Next, for each node $v \in V$, the cheapest access link type is selected in order to satisfy the upload and download flow of the node calculated according to the trees. The same procedure is repeated for every tree $t \in T$. When the largest possible access link does not provide sufficient capacity for allocated flows, a penalty function can be applied, i.e., an additional cost is included in the objective function. For more discussion on the penalty function approach refer to Sect. 2.2.2.

The crossover operator proposed according to encoding (4.2.6) combines two parent chromosomes $X_1$ and $X_2$ into one new individual $X_3$. The following procedure is replicated for every tree $t \in T$. The first gene of $X_1$ is set as the first gene of $X_3$. Then, the last gene from $X_2$ is set as the last gene of $X_3$. Next, the second gene of $X_1$ is set as the second gene of $X_3$, and the penultimate gene of $X_2$ is copied in the same place in $X_3$, etc. The process is continued until the sum of all genes included in $X_3$ exceeds the required number of connections defined as $(|V| - 1)$. If the chromosome is not feasible (sum of all genes is not equal to $(|V| - 1)$), it is repaired, i.e., some randomly selected genes are decremented.

The mutation operator simply decrements one randomly selected gene, and compensates for the reduced number of outgoing connections by incrementing a randomly chosen gene. The EA is easy to implement using the encoding scheme and operators presented above. For more details see [45].

### Neighborhood Search Methods

Many metaheuristic algorithms facilitate the concept of a neighborhood search, e.g., Local Search (LS), Tabu Search (TS), Simulated Annealing (SA), and Greedy Randomized Adaptive Search Procedure (GRASP) [46, 48, 49, 51–55]. The key issue in developing a neighborhood search heuristic is to define the solution encoding and the neighborhood generation.

In the context of the overlay network design problem (4.2.2), several encoding schemes can be proposed. Firstly, the approach proposed in (4.2.6) can be applied. To generate the neighborhood of (4.2.6), the value of $c_{vt}$ is decremented by 1 for every element $c_{vt}$ of solution $X$ that satisfies condition $c_{vt} > 0$. Next, all other elements of the same tree $t$ that holds $c_{wt} > |V| - 1$ are examined one by one and the value of $c_{wt}$ in incremented by 1. The same procedure is repeated for every tree $t \in T$. The size of the neighborhood space can be estimated as $(|V|(|V| - 1)|T|)$. To obtain values of flow variables $x_{vwtl}$ and access link capacity variables $y_{vk}$, the same procedures are used as described in the context of EA.

The second solution encoding represents for a parent node of a particular node $v$ and tree $t$ [56, 57]. Let $p_{vt}$ denote the index of the node which is the parent node of $v$ in tree $t$. Thus, the encoding scheme can be written as follows:

$$X = [p_{11}, p_{21}, \ldots, p_{|V|1}, p_{12}, p_{22}, \ldots, p_{|V|2}, \ldots, p_{1|T|}, p_{2|T|}, \ldots, p_{|V||T|}].$$

$$(4.2.7)$$

Note that since there are exactly $(|V| - 1)$ child nodes, the following condition must be satisfied for every tree $t \in T$: $\sum_{v \in V} p_{vt} = (|V| - 1)$. Encoding (4.2.7) directly defines flow variables $x_{vwtl}$, since using the information on parent nodes, a tree can be constructed and thus the level of each node can be obtained. A neighborhood of solution (4.2.7) is created by changing the parent node of one selected node. In consequence, the size of the neighborhood can be estimated as $((|V| - 1)^2|T|)$. However, if the new solution is not feasible (i.e., the multicast tree includes loops), the following repair mechanism is required. When a new neighbor solution is created (i.e., a new parent node of $v$ is selected) and the tree is disconnected, all descendants of node $v$ are directly connected to the previous parent node of $v$ and consequently the loop is avoided. Having defined feasible multicast flows, the access link capacity variables $y_{vk}$ can be acquired analogously to encoding (4.2.6).

### 4.2.3 Numerical Results

This section examines the performance of the heuristic algorithms proposed for solving the overlay network design problem with multicast flows [28, 29, 45]. We compare three methods: greedy algorithm (GA), Lagrangian relaxation (LR), and evolutionary algorithm (EA). Moreover, model (4.2.2) was implemented in CPLEX [58] to obtain optimal results. Following several preliminary experiments, the number of overlay nodes for experiments using CPLEX was selected as 20. Due to the low scalability of CPLEX, larger networks with 100 and 300 nodes were tested using heuristics only. To obtain real data of access links parameters, price lists of four ISPs were used: two operating in Poland and two operating in Germany. Each node was randomly assigned to one of the ISPs to create the set of access links. The values of download background traffic were selected at random in the range 512 kb/s and 1024 kb/s, and the values of upload background traffic were selected at random between 64 kb/s and 128 kb/s. The streaming rate $h = 360$ kb/s was divided proportionally to multicast trees and the number of trees was between 1 and 6. The number of levels was between 2 and 9. Numerical experiments were run on a PC with a

2 GHz processor and 4GB RAM. The EA was tuned to find the best values of input
tuning parameters; detailed results of the tuning process can be found in [45].

Table 4.1 reports results of algorithm comparison obtained for 20-node networks
as a function of the level number. The table includes optimal results yielded by
the CPLEX solver presented in the second column and the average gaps to optimal
results of heuristics GA, LR and EA included in columns 3–5. The algorithms provide
results which are on average 3.64, 2.86 and 1.33 % worse than optimal, respectively.
However, it should be stressed that the optimality gap of the heuristics increases with
the number of trees. This is because with more trees, the solution space increases
and consequently the heuristics become less efficient in searching the larger solution
space. The average execution times for CPLEX, GA, LR and EA were (in seconds)
393, 0.08, 12 and 120, respectively.

Table 4.2 presents a comparison of heuristic algorithms for larger networks with
100 and 300 nodes. For the smaller network, results of GA, LR and EA are reported,
while for the larger network only results of GA and EA are included, since the
LR algorithm is not able to provide results in 1 hour due to high complexity. The
main observation is that for lower numbers of trees, EA outperforms other methods.

**Table 4.1** Comparison of optimization algorithms for the overlay network design problem with
multicast flows for a 20-node network

| Number of trees | Average optimal cost | Average gap to optimal results | | |
|---|---|---|---|---|
| | | GA (%) | LR (%) | EA (%) |
| 1 | 747 | 2.79 | 2.74 | 0.09 |
| 2 | 647 | 5.16 | 2.39 | 0.00 |
| 3 | 627 | 2.45 | 2.14 | 0.85 |
| 4 | 613 | 4.16 | 4.16 | 4.36 |

**Table 4.2** Comparison of heuristic algorithms for the overlay network design problem with multicast flows for 100-node and 300-node networks

| Number of nodes | Number of trees | Average cost | | | Average gap to EA | |
|---|---|---|---|---|---|---|
| | | GA | LR | EA | GA (%) | LR (%) |
| 100 | 1 | 3775 | 3753 | 3612 | 4.27 | 3.68 |
| 100 | 2 | 3551 | 3492 | 3380 | 3.96 | 2.82 |
| 100 | 3 | 3427 | 3306 | 3306 | 3.04 | −0.02 |
| 100 | 4 | 3301 | 3280 | 3366 | −2.30 | −2.88 |
| 300 | 2 | 12184 | – | 10550 | 11.56 | – |
| 300 | 3 | 10465 | – | 10297 | 1.41 | – |
| 300 | 4 | 12517 | – | 10820 | 1.83 | – |
| 300 | 5 | 10740 | – | 10673 | 0.37 | – |
| 300 | 6 | 9970 | – | 10638 | −6.70 | – |

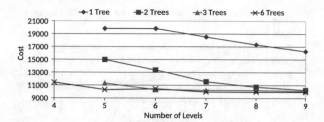

**Fig. 4.1** Overlay network cost as a function of the number of levels and multicast trees

However, as the number of trees increases, GA and LR prevail. Again, this trend is a consequence of the expansion of the solution space and indicates that the EA does not provide good scalability.

The next goal of was to evaluate various scenarios of network design for overlay multicasting. Experiments were conducted using the 300-node network and the GA algorithm. Figure 4.1 presents the overlay network cost as a function of the number of the tree levels and number of trees. A streaming rate of 360 kb/s was proportionally divided to one, two, three and six trees. Increasing the number of trees from one to three can significantly decrease the network cost, while the difference between three and six trees is not significant. The main benefit of using six trees is that a feasible solution exists for four levels, while the required level is five in the case of between one and three trees. The main trend clearly visible in Fig. 4.1 is that as the number of levels increases, the network cost decreases, although the gain declines with the increased number of trees. Accordingly, the key advantage of a high number of tree levels is the expansion of the system capacity measured as the overall streaming rate.

Figure 4.2 reports a scenario where two multicast trees transmit a streaming rate of 360 kb/s divided into (180, 180), (210, 150), (240, 120), and (270, 90). The largest gap between particular scenarios is approx. 20 %. Consequently, the way which the streaming rate is assigned to trees can have major influence on the overlay network cost.

Figure 4.3 shows another scenario related to splitting the stream, i.e., streaming rates of 240, 300 and 360 kb/s assigned to three trees equally. The first observation is

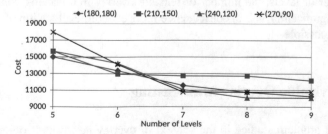

**Fig. 4.2** Overlay network cost as a function of the number of levels for two trees with a range of streaming rate allocations

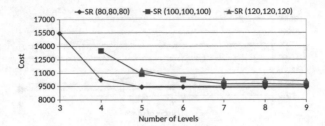

**Fig. 4.3** Overlay network cost as a function of the number of levels for three trees with a range of streaming rate allocations

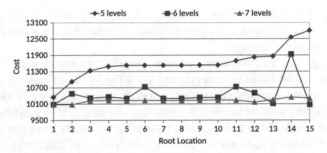

**Fig. 4.4** Overlay network cost as a function of the number of levels and location of the root node

that the higher the streaming rate, the more levels are needed to build feasible trees. Moreover, when the number of levels increases, the difference in the network cost between all three scenarios is reduced. Therefore, we can conclude that when the number of tree levels is not a critical issue, a relatively low additional investment in the overlay network significantly increases the overlay multicasting system capacity expressed as the overall streaming rate.

Finally, Fig. 4.4 plots the influence of the root location on the overlay network cost. The streaming rate of 360 kb/s is transmitted proportionally using three multicast trees. We randomly selected 15 locations of the root. The main observation is that as the number of levels increases, the impact of root location decreases; for five levels, the largest gap between results obtained for two different root locations is 19.19 %, while for seven levels the corresponding gap is only 2.61 %. This is because with a high number of levels, the number of feasible overlay multicasting configurations increases significantly, and in consequence more economic allocation of access link capacity is possible.

## 4.3   Survivable Overlay Multicasting

The second problem studied in the context of overlay networks is optimization of multicast flows with additional survivability requirements. The concept of overlay multicasting can be used to deliver critical information to end users, e.g., public

**Fig. 4.5** Example of survivable overlay multicasting with streaming server protection

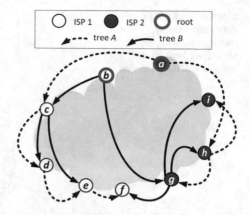

security notifications, surveillance, software updates, weather forecasts, hurricane warnings, traffic information, etc. To guarantee an uninterrupted overlay multicasting service, the approach of *survivable overlay multicasting* is proposed in [38]. Following the concept of DPP 1+1 protection used in optical networks [59], two (or more) failure disjoint multicast trees are established in the overlay network. All trees deliver the same information. In the event of a network failure, at least one of the trees should remain unaffected by the failure in order to enable uninterrupted transfer of critical content to all receivers. For ease of reference, the aforementioned protection method developed for multicast trees is referred to as *Dedicated Tree Protection* (DTP). To address specific features of overlay networks, the following failure scenarios are considered in the context of survivable overlay multicasting: streaming server failure, overlay link failure, uploading node failure and ISP (Internet Service Provider) link failure. Note that an idea similar to survivable overlay multicasting for reliable video broadcast in optical networks is presented in [60]. Moreover, papers [15, 61] analyze the resilience of P2P and overlay multicasting in the context of dynamic optimization.

Figures 4.5 and 4.6 present a simple example illustrating the concept of survivable overlay multicasting with DTP. The overlay network consist of nine nodes, namely, $a, b, c, d, e, f, g, h$, and $i$. Two multicast trees are established in the network; tree $A$ uses $a$ at the root node, while tree $B$ starts at node $b$. Nodes $c, d, e, f, g, h$, and $i$ are receivers which must be connected to both trees. However, these nodes belong to two ISPs, i.e., nodes $c, d, e$, and $f$ are assigned to ISP1, while nodes $g, h$, and $i$ belong to ISP2. Note that in Fig. 4.5 tree $A$ has four levels of uploading nodes (the longest path ($a, c, d, e$, and $f$) includes four hops), while tree $B$ has only two levels.

The example shown in Fig. 4.5 only provides protection against a streaming node failure, since the configuration of trees is vulnerable to other possible network failures. For instance, overlay link $(c, d)$ is included in both trees and when this link is broken, node $d$ is disconnected. Node $c$ is an uploading node in both trees and a failure of $c$ once again disconnects some network nodes from both trees. Lastly, the link connecting ISP 1 and ISP 2 is shared by both trees, i.e., overlay link $(a, c)$ of tree $A$ and overlay link $(b, g)$ of tree $B$ share the link between ISPs.

**Fig. 4.6** Example of
survivable overlay
multicasting with streaming
server, overlay link and
uploading node protection

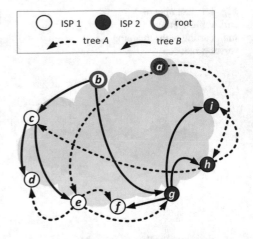

Figure 4.6 illustrates a configuration of trees which provides additional protection
against a single failure of an overlay link or an uploading node. Since there are only
two ISPs, it is impossible to construct a multicast configuration resilient to the ISP
link failure.

### 4.3.1 Formulations

The survivable overlay multicasting optimization problem concerns the flow allocation only, i.e., an existing overlay network with fixed access link capacities is
assumed. The optimization involves optimizing multicast flows with additional survivability constraints. We present two basic models ensuring resilience against a
streaming server failure and based on the flow and level formulations. Next, constraints to impose additional survivability requirements are introduced. The default
objective function is a routing cost, and two other functions are accounted for, namely,
maximum delay and throughput [38, 39, 57]. The notation and modeling is analogous
to Sect. 4.2.1. Note that the problem considered in this section is $\mathcal{NP}$-complete,
since it is equivalent to the Hop-Constrained Minimum Spanning Tree Problem [44].

**Flow Formulation**

The basic formulation of the survivable overlay multicasting problem addresses the
simplest survivability scenario, in which the system is protected against streaming
server failure only. Additional constraints to ensure other survivability scenarios of
overlay link failure, uploading node failure and ISP link failure are formulated below.

---

**OVR/M/FA/DTP/Cost/Flow**
**sets**

| | |
|---|---|
| $V$ | nodes (peers) |
| $L$ | levels |
| $T$ | disjoint trees |
| $R(t)$ | receivers of tree $t$ |

**constants**

| | |
|---|---|
| $d_v$ | download capacity of node $v$ (kb/s) |
| $u_v$ | upload capacity of node $v$ (kb/s) |
| $\zeta_{vw}$ | unit routing cost on link $(v, w)$ |
| $s_t$ | streaming node of tree $t$ |
| $h$ | tree streaming rate of (kb/s) |

**variables**

$x_{vwtr}$  =1, if in tree $t$ the streaming path from the root to receiver $r$ includes overlay link from node $v$ to node $w$ (no other overlay nodes in between); 0, otherwise (binary)

$x_{vwt}$  =1, if the overlay link from node $v$ to node $w$ (no other overlay nodes in between) is included in tree $t$; 0, otherwise (binary)

**objective**

$$\text{minimize} \quad F = \sum_{v \in V} \sum_{w \in W} \sum_{t \in T} \zeta_{vw} x_{vwt} \tag{4.3.1a}$$

**constraints**

$$\sum_{v \in V} x_{vwtr} - \sum_{v \in V} x_{wvtr} = \begin{cases} +1 & \text{if } v = r \\ -1 & \text{if } v = s_t, \\ 0 & \text{otherwise} \end{cases} \quad w \in V, t \in T, r \in R(t) \tag{4.3.1b}$$

$$x_{vwtr} \leq x_{vwt}, \quad v \in V, w \in V, t \in T, r \in R(t) \tag{4.3.1c}$$

$$\sum_{v \in V} \sum_{t \in T} x_{vwt} h \leq d_w, \quad w \in V \tag{4.3.1d}$$

$$\sum_{w \in V} \sum_{t \in T} x_{vwt} h \leq u_v, \quad v \in V \tag{4.3.1e}$$

$$\sum_{v \in V} \sum_{w \in V} x_{vwtr} \leq |L|, \quad t \in T, r \in R(t). \tag{4.3.1f}$$

The objective (4.3.1a) is to minimize the overall routing (streaming) cost considering all trees. Condition (4.3.1b) is the flow conservation constraint controlled for all receivers and trees. Inequality (4.3.1c) ensures the coupling between flow variables $x_{vwtr}$ and variables $x_{vwt}$ that define links included in every tree $t \in T$. Conditions (4.3.1d) and (4.3.1e) define the download and upload access link capacity constraints, respectively. Finally, constraint (4.3.1f) imposes the upper limit on the tree.

### Level Formulation

To write an optimization model of the survivable overlay multicasting problem based on the level formulation, model (4.2.2) presented in Sect. 4.2.1 must be modified slightly. Specifically, variable $x_{vwtl}$ is continuous here and denotes the flow allocated on overlay link $(v, w)$ in tree $t$ and node $v$ is located on level $l$.

---

### OVR/M/FA/DTP/Cost/Level
### variables (additional)

$x_{vwtl}$   flow on overlay link from node $v$ to node $w$ (no other overlay nodes in between) in multicast tree $t$ and node $v$ is located on level $l$ (continuous, non-negative)

### objective

$$\text{minimize} \quad F = \sum_{v \in V} \sum_{w \in W} \sum_{t \in T} \zeta_{vw} x_{vwt} \tag{4.3.2a}$$

### constraints

$$\sum_{v \in V} \sum_{l \in L} x_{vs_t tl} = 0, \quad t \in T \tag{4.3.2b}$$

$$\sum_{v \in V} \sum_{l \in L} x_{vrtl} = h, \quad t \in T, r \in R(t) \tag{4.3.2c}$$

$$\sum_{w \in V} x_{vwt1} = 0, \quad t \in T, v \in W \backslash \{s_t\}, \tag{4.3.2d}$$

$$x_{vwtl} \leq \sum_{u \in V} x_{uvt(l-1)}, \quad v \in V, w \in V, t \in T, l \in L \backslash \{1\} \tag{4.3.2e}$$

$$\sum_{v \in V} \sum_{t \in T} \sum_{l \in L} x_{vwtl} \leq d_w, \quad w \in V \tag{4.3.2f}$$

$$\sum_{w \in V} \sum_{t \in T} \sum_{l \in L} x_{vwtl} \leq u_v, \quad v \in V \tag{4.3.2g}$$

$$\sum_{l \in L} x_{vwtl} \leq h x_{vwt}, \quad v \in V, w \in V, t \in T \tag{4.3.2h}$$

$$x_{vwt} \leq \sum_{l \in L} x_{vwtl}, \quad v \in V, w \in V, t \in T \tag{4.3.2i}$$

$$\sum_{v \in V} x_{vwt} = 1, \quad w \in R(t), t \in T. \tag{4.3.2j}$$

The criterion function (4.3.2a) is the overall routing (streaming) cost. Condition (4.3.2b) imposes that for every tree $t \in T$ the root node of the tree cannot download any traffic related to tree $t$. Equation (4.3.2c) ensures that each receiving node must download the requested stream in every tree. Constraint (4.3.2d) meets the requirement that only the root node can be located on level $l = 1$. Constraint (4.3.2e) states that node $v$ cannot upload to any other node $w$ located on level $(l + 1)$ more than node $v$ downloads on level $l$. Conditions (4.3.2f) and (4.3.2g) set up the access link capacity constraints. Inequalities (4.3.2h) and (4.3.2i) bind variables $x_{vwtl}$ and $x_{vwt}$, respectively. The last condition (4.3.2j) guarantees that each receiving node has exactly one parent node in a particular tree, thus ensuring that all trees include exactly $|R(t)|$ links.

**Survivability Constraints**

Models (4.3.1) and (4.3.2) protect the overlay multicasting against a streaming server failure only. To protect the overlay multicasting against a single failure of the overlay link, uploading node or ISP link, additional constraints are required. Note that the following constraints are applicable to both (4.3.1) and (4.3.2) models. The first model ensures that the created trees are link disjoint.

**OVR/M/FA/DTP/Link Disjoint**
**constraints (additional)**

$$\sum_{t \in T} (x_{vwt} + x_{wvt}) \leq 1, \quad v \in V, w \in V. \tag{4.3.3a}$$

It is assumed that in the case of a failure of an overlay link between nodes $v$ and $w$, both directed links $(v, w)$ and $(w, v)$ are broken. This follows from the fact that a failure in an overlay network mainly impacts the transfer in both directions.

The next model accounts for the uploading node failure, i.e., additional constraints ensure that the trees are node disjoint. A new variable $x_{vt}$ denotes whether node $v$ uploads traffic in tree $t$.

**OVR/M/FA/DTP/Node Disjoint**
**constants (additional)**

$M$ large number

**variables (additional)**

$x_{vt} = 1$, if node $v$ uploads traffic in tree $t$; $0$, otherwise (binary)

**constraints (additional)**

$$\sum_{w \in V} x_{vwt} \leq M x_{vt}, \quad v \in V, t \in T \tag{4.3.4a}$$

$$\sum_{t \in T} x_{vt} \leq 1, \quad v \in V. \tag{4.3.4b}$$

The last model concerns the ISP link failure. As in the case of the overlay link failure, ISP link failure includes the failure of both directed links connecting a pair of ISPs. A new binary variable $z_{ijt}$ denotes whether there is at least one link between a node located in ISP $i$ and a node located in ISP $j$ in tree $t$.

**OVR/M/FA/DTP/ISP Disjoint**
**sets (additional)**

$I$        Internet Service Providers (ISPs)
$\Upsilon(i)$   nodes belonging to ISP $i$

**variables (additional)**

$z_{ijt}$   $=1$, if in tree $t$ there is at least one link between a node located in ISP $i$ and a
        node located in ISP $j$; 0, otherwise (binary)

**constraints (additional)**

$$\sum_{v\in\Upsilon(i)} \sum_{w\in\Upsilon(j)} (x_{vwt} + x_{wvt}) \le M z_{ijt}, \quad i \in I, j \in I, t \in T \tag{4.3.5a}$$

$$\sum_{t\in T} z_{ijt} \le 1, \quad i \in I, j \in I. \tag{4.3.5b}$$

**Maximum Delay and Throughput Objective Functions**

Models (4.3.1) and (4.3.2) use the routing cost as the optimization goal. This sub-section shows how to write corresponding optimization problems with two other objective function which arise naturally in the context of overlay multicasting, i.e., maximum delay and throughput.

The maximum delay function in the context of overlay multicasting is defined as the maximum transmission delay which occurs on a routing path established from the root (streaming server) to a receiver [39, 62–64]. To formulate the maximum delay function, we first assume that for each overlay link $(v, w)$ constant $\zeta_{vw}$ denotes a communication delay on this link given in milliseconds. For each receiving node $r \in R(t)$ and tree $t \in T$, the delay can be calculated as a sum of delays of all links included in the routing path from root node $s_t$ to node $r$ using formula $\sum_{v\in V} \sum_{w\in W} x_{vwtr} \zeta_{vw}$. The optimization goal is to minimize the maximum value of this delay over all receiving nodes and all trees. To formulate the survivable overlay multicasting problem with the maximum delay objective, only flow formulation (4.3.1) is applicable, since level formulation (4.3.2) does not include detailed information on each routing path between root and receiver.

**OVR/M/FA/DTP/Maximum Delay/Flow**
**constant (additional)**

$\zeta_{vw}$   delay on the overlay link from node $v$ to node $w$

**variables (additional)**

$x$   maximum delay (continuous, non-negative)

**objective**

$$\text{minimize} \quad F = x \tag{4.3.6a}$$

**constraints** (4.3.1b)–(4.3.1f) **and**

$$\sum_{v \in V} \sum_{w \in W} x_{vwtr} \zeta_{vw} \leq x, \quad t \in T, r \in R(t). \tag{4.3.6b}$$

The throughput function aims to maximize the aggregate receiving rate at each participating node [19, 38, 39]. To formulate the survivable overlay multicasting problem, we use the level formulation (4.3.2). However, note that flow formulation (4.3.1) may also be applied in this context, although supplementary flow variables are necessary, which further complicates the model. The main modifications compared to model (4.3.2) is that $h$ denoting the streaming rate of the tree is now used as a variable. Using this technique, all constraints of the model (4.3.2) remain unchanged, although it should be pointed out that the right-hand side of (4.3.2c) is now not a constant, but a variable.

**OVR/M/FA/DTP/Throughput/Level**
**variables (additional)**

$h$   streaming rate (continuous, non-negative)

**objective**

$$\text{maximize} \quad F = h \tag{4.3.7a}$$

**constraints** (4.3.2b)–(4.3.2j)

Note that additional survivability constraints defined in (4.3.3), (4.3.4) and (4.3.5) can be combined with the maximum delay model (4.3.6) and the throughput model (4.3.7).

### 4.3.2  Numerical Results

The CPLEX solver [58] is used to solve the optimization models proposed in Sect. 4.3.1 in the context of the survivable overlay multicasting optimization problem. As the goal was to obtain optimal results in a reasonable time, the size of overlay networks was limited to 20 nodes. Twelve test scenarios were generated at random

with the following assumptions: nodes either have symmetric access links (2048 or
4096 kb/s) or asymmetric access links (2048/256 or 4096/512 or 6144/768 kb/s),
each node is assigned to one of five ISPs, the link costs is in the range 3–109, and
two trees ($|T| = 2$) are used to stream with a bit-rate of $h = 256$ kb/s [38, 39].

The first goal of the numerical experiments was to compare flow (4.3.1) and level
(4.3.2). The CPLEX execution time was limited to 1 hour, so it was possible for the
solver to yield a solution without the optimality guarantee in some cases. The CPLEX
was run for both formulations and four failure scenarios referred to as SD (streaming
server disjoint), LD (overlay link disjoint), ND (uploading node link disjoint) and
ID (ISP link disjoint).

Table 4.3 presents results of the comparison of both models as a function of the
number of levels and survivability scenario. The third column denotes average cost
(value of objective function). Columns 4 and 5 include the average execution time
of each model is seconds. In turn, columns 6 and 7 present the number of cases
where a particular model reached the 1 hour time limit (i.e., the solution does not
have the optimality guarantee). The last column shows the average cost difference
between the level and flow formulation, i.e., a value greater than 0 denotes that the

**Table 4.3** Comparison of level and flow formulations of the survivable overlay multicasting optimization problem

| Number of levels | Failure scenario | Average cost | Execution time (s) | | 1h limits | | Cost difference (%) |
|---|---|---|---|---|---|---|---|
| | | | Flow | Level | Flow | Level | |
| Root nodes in separate ISPs | | | | | | | |
| 2 | SD | 1035 | 57.5 | 4.3 | 0 | 0 | 0.00 |
| 2 | LD | 1095 | 88.0 | 7.0 | 0 | 0 | 0.00 |
| 2 | ND | 1054 | 75.2 | 10.1 | 0 | 0 | 0.00 |
| 2 | ID | 1480 | 114.3 | 14.8 | 0 | 0 | 0.00 |
| 4 | SD | 671 | 736.2 | 1229.1 | 0 | 1 | 0.00 |
| 4 | LD | 707 | 1467.0 | 1192.2 | 4 | 2 | 0.75 |
| 4 | ND | 690 | 653.2 | 606.5 | 0 | 0 | 0.00 |
| 4 | ID | 838 | 2333.6 | 1386.9 | 3 | 0 | 0.59 |
| Root nodes in the same ISP | | | | | | | |
| 2 | SD | 1086 | 9.9 | 5.6 | 0 | 0 | 0.00 |
| 2 | LD | 1176 | 100.2 | 10.9 | 0 | 0 | 0.00 |
| 2 | ND | 1119 | 9.8 | 10.7 | 0 | 0 | 0.00 |
| 2 | ID | 2023 | 100.0 | 43.5 | 0 | 0 | 0.00 |
| 4 | SD | 784 | 341.1 | 872.3 | 0 | 2 | 0.00 |
| 4 | LD | 836 | 1244.8 | 883.9 | 4 | 1 | 1.05 |
| 4 | ND | 807 | 794.6 | 638.2 | 0 | 0 | 0.00 |
| 4 | ID | 984 | 3600.0 | 2391.2 | 12 | 4 | 14.69 |

level formulation provides a better (lower) value of the objective function. The last column is connected to the corresponding results in columns 6–7. Note that each row presents the average results of 12 test scenarios. Table 4.3 contains results run for two different cases of the root (streaming server) location, namely, roots are located in separate ISPs and both roots belong to the same ISP.

Results included in Table 4.3 confirm the complexity of the models analyzed in Sect. 1.2.3. In particular, the level formulation has a significantly shorter execution time than the flow formulation, especially for lower numbers of levels. On average, the level model is approx. 9.25 and 3.11 times faster than the flow model for separate and common IPSs scenarios, respectively. The corresponding values observed for four levels are 1.18 and 1.25, respectively. In addition, only in ten of 192 cases the level model does not finish the calculation within 1 hour, while the corresponding number for the flow model is 23 and for four levels the reported numbers were larger than for two levels. These observations are consistent with the fact that the level model minimizes the number of variables according to the level limit.

Comparing the performance for different numbers of levels, increasing the number of levels reduces the system cost by approx. 37 % taking into account all survivability scenarios. Moreover, the average execution time of both models increases with the increase of levels. This data can be explained by the fact that fewer levels implies a smaller solution space and, in consequence, a lower execution time.

Concerning the comparison between particular survivability scenarios, the SD case protecting the network against streaming server failure only yields the lowest cost, while the ID case protecting the network against ISP link failure returns the highest cost, with the average gap from 20 to 46 %. Two other scenarios, i.e., LD and ND, return only slightly worse performance than the SD case. In terms of execution time, the simplest SD scenario is usually the fastest model, while the additional survivability constraints added to models (LD, ND and ID) increases the execution time in most cases.

The next goal of the experiments was to compare all four failure scenarios and investigate how the additional survivability constraints impact the objective functions of routing cost, maximum delay and throughput. Table 4.4 presents the average percentage value of additional overheads of particular objective functions imposed by the fact that the basic SD model is enhanced with supplementary survivability requirements. Note that due to low scalability of the optimization model, in the case of the maximum delay function, the table includes results obtained for a smaller 15-node network. A general trend noticeable for all cases is that the LD and ND models provide results close to the reference SD model. A much larger gap is observed for the ID model.

Figure 4.7 shows the routing cost as a function of the number of levels and survivability scenario. The main observation is that the cost decreases as the number of levels increases. The largest gap is identified between two and three levels. Next, starting from five levels, the cost converges to a stable value for each curve. The differences between particular survivability scenarios are comparable to the trends presented in Table 4.3.

**Table 4.4** Overheads of using additional survivability constraints for the objective functions of routing cost, maximum delay and throughput

| Number of levels | Root nodes in separate ISPs | | | Root nodes in the same ISP | | |
|---|---|---|---|---|---|---|
| | LD (%) | ND (%) | ID (%) | LD (%) | ND (%) | ID |
| Routing cost | | | | | | |
| 2 | 7.03 | 3.18 | 30.97 | 7.83 | 2.84 | 46.54 |
| 4 | 7.90 | 5.79 | 22.80 | 7.76 | 2.95 | 19.86 |
| Maximum delay | | | | | | |
| 2 | 1.75 | 0.29 | 29.91 | 2.19 | 1.18 | 39.83 |
| 4 | 0.83 | 0.22 | 28.05 | 2.98 | 0.12 | 32.63 |
| Throughput | | | | | | |
| 2 | 0.03 | 4.30 | 2.79 | 0.00 | 4.31 | 18.40 |
| 4 | 0.03 | 4.30 | 2.78 | 0.03 | 4.30 | 18.39 |

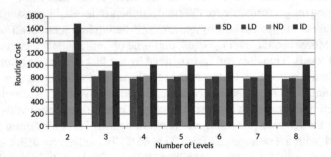

**Fig. 4.7** Routing cost as a function of number of levels and survivability scenario

Figure 4.8 presents the routing cost as a function of the streaming rate $h$ and survivability scenario. When the streaming rate decreases, the routing cost drops However, it should be noted that the streaming rate of 256 kb/s is a kind of a threshold, i.e., the routing cost is reduced dramatically at this point. This follows from the fact that the upload link capacity of some nodes is only 256 kb/s. In consequence, if the streaming rate is above 256 kb/s, these nodes cannot upload and other, generally more expensive overlay links are used in the trees. Again, the difference of costs observed for various survivability scenarios is largely consistent with the results reported in Table 4.3.

In [39, 57] two metaheuristic algorithms solving the survivable overlay multicasting problem, i.e., Evolutionary Algorithm and Tabu Search, are presented and evaluated. Papers [38, 39] report results of dynamic routing experiments run to evaluate how additional survivability constraints guaranteeing failure-disjoint trees impact the performance of overlay multicasting.

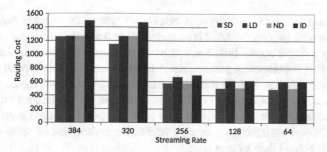

**Fig. 4.8** Routing cost as a function of the streaming rate and survivability scenario

## 4.4 Overlay Multicasting with Dual Homing Protection

This section concentrates on the optimization of overlay networks with multicast flows protected by the concept of *dual homing*. Research into the idea of multihoming began in the late 1980s, e.g., [65–68]. *Multihoming* can be defined as a network architecture in which a user is attached to more than a single node of the network. The main goal of using multihoming is to increase network survivability by a user—being connected to more than one network node—being equipped with several routing paths. In consequence, if a network failure impacts one of the available routing paths, the information can still be transmitted to/from the user. A situation where the number of connections (homes) that a user is equipped with is limited to two is called *dual homing*. Note that some works have focused on the application of the dual homing approach to protect multicast flows realized in lower network layers e.g., [69–72]. The idea of using dual homing protection (DHP) in the context of overlay multicasting was proposed in [73, 74] and further elaborated in [56, 75–77].

In general, the scenario considered in this section is similar to the concept of survivable overlay multicasting described in Sect. 4.3 and presented in Figs. 4.5 and 4.6. In particular, two disjoint multicast trees are used to transmit the same information in order to provide survivability guarantees. Here, the main modification following from the dual homing architecture is that it is assured that an access link can carry the flow of only one of the trees. In other words, a node having two access links (homes) uses one of the links to transmit the data of the first multicast tree, while the second link transfers the data of the second multicast tree. In consequence, a single failure of one access link affects only one of the trees, and the second tree still can provide the service.

### 4.4.1 Formulation

The general notation is the same as in Sect. 4.3. However, in order to model dual homing some new notation is required. More specifically, to model the fact that every node is connected to the overlay network by the use of two access links, a single *primal* node with two access links is modeled by two *virtual* nodes, i.e., $v$ and $\sigma(v)$, each having a single access link. Each primal node consisting of virtual nodes $v$ and $\sigma(v)$ has in fact four capacity parameters: constants $u_v$ and $d_v$ are respectively upload and download capacity of node $v$ and constants $u_{\sigma(v)}$ and $d_{\sigma(v)}$ are the corresponding parameters of the virtual node $\sigma(v)$. It is assumed that set $V$ includes virtual nodes, and thus the number of nodes in $V$ is two times higher than the number of primal nodes. Additional variable $x_{vt}$ denotes whether the access link of node $v$ is used to download or upload flow of tree $t$.

The model presented model ensures a basic protection against a single access link failure. Additional constraints to assure other survivability scenarios can be added in an analogous way, as in Sect. 4.3.1.

---

**OVR/M/FA/DHP/Cost/Flow**

**sets**

| | |
|---|---|
| $V$ | virtual nodes (peers) |
| $L$ | levels |
| $T$ | disjoint trees |
| $R(t)$ | receivers of tree $t$ |

**constants**

| | |
|---|---|
| $d_v$ | download capacity of node $v$ (kb/s) |
| $u_v$ | upload capacity of node $v$ (kb/s) |
| $\zeta_{vw}$ | unit routing cost on link $(v, w)$ |
| $s_t$ | streaming node of tree $t$ |
| $h$ | tree streaming rate of (kb/s) |
| $\sigma(v)$ | node associated with node $v$ (virtual node) |
| $M$ | large number |

**variables**

| | |
|---|---|
| $x_{vwtr}$ | =1, if in tree $t$ the streaming path from the root to receiver $r$ includes overlay link from node $v$ to node $w$ (no other overlay nodes in between); 0, otherwise (binary) |
| $x_{vwt}$ | =1, if overlay link from node $v$ to node $w$ (no other overlay nodes in between) is included in tree $t$; 0, otherwise (binary) |
| $x_{vt}$ | =1, if access link of node $v$ is used to download or upload flow of tree $t$; 0, otherwise (binary) |

**objective**

$$\text{minimize} \quad F = \sum_{v \in V} \sum_{w \in W} \sum_{t \in T} \zeta_{vw} x_{vwt} \tag{4.4.1a}$$

**constraints**

$$x_{vt} + x_{\sigma(v)t} = 1, \quad v \in V, t \in T \tag{4.4.1b}$$

$$\sum_{v \in V} \sum_{t \in T} x_{vwt} + \sum_{v \in V} \sum_{t \in T} x_{wvt} \leq M x_{wt}, \quad w \in V \tag{4.4.1c}$$

$$x_{v\sigma(v)t} = 0, \quad v \in V, t \in T \tag{4.4.1d}$$

$$\sum_{v \in V} x_{vwtr} - \sum_{v \in V} x_{wvtr} = \begin{cases} +x_{vt} & \text{if } v = r \\ -x_{vt} & \text{if } v = s_t, \quad w \in V, t \in T, r \in R(t) \\ 0 & \text{otherwise} \end{cases} \tag{4.4.1e}$$

$$x_{vwtr} \leq x_{vwt}, \quad v \in V, w \in V, t \in T, r \in R(t) \tag{4.4.1f}$$

$$\sum_{v \in V} \sum_{t \in T} x_{vwt} h \leq d_w, \quad w \in V \tag{4.4.1g}$$

$$\sum_{w \in V} \sum_{t \in T} x_{vwt} h \leq u_v, \quad v \in V \tag{4.4.1h}$$

$$\sum_{v \in V} \sum_{w \in V} x_{vwtr} \leq |L|, \quad t \in T, r \in R(t). \tag{4.4.1i}$$

The objective denotes (4.4.1a) the overall routing cost. Constraints (4.4.1b)–(4.4.1c) are used to define variable $x_{vt}$. In more detail, due to the dual homing protection, a single virtual node $v$ can belong (download or upload) only to one multicast tree, while at the same time node $\sigma(v)$ belongs to the second multicast tree. This requirement is assured by equality (4.4.1b). Inequality (4.4.1c) guarantees that variable $x_{vt}$ is switched on if node $v$ downloads or uploads in tree $t$. Constraint (4.4.1d) imposes that there cannot be a transmission within any primal node between nodes $v$ and $\sigma(v)$. Condition (4.4.1e) defines the flow conservation law for unicast paths from the root node to each receiver. The key modification compared to constraint (4.3.1b), is that in the right-hand side, i.e., variable $x_{vt}$ is used instead of 1, since due to the concept of virtual nodes, node $v \in V$ is connected to only one tree. Remaining constraints are the same as in the model (4.3.1).

Model (4.4.1) implements a basic survivability scenario when two streaming trees are disjoint with respect to dual homes (two disjoint access links connecting the same node to the overlay network). In order to account for additional survivability requirements on the overlay multicasting and to make the streaming trees overlay link disjoint, uploading node disjoint or ISP link disjoint, the additional constraints defined in (4.3.3), (4.3.4) or (4.3.5) must be added to the model (4.4.1), respectively. For more details, check [75–77]. Moreover, once again the considered models are $\mathcal{NP}$-complete, since they can be reduced to the Hop-Constrained Minimum Spanning Tree Problem [44].

Note that on the base of models (4.3.6) and (4.3.7), the DHP overlay multicasting problem with maximum delay and throughput objective functions respectively, can be formulated. For more details, see [56]. Moreover, references [76, 77] include ILP formulations and corresponding results related to the overlay network design problem with dual homing protection. In turn, papers [56, 74–76] present and evaluate metaheuristic algorithms based on TS and SA approaches which are proposed to solve various optimization problems for overlay multicasting with dual homing protection.

## 4.4.2  Numerical Results

The first goal of the numerical experiments was to compare the performance of two ILP formulations of the multicast flow allocation problem in an overlay network with dual homing protection—namely, flow formulation (4.4.1) and the corresponding level formulation [74]. To solve ILP models in an optimal way, the CPLEX solver [58] was applied. The tested overlay networks were generated at random with the following assumptions: 20 primal nodes; 2 disjoint trees; link costs in ranges 1–20, 1–50, 1–100, 1–200; access link capacity either symmetric (1024/1024 or 512/512 kb/s) or asymmetric (1024/512 or 512/384 or 512/256 or 512/128 kb/s); streaming rate 128 kb/s, levels in rage 2–20. The execution time of CPLEX was limited to 3600 s.

Table 4.5 summarizes results related to the comparison of both optimization models in terms of the execution time. Note that each presented value is an average of 5 experiments run on different networks and the tested scenarios are divided into 4 groups according to the values of link costs. The presented running times belong to three categories. Firstly, term 'INF' indicates that CPLEX using a particular formulation was not able to find feasible results within the 3600 s time limit. Secondly, value '3600' denotes a case when CPLEX found a feasible solution but without optimality guarantees. Thirdly, a value smaller than 3600 implies that CPLEX yielded an optimal solution with an execution time smaller than 1 h. The results clearly show that the level model outperforms the flow model. However, the gap between both

**Table 4.5**  Comparison of ILP formulations performance for the multicast flow allocation problem in overlay network with dual homing protection—average execution time in seconds

| Number of levels | Link cost 1–20 | | Link cost 1–50 | | Link cost 1–100 | | Link cost 1–200 | |
|---|---|---|---|---|---|---|---|---|
|  | Level | Flow | Level | Flow | Level | Flow | Level | Flow |
| 2 | 3 | INF | 1 | INF | 1 | 3600 | 3 | 3600 |
| 3 | 3 | INF | 8 | INF | 5 | 3600 | 7 | 3600 |
| 4 | 8 | INF | 13 | INF | 10 | 3600 | 14 | 3600 |
| 5 | 20 | INF | 28 | 3600 | 19 | 3600 | 19 | 3400 |
| 10 | 82 | 1439 | 171 | 3600 | 276 | 1009 | 201 | 1221 |
| 15 | 361 | 1581 | 3600 | 3168 | 442 | 1155 | 556 | 1316 |
| 20 | 589 | 1225 | 3600 | 2414 | 1150 | 913 | 971 | 962 |

formulations decreases with the increase of the level number and the larger range of link costs. The flow model for a small value of the level number was either not able to yield a feasible solution or was not able to find an optimal solution. On the contrary, the level model succeeds finding the optimum value for almost every case. However, the execution time of the level formulation grows with the increase of the levels. The general conclusion is consistent with the results presented in Sect. 4.3.2 in the context of the survivable overlay multicasting with DTP, i.e., in the case of a smaller number of levels, the level model provides much better performance than the flow model. Moreover, when the number of allowed levels becomes larger, the flow model can be more efficient.

The second goal of the experiments was to examine how additional survivability constraints—namely, overlay link disjoint (LD), uploading node disjoint (ND) and ISP link disjoint (ID)—impact the performance of overlay multicasting protected by the dual homing approach [56, 77]. Overlay networks were generated at random assuming: link costs in range 1–50; streaming rate of 256 kb/s; level limit L in range from 2 to 5; asymmetric link capacity nodes (1024/512, 2048/512, 6144/512, 10240/1024, 20480/1024, 51200/2048 kb/s); 5 ISPs. Three types of optimization problems were examined: flow allocation with routing cost, flow allocation with maximum delay and network design with capacity cost. Table 4.6 reports the overhead of using additional survivability constraints. The largest gaps are observed for the flow allocation problem with routing cost function, while the smallest ones are obtained in the case of the network design problem with capacity cost. The reason for this lies in the fact that the additional survivability constraints directly affect the multicast trees' configuration, and accordingly the routing cost function is also affected to a large extent. The maximum delay function in the lower extent depends on the detailed configuration of the trees. Finally, in the context of the network design problem, a change in the configuration of the multicast trees influences the objective function only in an indirect way. Moreover, the fact that link capacity is allocated in some modular units, not in a continuous way, must be accounted for, i.e., in many case a

**Table 4.6** Overhead of using additional survivability constraints for objective function: routing cost, maximum delay and capacity cost

| Optimization problem | Objective function | Number of nodes | Failure scenario | | |
|---|---|---|---|---|---|
| | | | LD (%) | ND (%) | ID (%) |
| Flow allocation | Routing cost | 10 | 22.8 | 29.0 | 33.5 |
| Flow allocation | Routing cost | 20 | 15.0 | 27.5 | 25.3 |
| Flow allocation | Maximum delay | 10 | 12.4 | 0.1 | 22.7 |
| Network design | Capacity cost | 10 | 2.7 | 2.9 | 4.1 |

change of tree configuration due to the additional survivability constraint does not change the assigned link capacities at all, since residual capacity is facilitated to establish extra links.

For more results on dual homing protection including comprehensive evaluation of heuristic algorithms the reader is referred to [56, 73–77].

## 4.5   Overlay Computing System for Machine Learning Tasks

In recent years, due to the tremendous growth in popularity of various IT services commonly applied in both industry and academia, many institutions have begun to collect huge amounts of data, which cannot by efficiently analyzed by a human being without the support of dedicated algorithms. In fact, traditional methods of data analysis are not sufficient to answer the new challenges and increasing competition, since they do not allow the data to be used professionally for smart decisions, for which the complex knowledge hidden in data is necessary. Therefore, a rapid progress in the development of machine learning approaches, especially knowledge data discovery and data mining methods, has been observed recently. Moreover, the amount of currently collected data means that time-efficient parallel machine learning methods implemented in a distributed network computing environment are necessary. For this reason, this section is related to the optimization of overlay computing systems. In particular, we focus on a k-nearest neighbors pattern recognition algorithm as a representative example of a machine learning method, which is often adapted to distributed computing. Note that the various machine learning approaches are commonly applied to a wide range of practical tasks, e.g., credit approval, fraud detection, prediction of customer behavior, designing of IT security systems like IPS/IDS (Intrusion Prevention System/Intrusion Detection System), medical diagnosis [78–81].

The main goal of a classification task—known also as pattern recognition—is to place an object included in the database into a predefined category, according to the features describing the object. Various approaches have been proposed to for the construction of efficient classifiers, e.g., neural networks, statistical learning, and symbolic learning. The *k-nearest neighbors rule* proposed in [82] is one of the most fundamental and simple classification algorithms. The minimal distance classification is attractive from both a theoretical and practical point of view. The main idea behind the k-nearest neighbors algorithm is to classify an object by a majority vote of its neighbors included in the same database (data set used for classification). Typically, a Euclidean distance formula is used to calculate the distance metric between objects. The computational load of a single object classification with the use of the k-nearest neighbors algorithm is $O(n\,d)$, where $d$ denotes the dimension of the features vector and $n$ denotes the number of training samples. Consequently, the process of classification is time consuming, especially if $n$ is large. However, the potential advantage of the k-nearest neighbors algorithm is that the processing can be parallelized and processed in a distributed computing environment, since the original database can be split into $k$ partitions stored and processed in various sites. In more detail, for each partition of data sets, a decision about the object classification is made independently. Next, combining the aforementioned $k$ decisions (one decision for each partition), the final decision is determined using a fusion method such as majority voting or weighted voting [79, 80]. For more information on various aspects of machine learning check [78, 81, 83] and the references therein.

The general architecture of an overlay computing system design for performing machine learning tasks was proposed in [84, 85] and further applied in the context of various optimization problems in papers [79, 80, 86–88]. Thereafter, survivability aspects of overlay computing systems have been examined in [89–92] (for more details see Sect. 4.6).

### 4.5.1 Formulation

The optimization model presented below was formulated for the first time by Walkowiak et al. in [80]. Let set $R$ include all databases that are available in the considered computing system. Each $r \in R$ represents a separate database including classification data (training samples) used in the k-nearest neighbors method. Moreover, each database is divided into uniform units including a particular number of individual training samples. It is assumed that each unique unit requires the same processing power to be analyzed. Let $n_r$ be the number of units in database $r \in R$. Grouping various units in separate sets, the original database can be divided into several partitions assigned to different computing nodes for processing.

The considered computing system works on the top of an overlay network. Overlay nodes included in set $V$ represent some computing resources (e.g., single machine, cluster, data center) located at the same physical location. Let $c_v$ denote a limit on the maximum number of units which node $v$ can process. This limit accounts for various constraints of each computing node, e.g., processing power, storage space, access link capacity and others. For each node $v \in V$, constant $p_v$ denotes a processing rate given in units/millisecond. This limit represents the number of units which node $v$ can process in one millisecond. In addition, the model assumes that the overall number of computing nodes involved in a particular database is limited by constant $S$. In other words, constant $S$ denotes the maximum number of partitions related to a single database. For instance, $S = 4$ means that each database can be split to maximum of 4 computing nodes.

Set $D$ contains demands (clients). The workflow of the system is as follows. Client $d \in D$ generates an object classification request related to one (or more) of computational databases $r \in R$ and expects to receive the classification decision as quickly as possible. For this purpose, client $d$ sends the request to all nodes involved in database $r$. Next, all nodes included in database $r$ run the k-nearest neighbors algorithm to obtain the local decision according to the partition of database stored at a particular node. Finally, the local decisions are sent back to the client, which makes the final decision.

Let binary constant $b_{rd}$ denote, if demand $d$ generates requests related to database $r$. Every client $d \in D$ has up-to-date knowledge about which computing nodes serve particular databases. Note that this information is provided by a special central service available for each demand. Let $t_{vd}$ denote a network delay between node $v$ and the node of client $d$ given in milliseconds. For the sake of simplicity, it is assumed that delay $t_{vd}$ is symmetrical (the same in both directions). Nevertheless, the described model can easily be adapted to address asymmetric delays. The network delay can be estimated by special networking mechanisms, e.g., using the ICMP protocol.

The main decision variable is $z_{rv}$, which denotes the number of units of database $r$ assigned to node $v$. Accordingly, the processing time of database $r$ at node $v$ is $z_{rv}/p_v$. In addition, an auxiliary binary variable $u_{rv}$ denotes whether node $v$ is involved in database $r$, i.e., if at least one unit of database $r$ is located at node $v$. Using variables $z_{rv}$ and $u_{rv}$, variable $w_{rvd} = 2u_{rv}t_{vd} + z_{rv}/p_v$ denotes the response time related to database $r$, computing node $v$ and demand $d$. Notice that $w_{rvd}$ includes the network delay between the demand node $d$ and computing node $v$ (in both directions) as well as the processing time related to database $r$ calculated at node $v$.

Let variable $w_{rd}$ denote the overall decision time regarding requests related to demand $d$ and database $r$. Since $w_{rd}$ must account for all computing nodes involved in database $r$ (the final decision can be made only when the client collects all responses), variable $w_{rd}$ is defined as the maximum value of $w_{rvd}$ over all network nodes. Finally, let variable $w$ represent the maximum response time of the computing system accounting for all demands and databases. Variable $w$ comprises two elements: (i) network delay related to overall time required to send all requests from the client node to processing nodes involved in the database and corresponding replies, (ii) the processing time needed to make the local classification decision at a particular processing node. The goal of the optimization is to minimize $w$.

Figure 4.9 illustrates the workflow of the system and the maximum response time objective function. In particular, there are 3 computing nodes involved in the considered database. Step 1 (the number of steps are shown on the figure in circles) is generated by the client node of demand $d$, i.e., a query is sent to each computing node $r$ related to the considered database. The delay of this operation is $t_{vd}$ for each computing node $v$. Afterward, each computing node processes the query (step 2) what takes $z_{rv}/p_v$ and returns to the client node the decision (step 3 and delay $t_{vd}$). When, the client node collects all replies from computing nodes, the final decision is made in a very short time which is a constant, and so is not considered in the model (step 4).

Note that in order to make the optimization problem feasible, the computing system is dimensioned according to the predicted load. In more detail, each new request is processed almost immediately, without the need to queue the request. Therefore, the optimization objective (i.e., maximum response time) does not include any queuing delay. This can be accomplished by the parallel processing of arriving requests and by the overprovisioning of network resources.

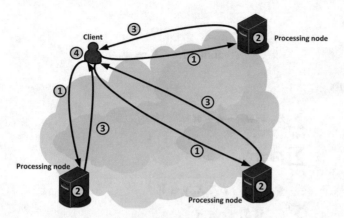

**Fig. 4.9** Workflow of an overlay computing system processing requests related to k-nearest neighbors method

## OVR/A/LD/Response Time

### sets

| | |
|---|---|
| $V$ | computing nodes |
| $D$ | demands (clients) |
| $R$ | databases (data sets) |
| $R(d)$ | databases, for which demand $d$ generates requests |

### constants

| | |
|---|---|
| $c_v$ | capacity of node $v$ |
| $p_v$ | processing rate of node $v$ (number of units that $v$ can process in 1 ms) |
| $n_r$ | size of the database $r$ (number of database units) |
| $t_{vd}$ | network delay between computing node $v$ and client node of demand $d$ (ms) |
| $b_{rd}$ | $=1$, if demand $d$ generates requests related to database $r$; 0, otherwise |
| $S$ | split limit (maximum number of computing nodes involved in one database) |
| $M$ | large number |

### variables

| | |
|---|---|
| $z_{rv}$ | number of units of database $r$ assigned to node $v$ (integer, non-negative) |
| $u_{rv}$ | $=1$, if database $r$ uses computing node $v$; 0, otherwise (binary) |
| $w_{rvd}$ | overall response time regarding requests related to database $r$, computing node $v$ and demand $d$ (continuous) |
| $w_{rd}$ | overall decision time regarding requests related to demand $d$ and database $r$ (continuous) |
| $w$ | maximum response time of the system (continuous) |

**objective**

$$\text{minimize} \quad F = w \tag{4.5.1a}$$

**constraints**

$$\sum_{v \in V} z_{rv} = n_r, \quad r \in R \tag{4.5.1b}$$

$$\sum_{r \in R} z_{rv} \leq c_v, \quad v \in V \tag{4.5.1c}$$

$$z_{rv} \leq M u_{rv}, \quad r \in R, v \in V \tag{4.5.1d}$$

$$\sum_{v \in V} u_{rv} \leq S, \quad r \in R \tag{4.5.1e}$$

$$w_{rvd} = 2u_{rv}t_{vd} + z_{rv}/p_v \quad d \in D, r \in R(d), v \in V \tag{4.5.1f}$$

$$w_{rvd} \leq w_{rd} \quad d \in D, r \in R(d), v \in V \tag{4.5.1g}$$

$$w_{rd} \leq w \quad d \in D, r \in R(d). \tag{4.5.1h}$$

The objective (4.5.1a) is to minimize the maximum response time. Condition (4.5.1b) assures that for every database $r \in R$ all units are assigned to computing nodes. Inequality (4.5.1c) sets an upper limit to the number of units assigned to a computing node. Constraint (4.5.1d) defines variable $u_{rv}$. Condition (4.5.1e) is the model to control the split limit. Constraints (4.5.1f)–(4.5.1h) are used to express the maximum response time. Model (4.5.1a)–(4.5.1h) is $\mathcal{NP}$-complete, since it is equivalent to the Multidimensional Knapsack problem proved to be $\mathcal{NP}$-complete in [93].

Note that in [79], a similar optimization model was presented and solved by an evolutionary algorithm. However, a centralized service was assumed to control the computing system. In turn, papers [84–87] concerned comparable optimization problems of overlay computing systems, but with the objective of minimizing system cost as well as system throughput. Corresponding ILP formulations were presented with a heuristic algorithm based on the GRASP method and the results of numerical experiments. Moreover, the authors of [88] considered a problem of joint scheduling and access link capacity design in overlay computing systems. In addition to the optimization model formulation, various heuristic algorithms (greedy, evolutionary algorithm and tabu search) were proposed and evaluated.

## 4.6  Survivable Overlay Computing System

The previous Sect. 4.5 focused on the optimization of an overlay computing system assuming a quite simple architecture, since access link capacity constraints were not included in the optimization model. This section addresses a more comprehensive scenario of a survivable overlay computing system which embraces access link capacity constraints, i.e. survivability aspects are considered and each computing task is assigned to two disjoint computing nodes [89–92].

The main assumptions of the overlay computing system considered in this section are made on the basis of previous works on this topic e.g., [5, 22, 23, 27], and real computing systems like Seti@Home using the BOINC framework [24]. The goal of the considered system is to process various computational projects and minimize the system cost. Each project is divided into uniform tasks, and every task requires the same amount of processing power to be calculated. Each project has one source node which provides the input data, and one or more destination nodes which receive the computational results. Each task must be assigned to an overlay node to be processed and to obtain the output data sent to the destination nodes. The workflow of the system can be described in the following way. The input data of a task is sent from the source node to a computing node, which processes the data. Then, the output data (results of computations) is transmitted from the computing node to the destination nodes. It is assumed that computational projects are long-lived and established for relatively long time periods. Such computing systems are applied in the context of the following computational tasks: weather forecasting, scientific experiments (e.g., the Large Hadrons Collider), data classification for IT security (e.g., anti-spam filters, IPS/IDS), data mining (e.g., sales forecasts). Therefore, the assumed model does not consider the time dependency of the projects (starting time, completion time, etc.). The input and output data associated with each task is continuously transmitted in the network, and can be expressed in bits per second. In consequence, computational and network resources can be allocated in the system according to the offline optimization approach.

Overlay nodes are able to process the computational tasks, since they are equipped with various devices which can be applied for computing (e.g., computers, clusters, data centers, etc.). Each node has a processing power limit defined as the maximum number of uniform computational tasks that the node can calculate. In addition, each node is connected to the overlay network by an access link with a specified capacity, which is used only to transfer the input and output data of task processing. In particular, each node $v$ downloads two types of data: task input data (if node $v$ is selected for processing of the considered task), and task output data (if $v$ is the destination node of the considered task). Correspondingly, each node $v$ uploads two types of data: task input data (if $v$ is the source node of the considered task), and output data (if node $v$ is selected for processing of the considered task).

To illustrate the workflow of the described overlay computing system a simple example is shown in Fig. 4.10. Briefly, the overlay computing system includes five nodes denoted as $a, b, c, d, e$. The system processes one project with three tasks. Node $a$ is the source node of all tasks, while nodes $b$ and $e$ are the destination nodes. Node $a$ provides the input data related to particular tasks (green rectangles labeled i1, i2 and i3). Blue rectangles labeled p1, p2, and p3 denote the nodes where a particular task is processed, i.e., node $b$ processes task 1, node $a$ processes task 2, and node $d$ processes task 3. Finally, red rectangles labeled o1, o2, and o3 denote results of computations related to tasks 1, 2 and 3, respectively. The allocation of tasks for

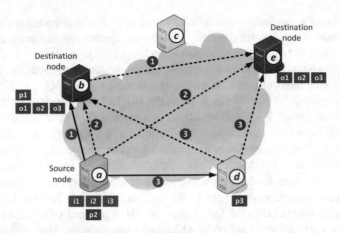

**Fig. 4.10** Example of an overlay computing system

processing results in network flows between nodes. Solid lines represent the flow of input data and a green circle with a number indicates the number of the task the input data is related to. For instance, since node *b* computes task 2, the input data of this task is transmitted from node *a* to node *b*. In turn, dotted lines represent the flow of output data. Once more, the numbers in red circles indicate the number of the task the output data is related to. For example, since node *d* processes task 3, this node uploads the output data to nodes *b* and *e* (destination nodes). Notice that the input data of task 2 is not uploaded from node *a*, since node *a* processes the task by itself. In a similar way, the output data of task 1 calculated in node *b* is uploaded only to node *e*.

Overlay systems are subject to various unintentional failures (e.g., natural disasters, software errors, human errors) and intentional failures (e.g., sabotage) [59]. Such failures can impact the underlying network infrastructure connecting computing nodes, as well as the access links connecting tthe overlay nodes to the network. In addition, elements of the systems responsible for processing are also subject to various breakdowns. Consequently, if the overlay computing system is utilized to process computational projects of large importance, some additional survivability mechanisms are required to guarantee the computational tasks' completion and delivery of all required results.

To protect the overlay computing system against a failure that results in a situation where the results of computational projects are not delivered to all destination nodes, a Dedicated Node Protection (DNP) mechanism is proposed. Specifically, for each computational task, two computing nodes are assigned: a primary (working) node and a completely reserved backup node. Both nodes simultaneously process the same input data, and send the output data (results) to all destination nodes. The idea of DNP is very similar to the 1+1 DPP method developed in the context of optical networks [59]. The main advantage of the DNP approach is a very fast reaction time after a failure. However, the main disadvantage is a relatively large cost following

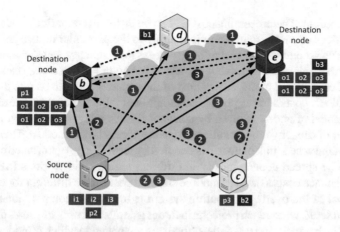

**Fig. 4.11** Example of a survivable overlay computing system

from the system redundancy. The architecture of a survivable overlay computing system with DNP protection is very similar to the overlay computing system introduced above in this section. But the need to process an additional copy of every task increases the network traffic and requires more computational resources.

Figure 4.11 shows a survivable overlay computing system with DNP protection using the same example as in Fig. 4.10. To account for additional survivability requirements, each of three tasks is processed at two separate nodes, i.e., primary tasks are marked with light blue rectangles p1, p2, and p3, while backup tasks are labelled with dark blue rectangles b1, b2 and b3, respectively. For instance, task 1 is processed at nodes $b$ and $d$. As pointed out above, the duplication of task processing generates extra network traffic compared to the example included in Fig. 4.10. For example, node $a$ uploads two copies of the input data related to tasks 1 and 3, and one copy related to task 2.

### 4.6.1 Formulation

Set $V$ includes overlay nodes which have the capability to process tasks as well as being the source and/or destinations of the computational projects. Since according to [30–36] capacity constraints of access links are typically sufficient in overlay networks and the underlying network is assumed to be overprovisioned, the only network capacity constraints in the model refer to access links.

Each node $v \in V$ is assigned with a maximum processing rate $p_v$, i.e., $p_v$ denotes the number of uniform computational tasks that node $v$ can calculate. Two types of cost are assigned to the nodes. Firstly, let $\psi_v$ denote the cost related to the processing of one uniform task at node $v$. For example, $\psi_v$ can be interpreted as the OPEX cost necessary to process the uniform computational tasks, including power consumption, maintenance, etc. The processing cost $\psi_v$ can be different for various nodes, due the fact that prices of energy and IT services apply to the particular node's geographical location. The second type of cost refers to access link capacity. Let $\xi_v$ denote the cost of one access link capacity module allocated for node $v$. Constant $\xi_v$ can represent the OPEX cost related to the leasing cost of the capacity module paid to the ISP and all other OPEX costs such as power consumption, maintenance, etc. The size of one capacity model of node $v$ is given by constant $m_v$. An integer variable $y_v$ represents the number of capacity modules allocated to the access link of node $v$. Therefore, the capacity of the access link of node $v$ is denoted by term $m_v z_v$. Again, since computing nodes can be spread geographically in different regions with various ISPs, energy and maintenance costs as well as module size and cost can be different for each node.

The goal of the overlay computing system is to process computational projects included in set $R$, where each project $r$ includes a number of uniform tasks denoted by set $K(r)$. Each project and task is described by a source node which provides the input data, and destination nodes which receive the output data (results of computations). Binary constants $s_{kv}$ and $d_{kv}$ denote, if node $v$ is the source or destination node of task $k$, respectively. Input and output data transmission rates of task $k$ are defined as constants $h_k$ and $g_k$, respectively.

Two binary variables $z_{kv}$ and $b_{kv}$ determine the allocation of task $k$ to primary and backup node, respectively. However, by using constant $\alpha_k$ it is possible to indicate which projects (tasks) require protection. In particular, only if task $k$ is to be protected (i.e., $\alpha_k = 1$), a backup computing node is assigned to process this task.

The goal of optimization is to minimize the cost of the computing system, including expenses related to network resources (access links) and processing (computations). It should be stressed that the cost is perceived from the perspective of the overlay computing system, not from the perspective of the underlying network operator. Consequently, the part of the cost related to network resources is limited only to access links and does not account for costs related to the backbone network underlying the overlay computing system.

## OVR/A/NDL/DNP/Cost
### sets

| | |
|---|---|
| $V$ | computing nodes |
| $R$ | projects |
| $K$ | tasks |
| $K(r)$ | tasks included in project $r$ |

### constants

| | |
|---|---|
| $p_v$ | processing rate of node $v$ (number of units which $v$ can process in 1 second) |
| $m_v$ | size of the capacity module for node $v$ (Mb/s) |
| $\psi_v$ | cost of processing one uniform task for node $v$ |
| $\xi_v$ | cost of one access link capacity module for node $v$ |
| $h_k$ | transmission rate of input data of task $k$ (Mb/s) |
| $g_k$ | transmission rate of output data of task $k$ (Mb/s) |

$n_k$   number of destination nodes for task $k$
$s_{kv}$  =1, if node $v$ is the source node of task $k$
$t_{kv}$  =1, if node $v$ is the destination node of task $k$
$\alpha_k$  =1, if task $k$ requires protection; 0, otherwise
$S$   split limit (maximum number of computing nodes involved in one project)

**variables**

$z_{kv}$  =1, if task $k$ is allocated to primary computing node $v$; 0, otherwise (binary)
$b_{kv}$  =1, if task $k$ is allocated to backup computing node $v$; 0, otherwise (binary)
$y_v$   capacity of node $v$ access link expressed as the number of capacity modules (integer, non-negative)
$u_{rv}$  =1, if project $r$ uses computing node $v$; 0, otherwise (binary)

**objective**

$$\text{minimize} \quad F = \sum_{v \in V} \xi_v y_v + \sum_{k \in K} \sum_{v \in V} \psi_v (z_{kv} + b_{kv}) \tag{4.6.1a}$$

**constraints**

$$\sum_{v \in V} z_{kv} = 1, \quad k \in K \tag{4.6.1b}$$

$$\sum_{v \in V} b_{kv} = \alpha_k, \quad k \in K \tag{4.6.1c}$$

$$(z_{kv} + b_{kv}) \leq 1, \quad k \in K, v \in V \tag{4.6.1d}$$

$$\sum_{k \in K} (z_{kv} + b_{kv}) \leq p_v, \quad v \in V \tag{4.6.1e}$$

$$\sum_{k \in K} (1 - s_{kv}) h_k (z_{kv} + b_{kv}) + \sum_{k \in K} t_{kv} g_k (1 - z_{kv} + \alpha_k - b_{kv}) \leq m_v y_v, \ v \in V \tag{4.6.1f}$$

$$\sum_{k \in K} s_{kv} h_k (1 - z_{kv} + \alpha_k - b_{kv}) + \sum_{k \in K} (n_k - t_{kv}) g_k (z_{kv} + b_{kv}) \leq m_v y_v, \ v \in V \tag{4.6.1g}$$

$$z_{kv} \leq u_{rv}, \quad r \in R, k \in K(r), v \in V \tag{4.6.1h}$$

$$b_{kv} \leq u_{rv}, \quad r \in R, k \in K(r), v \in V \tag{4.6.1i}$$

$$\sum_{v \in V} u_{rv} \leq S, \quad r \in R. \tag{4.6.1j}$$

The optimization goal is to minimize the cost of the computing system (4.6.1a). The objective function includes two elements: cost of access links and cost of task processing at primary and backup computing nodes. Equality (4.6.1b) guarantees that exactly one node is selected as the primary computing node for a particular task. In an analogous way, equality (4.6.1c) assures that if task $k$ is to be protected ($\alpha_k = 1$), then exactly one backup node has to be assigned for this task. The next constraint (4.6.1d) implements the DNP protection method and ensures that primary and backup computing nodes are disjoint. Since each computing node $v \in V$ has a processing limit, constraint (4.6.1e) ensures that node $v$ cannot be assigned with more tasks to calculate than it can process.

Inequality (4.6.1f) defines the download capacity constraint for an access link. In more detail, the left-hand side of (4.6.1f) represents the flow incoming to node $v$ and comprises two elements. The first one denotes the flow related to transmission of input data for computations, i.e., node $v$ that is selected as a primary node of task $k$ ($z_{kv} = 1$), or a backup node of task $k$ ($b_{kv} = 1$), must download the input data with transmission rate $h_r$. Only if the considered node $v$ is the source node of task $k$ ($s_{kv} = 1$) is there no need to send the input data. The second element on the left-hand side of (4.6.1f) denotes the output data transmission, i.e., each destination node $v$ of task $k$ ($t_{kv} = 1$) must download the output data of this task with rate of $g_k$ from both primary and backup computing nodes. But if the considered node $v$ is selected as the primary node of task $k$ ($z_{kv} = 1$), data does not have to be sent to the primary node. In the same way, if task $k$ is protected ($\alpha_k = 1$) and node $v$ is selected as a backup node of task $k$ ($b_{kv} = 1$), there is no need to download data to the backup node. In turn, the right-hand side of (4.6.1f) denotes the access link capacity.

Condition (4.6.1g) formulates the upload access link capacity constraints. Analogous to (4.6.1f), the left-hand side of (4.6.1g) defines the flow leaving node $v$ and again includes two terms. The former one denotes the flow of input data sent from source node $v$ of task $k$ ($s_{kv} = 1$) to primary and backup computing nodes. However, if the considered node $v$ is selected as either the primary computing node ($z_{kv} = 1$), or the backup computing node ($b_{kv} = 1$) of task $k$, the transmission is not established. The latter term represents the flow of output data to all destination nodes of task $k$, i.e., each node $v$ that is selected as either the primary node of task $k$ ($z_{kv} = 1$), or the backup node of task $k$ ($b_{kv} = 1$), must upload the output data to all $n_k$ destination nodes of task $k$. If node $v$ is one of destination nodes of task $k$ ($t_{kv} = 1$), then the flow is decreased correspondingly.

Finally, constraints (4.6.1h) and (4.6.1i) define variable $u_{rv}$ which is next used in (4.6.1j) to set the limit on the number of computing nodes involved in one computational project. Note that the problem (4.6.1) is $\mathcal{NP}$-complete, since it is equivalent to the network design problem with modular link capacities [46].

## 4.6.2  Cut Inequalities

Recall that cut inequalities are added to the original optimization problem to improve the performance of the branch-and-bound algorithm. This section introduces cut inequalities for problem (4.6.1) [92]. It is assumed that the proposed cuts are applied in the cut-and-branch variant of the branch-and-bound algorithm, i.e., cut inequalities are added to the root node of the solution tree and are used throughout the whole branch-and-bound tree. For a more comprehensive discussion on cut inequalities, refer to Sect. 2.6.2.

The first proposed cut is a constraint that formulates a lower bound of variable $y_v$. Notice that if node $v$ is a destination node of task $k$ ($t_{kv} = 1$), it must download the output data (results of computations) related to task $k$. Node $v$ downloads this data in two ways, i.e., (i) if node $v$ is assigned to process task $k$, the data related to task $k$ is downloaded as the input data (which is later processed in this node to obtain the output data), or (ii) as the output data from another node that processes task $k$. However, if $v$ is the source node of task $k$ ($s_{kv} = 1$), the considered node does not download the data related to task $k$. Firstly, a case when task $k$ is not protected ($\alpha_k = 0$) is analyzed. The download capacity of node $v$ related to the processing of task $k$ must then exceed the following value $t_{kv}(1 - s_{kv}) \min(h_k, g_k)$. If task $k$ is protected ($\alpha_k = 1$), node $v$ receives the results of computations twice. Nevertheless, on the base of constraint (4.6.1c), the same node cannot serve as both a primary and backup computing node of task $k$. Consequently, two cases are possible, either node $v$ downloads both the input and output data of task $k$ (i.e., node $v$ is the primary or backup node of task $k$) or node $v$ downloads the output data of task $k$ twice (node $v$ is neither primary nor backup node of task $k$). In consequence, the download capacity of node $v$ related to the processing of protected task $k$ must exceed the value of $t_{kv}(1 - s_{kv}) \min(h_k + g_k, 2g_k)$.

Similarly, the upload capacity of node $v$ related to the processing of task $k$ can be analyzed. More specifically, if node $v$ is the source node of task $k$, ($s_{kv} = 1$), the data related to task $k$ is sent to all destination nodes of this task. Again, data can be transferred either as the input data (rate $h_k$) to another processing node, or node $v$ performs task $k$ by itself and sends the output data (rate $g_k$) to ($n_k - t_{kv}$) nodes (all destination nodes excluding node $v$). In turn, if task $k$ is not protected ($\alpha_k = 0$), the upload capacity of node $v$ related to task $k$ must exceed $s_{kv} \min(h_k, (n_k - t_{kv})g_k)$. If task $k$ is protected ($\alpha_k = 1$), again due to constraint (4.6.1c) node $v$ cannot serve as both a primary and a backup node. Therefore, two cases must be examined. Firstly, node $v$ uploads input data (rate $h_k$) twice to the primary and backup nodes. Secondly, node $v$ processes task $k$ and sends the output data (rate $g_k$) to the destination nodes as well as sending the data related to the backup task as input data (rate $h_k$). Thus, the upload capacity of node $v$ related to task $k$ must be above $s_{kv} \min(2h_k, h_k + (n_k - t_{kv})g_k)$.

Let $d_{kv}$ and $u_{kv}$ denote the lower bounds of a download and upload flow related to node $v$ and task $k$, respectively, formulated as follows:

$$d_{kv} = \begin{cases} t_{kv}(1 - s_{kv}) \min(h_k, g_k) & \text{if } \alpha_k = 0 \\ t_{kv}(1 - s_{kv}) \min(h_k + g_k, 2g_k) & \text{if } \alpha_k = 1 \end{cases} \tag{4.6.2a}$$

$$u_{kv} = \begin{cases} s_{kv} \min(h_k, (n_k - t_{kv})g_k) & \text{if } \alpha_k = 0 \\ s_{kv} \min(2h_k, h_k + (n_k - t_{kv})g_k) & \text{if } \alpha_k = 1. \end{cases} \tag{4.6.2b}$$

Combining definitions (4.6.2) with the MIR approach [94, 95], the following cuts can be written:

$$\left\lceil \sum_{k \in K} d_{kv}/m_v \right\rceil \leq y_v \tag{4.6.3a}$$

$$\left\lceil \sum_{k \in K} d_{kv}/m_v \right\rceil \leq y_v. \tag{4.6.3b}$$

Moreover, cuts based on the Cover Inequality (CI) approach [96] can be formulated using constraints (4.6.1e) and (4.6.1j). To limit the number of possible cover inequalities, first the linear relaxation of the model (4.6.1) is solved. Afterwards, in the obtained solution, variables $z_{rv}$, $b_{rv}$ and $y_v$ that are not integers are identified, and for these variables the CI approach is applied in the context of constraints (4.6.1e) and (4.6.1j).

### 4.6.3   Algorithms

Since problem (4.6.1) is $\mathcal{NP}$-complete, to tackle larger problem instances three algorithms are described in this section, namely, greedy method [91], tabu search method [89, 90] and evolutionary algorithm [90]. Note that the algorithms are formulated for the problem (4.6.1a)–(4.6.1g), i.e., the split constraint is not considered.

**Greedy Algorithm**

Algorithm GA/OVR/M/NDL/DNP processes all tasks included in set $K$ in two runs; first, all tasks are assigned one-by-one to the primary nodes, next, the same operation is repeated for the backup nodes. Only feasible nodes (with available processing resources) are considered as candidate nodes to assign the current task $k$. A special metric $c_{kv}$ is used to evaluate the cost of allocation of task $k$ to node $v$, i.e., the algorithm selects an allocation pair $v$ and $k$ that provides the lowest cost of this metric.

In particular, $c_{kv}$ is defined as a weighted sum of two elements. Firstly, an average cost of allocation of task $k$ to node $v$ is calculated considering both the processing cost of node $v$ and the access link cost of node $v$. More specifically, allocation of task $k$ to node $v$ results in an additional flow of input data from the source node of task $k$ to node $v$, and next additional flow of the output data from node $v$ to all destination nodes of task $k$. For instance, the extra cost related to the download capacity is defined as $\xi_v h_k/m_v$, the extra cost related to the upload capacity is defined as $\xi_v(n_k - t_{kv})g_k/m_v$ and the additional processing cost is simply given by $\psi_v$. The second element of metric $c_{kv}$ is calculated in another way. The key idea is to check if the allocation of task $k$ to node $v$ causes new capacity module(s) to be allocated to some nodes in order to supply enough capacity to download/upload extra data. The motivation behind this idea is the observation that, in some cases, allocation of task

$k$ to node $v$ does not trigger the need to increase capacity of access links, since the already allocated capacity resources are sufficient to send all the required data. To find the best trade-off between both elements included in metric $c_{kv}$, the GA method uses a tuning parameter which weights the influence of both elements.

**Tabu Search**

The first step to implement a TS/OVR/M/NDL/DNP algorithm based on the Tabu Search approach is to define the solution encoding which next allows the construction of a move operation to search the neighborhood space. In the context of problem (4.6.1), the encoding is defined as two sets $Z$ and $B$ including $|K|$ elements, where each element $z_k \in Z$ and $b_k \in B$ denote the primary and backup node selected to process task $k$:

$$(Z, B) = [z_1, z_2, \ldots, z_{|K|}], [b_1, b_2, \ldots, b_{|K|}]. \tag{4.6.4}$$

Two solutions are considered as neighbors, if they differ at exactly one position considering both vectors $Z$ and $B$. The move operation which allows the generation of a neighborhood solution $(Z, B)$ is defined as a triple $(k, v, w)$, which denotes the reallocation of task $k$ from primary (backup) node $v$ to primary (backup) node $w$. However, to assure constraint (4.6.1c), it is not allowed to generate a neighbor solution in which both primary and backup nodes of a particular task $k$ are located at the same node $v$, i.e., condition $z_k \neq b_k$ must be satisfied for every $k \in K$. Having solution $(Z, B)$, the values of variables $y_v$ are calculated in order to find the minimum possible values that satisfy link capacity constraints (4.6.1f) and (4.6.1g).

All constraints of the problem—excluding (4.6.1e)—are assured directly by the solution encoding (4.6.4) and construction of the move operation. To address the node processing limit constraint (4.6.1d), the penalty function approach is used, i.e., a function that evaluates the value of solution $(Z, B)$ includes the problem objective function (4.6.1a) and the additional term that represents the potential overrun of constraint (4.6.1d). More details on the TS/OVR/M/NDL/DNP method can be found in [89, 90].

**Evolutionary Algorithm**

The EA/OVR/M/NDL/DNP algorithm is based on the same solution encoding (4.6.4) as the TS/OVR/M/NDL/DNP method, i.e., chromosome $(Z, B)$ includes individual genes $(z_k, b_k)$ that represent the primary and backup nodes selected for task $k \in K$. In order to obtain values of link capacity variables $y_v$, the same procedure as in the case of the TS algorithm is to applied.

A crossover operation in points $i$ and $j$ ($i < j$) assumes an exchange of all tasks from $i$ to $j$ between two individuals for both primary and backup nodes. Note that such a crossover operation assures that the survivability constraint (4.6.1c) is satisfied in the children chromosomes. A mutation operation is defined as a random modification of some genes, i.e., some primary and/or backup nodes are changed for randomly selected tasks. However, new solutions generated by the mutation operation must be controlled to assure the survivability constraint (4.6.1c).

To tackle the processing limit constraint (4.6.1c), both penalty function and repair function approaches are utilized. The repair function simply reallocates some tasks from nodes exceeding the processing limit to the nodes that still have residual processing power. The number of tasks to be moved is always selected as the lowest required to repair a particular chromosome. The repair function is executed for every individual in the population with a given frequency defined as a tuning parameter of the EA algorithm. A more comprehensive description of EA is presented in [90].

### 4.6.4   Numerical Results

The goal of the numerical results [89–92] was threefold: to evaluate performance of cut inequalities, compare heuristic algorithms and examine the impact of protection levels on the system cost. To obtain the optimal results, the CPLEX solver was used [58]. Two types of overlay computing systems were created at random: small systems embracing 30 nodes and large systems containing 200 nodes. In particular, for both small and large systems, six sets of systems and six sets of computing tasks were generated randomly according to ranges of particular parameters reported in Table 4.7. The size of one capacity module was equal to 100 Mb/s. Additionally, several configurations related to protection requirements were created with the following values of the protected tasks percentage (PTP): $0, 10, 20, \ldots, 100\%$. For instance PTP $= 20\%$ indicates that on average 20% of tasks are protected (i.e., assigned to two disjoint computing nodes), while the remaining 80% are not protected. The overall number of individual cases for each size of the system was 396 (i.e., $6 \times 6 \times 11$). The preliminary experiments showed that CPLEX with default setting of the optimality gap (i.e., 0.0001) was not able find optimal results for 30-node systems. Therefore, the optimality gap of CPLEX was set to 0.01. In turn, for 200-node systems, CPLEX was not able to yield a feasible result.

The first goal of the experiments was to evaluate cut inequalities proposed in Sect. 4.6.2. Table 4.8 presents results regarding the execution time of CPLEX as well as the number of branch-and-bound nodes obtained for 30-node systems. Four cases

**Table 4.7**  Parameters of tested systems

| Parameter name | Small systems | Large systems |
|---|---|---|
| Number of computing nodes | 30 | 200 |
| Number of tasks | 300–700 | 1200–4800 |
| Cost of one access link capacity module | 120–400 | 120–400 |
| Processing cost of one task | 50–150 | 50-150 |
| Processing limit per node | 10–40 | 10–40 |
| Number of destination nodes | 1–4 | 1–8 |
| Input and output data rates | 5–15 | 5–10 |

**Table 4.8** Performance of additional cut inequalities as a function of PTP

| PTP (%) | Av. execution time in seconds | | | | Av. number of BB nodes | | | |
|---|---|---|---|---|---|---|---|---|
| | NoCut | MIR | CI | AllCuts | NoCut | MIR | CI | AllCuts |
| 0 | 58 | 109 | 365 | 108 | 514 | 521 | 3079 | 536 |
| 10 | 225 | 417 | 572 | 388 | 2365 | 2360 | 3043 | 2510 |
| 20 | 86 | 163 | 2019 | 506 | 521 | 521 | 16725 | 2636 |
| 30 | 458 | 888 | 833 | 574 | 4533 | 4564 | 4529 | 2516 |
| 40 | 116 | 221 | 486 | 257 | 470 | 471 | 1316 | 495 |
| 50 | 507 | 979 | 2480 | 3309 | 2881 | 2900 | 12606 | 15383 |
| 60 | 1488 | 2913 | 1118 | 453 | 6189 | 6191 | 4047 | 879 |
| 70 | 4119 | 8203 | 6057 | 3685 | 20551 | 20570 | 22512 | 8443 |
| 80 | 1256 | 1213 | 3807 | 1589 | 4299 | 4299 | 17587 | 8586 |
| 90 | 11530 | 11324 | 4500 | 2056 | 74287 | 74307 | 20705 | 9516 |
| Sum | 19844 | 26430 | 22237 | 12925 | 116607 | 116702 | 106148 | 51498 |

are reported, namely, branch-and-bound without additional cut inequalities (NoCut), branch-and-bound only with MIR cut inequalities (MIR), branch-and-bound only with CI cut inequalities (CI) and branch-and-bound with both CI and MIR cut inequalities (AllCuts). It is easy to see that the application of both additional cuts (MIR and CI) provides the best performance, since the execution time is reduced by about 35 % compared to the case without additional cuts. On the other hand, the number of branch-and-bound nodes is reduced by about 55 %. Nevertheless, a detailed analysis of results obtained for individual cases indicates that the trends are not stable, i.e., for some cases, additional cuts do not improve the performance of the CPLEX solver.

The next goal of the numerical experiments was to compare heuristic algorithms GA, TS and EA proposed to solve the survivable overlay computing system optimization. Note that complete information on tuning of TS and EA algorithms can be found in [89, 90]. Table 4.9 reports a comparison of two types of system, i.e., 30-node and 200-node. For the former case, the average optimality gap with corresponding lengths of 95 % confidence intervals are presented. In the case of larger systems, the table includes the average gap to GA method with corresponding lengths of 95 % confidence intervals. Note that the average execution time for small systems was (in seconds): 29, 0.4, 2.5 and 150 for CPLEX, GA, TS and EA methods, respectively. The average execution time for large systems was (in seconds) 1, 6 and 1020 for AE, TS and EA, respectively. Combining information on the average quality of results reported in Table 4.9 and execution times, the TS algorithm seems to be the best optimization method for the considered problem.

Finally, Fig. 4.12 reports the average values of the relative cost for 200-node systems as a function of the PTP parameter. In more detail, the result obtained for the case without protection (PTP = 0 %) is the reference result that is next applied to calculate the relative cost yielded for larger values of PTP. Moreover, the cost is divided into two categories: capacity cost and processing cost (see objective

**Table 4.9** Comparison of algorithms—average gap to reference results and corresponding lengths of 95 % confidence intervals

| Algorithm | 30-nodes systems | | 200-nodes systems | |
|---|---|---|---|---|
| | Av. gap to opt. (%) | Conf. interval (%) | Av. gap to GA (%) | Conf. interval (%) |
| GA | 10.17 | 0.25 | – | – |
| TS | 5.77 | 0.11 | −8.10 | 0.20 |
| EA | 7.16 | 0.11 | −4.14 | 0.11 |

**Fig. 4.12** Relative capacity and processing costs as a function of the protected tasks percentage for 200-nodes systems

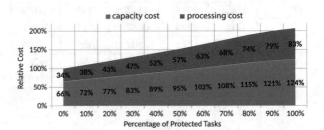

function (4.6.1a)). The numbers shown in the figure denote the percentage share of a particular type of cost assuming that the overall cost obtained for PTP = 0 % is 100 %. The general trend is that both types of cost increase almost linearly with the PTP parameter. However, the processing cost rises slightly faster when compared to the capacity cost. The reason for this lies in the fact that relatively cheaper nodes (i.e., nodes with lower values of costs related to capacity and processing) are saturated before selecting more expensive nodes. Consequently, when the number of tasks to be protected increases, some tasks have to be allocated to more costly nodes. Notice that the average cost of the system with full protection (PTP = 100 %) is about 106 % larger when compared to the case when all tasks are not protected. For more results and discussions on the performance of survivable computing systems refer to [89–92].

## 4.7  Optimization of Peer-to-Peer Flows

The concept of Peer-to-Peer (P2P) systems is a popular networking approach applied mostly to provide efficient delivery of various content. Numerous services including file-sharing, distributed computing, Internet based telephony, Internet television, have been successfully developed using various P2P mechanisms. Deployment of P2P systems has triggered a broad range of research challenges, but in this section we focus only on one aspect of P2P systems, namely the modeling and optimization of P2P flows. This problem is a natural consequence of the requirement to optimize performance of P2P systems from the network perspective. This section presents only

ILP formulations. However, note that various heuristic and metaheuristic algorithms have been proposed and tested in the context of P2P flow optimization, e.g., see [9, 26, 97–99].

## 4.7.1 Formulations

The optimization models of the flow allocation in a P2P system presented in this section were for the formulated first time in [99, 100] and next applied in the context of various optimization problems related to P2P systems, e.g., [26, 97, 101].

The main modeling assumptions are made according to analysis of real P2P systems such as BitTorrent and the previous works on the optimization of overlay and P2P systems, e.g., [2, 5, 7, 10, 20, 30–32, 34, 102–107]. The key challenge compared to previous optimization models presented and discussed in this book, is the fact that a P2P system changes over time. Up to now, it was assumed that the information to be transferred in the network is statically provisioned for requesting nodes. Here, in the context of P2P systems, along with the running of the system, nodes (peers) download more information, which is next made available for requesting nodes, which influences the traffic patterns. Therefore, a time scale of the system must be included in the model.

The P2P system consists of set $V$ including nodes. Each node is connected to an overlay network by an accesses link defined by upload capacity $u_v$ and download capacity $d_v$. To model the time scale of the considered P2P system, a concept *time slots* is applied [31, 32, 99, 100, 102, 103]. In particular, the P2P system operates in subsequent iterations and the length of one iteration is equal to one time slot. Set $T$ contains all available time slots. For the sake of simplicity, it is assumed that each time slot has the same length. However, the model presented below can be modified to address some other scenarios (for more details see [100]). Furthermore, at the end of each time slot, the indexing service is updated and all operations completed in the particular time slot are recorded. Thus, at the beginning of the next time slot, the detailed information on current availability of blocks at network nodes is delivered to all requesting nodes. For instance, if block $b$ was downloaded by node $v$ in iteration $t$, in the next time slot $t + 1$, all other peers can try to get this block from node $v$. According to [20], a P2P system modeled as described above is called *synchronous*. Note that the considered model is not limited to a single particular implementation of the indexing service, since the system can work with an indexing system organized either in a centralized manner (e.g., similar to tracker used in BitTorrent), or in a decentralized manner (e.g., DHT).

Data to be delivered in the system is divided into blocks (pieces) and set $B$ includes all blocks to be transferred. It is assumed that a transfer of one block is completed within one time slot. However, the model presented below can easily be modified to include some more heterogeneous scenarios, e.g., the transfer of one block takes more than one time slot. For each block $b \in B$, constant $h_b$ denotes a bit-rate required to deliver this block within one time slot. Before the P2P system begins the operation, some nodes called *seeds* hold the blocks, denoted by constant $g_{bv}$ which equals 1 if block $b$ is located at node $v$ before the P2P transfer starts, and 0 otherwise. There is one binary variable in the problem, namely, $x_{bvwt}$ denotes if block $b$ is transferred from node $v$ to node $w$ in time slot $t$.

---

**OVR/P2P/FA/Cost**

**sets**

   $V$  nodes (peers)
   $T$  time slots (iterations)
   $B$  blocks

**constants**

   $d_v$    download capacity of node $v$ (kb/s)
   $u_v$    upload capacity of node $v$ (kb/s)
   $h_b$    bit-rate required to transfer block $b$ in one time slot (kb/s)
   $g_{bv}$   =1, if block $b$ is located at node $v$ before the P2P transfer starts; 0, otherwise
   $\zeta_{vw}$   cost of block transfer from node $v$ to node $w$
   $M$   large number

**variables**

   $x_{bvwt}$  =1, if block $b$ is transferred from node $v$ to node $w$ in time slot $t$; 0, otherwise (binary)

**objective**

$$\text{minimize} \quad F = \sum_{b \in B} \sum_{v \in V} \sum_{w \in V} \sum_{t \in T} \zeta_{vw} x_{bvwt} \tag{4.7.1a}$$

**constraints**

$$g_{bw} + \sum_{v \in V} \sum_{t \in T} x_{bvwt} = 1, \quad b \in B, w \in V \tag{4.7.1b}$$

$$\sum_{b \in B} \sum_{w \in V} x_{bvwt} h_b \le u_v, \quad v \in V, t \in T \tag{4.7.1c}$$

$$\sum_{b \in B} \sum_{v \in V} x_{bvwt} h_b \le d_w, \quad w \in V, t \in T \tag{4.7.1d}$$

$$\sum_{w \in V} x_{bvwt} \le M \left( g_{bv} + \sum_{i \in T : i < t} \sum_{s \in V} x_{bsvt} \right), \quad b \in B, v \in V, t \in T. \tag{4.7.1e}$$

The objective function (4.7.1a) is to minimize the overall routing cost required to deliver all blocks to all requesting nodes using the P2P approach. Equality (4.7.1b) assures that each node receives requested blocks within all available time slots. There are two possible situations to fulfill this condition. In essence, either node $v$ is the seed of block $b$ ($g_{bw} = 1$) or block $b$ is delivered to node $w$ in one of possible time slots $t \in T$ ($\sum_{v \in V} \sum_{t \in T} x_{bvwt} = 1$). It is assumed in (4.7.1b) that every node $w \in V$ requests all blocks included in set $B$. But this constraint can easily be adjusted to address the requirement that a particular node needs to download only selected blocks by using the right-hand side of (4.7.1b)—instead of 1—a binary constant $r_{bw}$ denoting if node $w$ requests block $b$. Inequalities (4.7.1c) and (4.7.1d) denote the upload and download node capacities, respectively. Finally, (4.7.1e) defines a *possession* constraint [100, 104] assuring that block $b$ can be transferred from node $v$ to node $w$ in time slot $t$, only if node $v$ holds block $b$ before time slot $t$ begins [100, 104]. To be more specific, node $v$ can possess block $b$ in time slot $t$ due to two situations. Firstly, node $v$ can be simply the seed node of block $b$ ($g_{bv} = 1$). Secondly, node $v$ downloaded block $b$ from any node $s \in V$ in any time slot $i \in T$ preceding time slot $t$ ($i < t$). Problem (4.7.1) is $\mathcal{NP}$-complete, since an equivalent Minimum Broadcast Time problem is also is $\mathcal{NP}$-complete [108].

Another objective function which can be used in the context of P2P flows is to minimize the overall download time, i.e. the time in which all requesting peers download all requested blocks [6, 31, 32, 34, 100, 103, 106, 109, 110]. To formulate a model with the time objective, a new variable $z_t$ is introduced. In particular, $z_t$ denotes if in time slot $t$ there is at least one transfer between nodes. The below model (4.7.2) formulates the P2P flow allocation problem with the objective function of the download time.

**OVR/P2P/FA/Time**
**variables (additional)**

$z_t = 1$, if in time slot $t$, there is at least one transfer; 0, otherwise (binary)

**objective**

$$\text{minimize} \quad F = \sum_{t \in T} z_t \tag{4.7.2a}$$

**constraints** (4.7.1b)–(4.7.1e) **and**

$$\sum_{b \in B} \sum_{v \in V} \sum_{w \in V} x_{bvwt} \leq M z_t, \quad t \in T. \tag{4.7.2b}$$

An alternative criterion function is throughput defined as the number of blocks (which can be interpreted as the size of a file) which can be delivered to all nodes within given time slots. Let binary variable $u_{bv}$ denote, if block $b$ is delivered to node $v$ during the system operation. In turn, binary variable $u_b$ equals one, if block $b$ is delivered to all requesting nodes. Model (4.7.3) defines the P2P flow allocation problem with the objective related to system throughput.

**OVR/P2P/FA/Throughput**
**variables (additional)**

$u_{bv}$ =1, if block $b$ is delivered to node $v$; 0, otherwise (binary)
$u_b$ =1, if block $b$ is delivered to all requesting nodes; 0, otherwise (binary)

**objective**

$$\text{maximize} \quad F = \sum_{b \in B} u_b \qquad (4.7.3a)$$

**constraints** (4.7.1c)–(4.7.1e) **and**

$$g_{bw} + \sum_{v \in V} \sum_{t \in T} x_{bvwt} = u_{bw}, \quad b \in B, w \in V \qquad (4.7.3b)$$

$$|V|u_b \leq \sum_{v \in V} u_{bv}, \quad b \in B. \qquad (4.7.3c)$$

Note that constraint (4.7.3b) is a modified version of the condition (4.7.1b), since here it is not required that all blocks $b \in B$ must be delivered to every node. In turn, constraint (4.7.3c) defines variable $u_b$, i.e., only when all nodes download block $b$ (right-hand side of (4.7.3c) equals $|V|$), variable $u_b$ can be switched to 1.

### 4.7.2  Additional Constraints

Model (4.7.1) is a basic and generic formulation of the flow allocation problem in P2P systems. In the following, we will present some additional constraints that can enhance model (4.7.1), in order to account for requirements following from real P2P systems.

**Stochastic P2P Systems**

It is quite common in P2P systems such as BitTorrent that nodes (peers) join and/or leave the system frequently, which can make the system stochastic and dynamic [2, 5, 7, 10]. Offline optimization models considered in this book are deterministic. However note that the basic P2P model formulated in (4.7.1) assumes that the time scale of the system is divided into time slots. Consequently, the stochastic nature of a P2P system can be modeled using a special constant $a_{vt}$ which denotes the availability of node $v$ in time slot $t$. Specifically, let $a_{vt} = 1$, if node $v$ is available in time slot $t$ and 0 otherwise. The formulated model (4.7.4) shows how to incorporate the information on node availability into the node capacity constraint.

**OVR/P2P/FA/Stochastic**
**constants (additional)**

$a_{vt}$ =1, if node $v$ is available in time slot $t$; 0, otherwise

**constraints**

$$\sum_{b \in B} \sum_{w \in V} x_{bvwt} h_b \leq a_{vt} u_v, \quad v \in V, t \in T \tag{4.7.4a}$$

$$\sum_{b \in B} \sum_{v \in V} x_{bvwt} h_b \leq a_{wt} d_w, \quad w \in V, t \in T. \tag{4.7.4b}$$

Constraints (4.7.4a) and (4.7.4b) are adapted versions of constraints (4.7.1c) and (4.7.1d). The only modification is visible in the right-hand side of the new constraints, i.e., the available upload and download capacity is multiplied by constant $a_{vt}$ to denote whether particular node $v$ is available in time slot $t$.

**Fairness**

An important aspect of P2P systems is fairness [2, 34, 109]. In essence, the resource contribution in P2P systems should be fair, i.e., the average resource contribution of a single node should be within defined bounds according to average statistics observed in the system. Consequently, a peer should not be forced to upload much more than it downloads. Fairness can be regarded as a type of incentive for nodes to participate in the system, especially in situations where there is a shortage of upload capacity or ISPs charge users based on upload capacity usage. To impose fairness in a P2P system, a following constraint can be used. Let $\Psi_v$ denote fairness of node $v$ defined as the maximum proportion between the amount of uploaded and downloaded information. For instance, $\Psi_v = 1.5$ means that node $v$ uploads at most 1.5 times more information than it has downloaded during the whole period of the P2P system's operation. Note that peers which are seeds can be excluded from the fairness constraint.

One of requirements in enforcing fairness in the P2P system is to avoid nodes which are free riders or selfish peers. *Free riders* are defined as nodes which use resources of the P2P system and download files without contributing to the system. In turn, a *selfish* peer uploads much less than it downloads, or stops uploading when all blocks are collected [2, 5, 6, 34, 109, 111]. To tackle the problem of free riders and selfish peers, constant $\Omega_v$ is defined to represent the minimum proportion between upload and download for node $v$. For instance, $\Omega_v = 0.5$ indicates that node $v$ must upload at least 50% of downloaded information considering all time slots. Note that $\Omega_v = 0$ allows node $v$ to be a free rider.

**OVR/P2P/FA/Stochastic**
**constants (additional)**

$\Psi_v$ fairness of node $v$
$\Omega_v$ minimum proportion between upload and download for node $v$

**constraints**

$$\sum_{b \in B} \sum_{w \in V} \sum_{t \in T} x_{bvwt} \leq \Psi_v \sum_{b \in B} \sum_{w \in V} \sum_{t \in T} x_{bwvt}, \quad v \in V \tag{4.7.5a}$$

$$\sum_{b\in B}\sum_{w\in V}\sum_{t\in T}x_{bvwt} \geq \Omega_v\sum_{b\in B}\sum_{w\in V}\sum_{t\in T}x_{bwvt}, \quad v \in V. \tag{4.7.5b}$$

Constraint (4.7.5a) assures that nodes use the P2P system with defined fairness, while (4.7.5b) limits selfish behavior of peers. Note that the sum on the left-hand side of both inequalities denotes the number of blocks uploaded by the peer during all considered time slots. In turn, the sum on the right-hand side of both constraints represents the number of blocks downloaded by node $v$.

**Neighbors**

In some P2P systems, a new node which joins the system is provided with a list of peers it can connect to and next cooperate to exchange blocks (e.g., BitTorrent). Such peers are called *neighbors* [10, 106, 112]. To model this feature, binary constant $e_{wv}$ is used to denote whether peers $v$ and $w$ are neighbors or not.

**OVR/P2P/FA/Neighbors**
**constants (additional)**

$e_{vw}$ =1, if nodes $v$ and $w$ are neighbors

**constraints**

$$\sum_{b\in B}\sum_{t\in T}x_{bvwt} \leq Me_{vw}, \quad v \in V, w \in V. \tag{4.7.6a}$$

Constraint (4.7.6a) imposes that a transfer between nodes $v$ and $w$ is only allowed if they are neighbors.

# References

1. Aoyama, T.: A new generation network: beyond the internet and NGN. IEEE Commun. Mag. **47**(5), 82–87 (2009)
2. Buford, J., Yu, H., Lua, E.: P2P Networking and Applications. Morgan Kaufmann Publishers Inc., San Francisco (2008)
3. Miller, M.: Discovering P2P. SYBEX Inc., Alameda (2001)
4. Perros, H.: Connection-Oriented Networks: SONET/SDH, ATM, MPLS and Optical Networks. Wiley, New York (2005)
5. Shen, X., Yu, H., Buford, J., Akon, M.: Handbook of Peer-to-Peer Networking, 1st edn. Springer Publishing Company, Incorporated (2009)
6. Steinmetz, R., Wehrle, K. (ed.): Peer-to-Peer Systems and Applications. Lecture Notes in Computer Science, vol. 3485. Springer, Berlin (2005)
7. Tarkoma, S.: Overlay Networks: Toward Information Networking, 1st edn. Auerbach Publications, Boston (2010)
8. Chowdhury, N.M.M.K., Boutaba, R.: Network virtualization: state of the art and research challenges. IEEE Commun. Mag. **47**(7), 20–26 (2009)
9. Walkowiak, K.: Modeling and Optimization of Computer Networks. Wroclaw University of Technology, Wroclaw (2011)

10. Cohen, B.: Incentives build robustness in bittorrent. Technical report, bittorrent.org (2003)
11. Cisco. Cisco Visual Networking Index: Forecast and Methodology, pp. 2014–2019, 2015. White Paper. http://www.cisco.com/c/en/us/solutions/service-provider/visual-networking-index-vni/index.html
12. Banerjee, S., Lee, S., Bhattacharjee, B., Srinivasan, A.: Resilient multicast using overlays. IEEE/ACM Trans. Netw. **14**(2), 237–248 (2006)
13. Cha, M., Rodriguez, P., Moon, S., Crowcroft, J.: On next-generation telco-managed P2P tv architectures. In: Proceedings of the 7th International Conference on Peer-to-Peer Systems (IPTPS 2008), IPTPS'08, pp. 5–5. Berkeley (2008) (USENIX Association)
14. Cui, Y., Xue, Y., Nahrstedt, K.: Optimal resource allocation in overlay multicast. IEEE Trans. Parallel Distrib. Syst. **17**(8), 808–823 (2006)
15. Fei, Zongming, Yang, Mengkun: A proactive tree recovery mechanism for resilient overlay multicast. IEEE/ACM Trans. Netw. **15**(1), 173–186 (2007)
16. Ganjam, A., Zhang, Hui: Internet multicast video delivery. Proc. IEEE **93**(1), 159–170 (2005)
17. Picconi, F., Massoulie, L.: Is there a future for mesh-based live video streaming? In: Proceedings of the Eighth International Conference on Peer-to-Peer Computing (P2P 2008), pp. 289–298 (2008)
18. Sentinelli, A., Marfia, G., Gerla, M., Kleinrock, L., Tewari, S.: Will IPTV ride the peer-to-peer stream? (peer-to-peer multimedia streaming). IEEE Commun. Mag. **45**(6), 86–92 (2007)
19. Chuan, W., Baochun, L.: Optimal rate allocation in overlay content distribution. In: Akyildiz, I.F., Sivakumar, R., Ekici, E., Cavalcantede Oliveira, J., McNair, J. (eds.) NETWORKING 2007. Ad Hoc and Sensor Networks, Wireless Networks, Next Generation Internet. Lecture Notes in Computer Science, vol. 4479, pp. 678–690. Springer, Berlin (2007)
20. Wu, C., Li, B.: On Meeting P2P Streaming Bandwidth Demand with Limited Supplies (2008)
21. Krauter, K., Buyya, R., Maheswaran, M.: A taxonomy and survey of grid resource management systems for distributed computing. Softw.: Pract. Exp. **32**(2), 135–164 (2002)
22. Nabrzyski, J., Schopf, J., Weglarz, J.: Grid Resource Management—State of the Art and Future Trends. Kluwer Academic Publishers, Norwell (2004)
23. Travostino, F., Mambretti, J., Karmous-Edwards, G.: Grid Networks: Enabling Grids with Advanced Communication Technology. Wiley, New York (2006)
24. Anderson, D., Cobb, J., Korpela, E., Lebofsky, M., Werthimer, D.: Seti@home: an experiment in public-resource computing. Commun. ACM **45**(11), 56–61 (2002)
25. Chmaj, G., Walkowiak, K.: A P2P computing system for overlay networks. Futur. Gener. Comput. Syst. **29**(1), 242–249 (2013)
26. Chmaj, G., Walkowiak, K., Tarnawski, M., Kucharzak, M.: Heuristic algorithms for optimization of task allocation and result distribution in peer-to-peer computing systems. Appl. Math. Comput. Sci. **22**(3), 733–748 (2012)
27. Milojicic, D., Kalogeraki, V., Lukose, R., Nagaraja, K., Pruyne, J., Richard, B., Rollins, S., Xu, Z.: Peer-to-Peer Computing (2002)
28. Walkowiak, K.: Network design problem for P2P multicasting. In: Proceedings of the International Network Optimization Conferenc (INOC 2009) (2009)
29. Walkowiak, K.: P2P multicasting network design problem—heuristic approach. In: Proceedings of the 2010 IEEE GLOBECOM Workshops, pp. 1508–1512 (2010)
30. Akbari, B., Rabiee, H., Ghanbari, M.: An optimal discrete rate allocation for overlay video multicasting. Comput. Commun. **31**(3), 551–562 (2008)
31. Mundinger, J., Weber, R.: Effcient file dissemination using peer-to-peer technology. Technical Report 2004–01, Statistical Laboratory Research Reports, Cambridge (2004)
32. Mundinger, J., Weber, R., Weiss, G.: Optimal scheduling of peer-to-peer file dissemination. J. Sched. **11**(2), 105–120 (2008)
33. Shi, S., Turner, J., Waldvogel, M.: Dimensioning server access bandwidth and multicast routing in overlay networks. In: Proceedings of the 11th International Workshop on Network and Operating Systems Support for Digital Audio and Video (NOSSDAV 2001), NOSSDAV '01, pp. 83–91. ACM, New York (2001)
34. Wu, G., Chiueh, T.: Peer to peer file download and streaming. Technical Report tr-185 (2005)

35. Zhu, Ying, Li, Baochun: Overlay networks with linear capacity constraints. IEEE Trans. Parallel Distrib. Syst. **19**(2), 159–173 (2008)
36. Zhu, Y., Pu, K.Q.: Adaptive multicast tree construction for elastic data streams. In: Proceedings of the IEEE Global Telecommunications Conference (GLOBECOM 2008), pp. 1–5 (2008)
37. Birrer, S., Bustamante, F.E.: Resilience in overlay multicast protocols. In: Proceedings of the 14th IEEE International Symposium on Modeling, Analysis, and Simulation of Computer and Telecommunication Systems (MASCOTS 2006), pp. 363–372 (2006)
38. Walkowiak, K.: Survivability of P2P multicasting. In: Proceedings of the 7th International Workshop on Design of Reliable Communication Networks (DRCN 2009), pp. 92–99 (2009)
39. Walkowiak, K., Przewozniczek, M.: Modeling and optimization of survivable P2P multicasting. Comput. Commun. **34**(12), 1410–1424 (2011)
40. Small, T., Li, B., Liang, B.: Outreach: peer-to-peer topology construction towards minimized server bandwidth costs. IEEE J. Sel. Areas Commun. **25**(1), 35–45 (2007)
41. Venkataraman, V., Francisy, P., Calandrinoz, J.: Chunkyspread: multitree unstructured peer-to-peer multicast. In: Proceedings of the Fifth International Workshop on Peer-to-Peer Systems (IPTPS 2006) (2006)
42. Zhu, Ying, Li, Baochun, Guo, Jiang: Multicast with network coding in application-layer overlay networks. IEEE J. Sel. Areas Commun. **22**(1), 107–120 (2004)
43. Dahl, G., Gouveia, L., Requejo, C.: On formulations and methods for the hop-constrained minimum spanning tree problem. In: Resende, M., Pardalos, P. (eds.) Handbook of Optimization in Telecommunications, pp. 493–515. Springer, US (2006)
44. Gouveia, L., Simonetti, L., Uchoa, E.: Modeling hop-constrained and diameter-constrained minimum spanning tree problems as steiner tree problems over layered graphs. Math. Program. **128**(1–2), 123–148 (2011)
45. Wisniewski, M., Walkowiak, K.: Evolutionary algorithm for P2P multicasting network design problem. In: Corchado, E., Kurzynski, M., Wozniak, M. (eds.) Hybrid Artificial Intelligent Systems. Lecture Notes in Computer Science, vol. 6678, pp. 182–189. Springer, Berlin Heidelberg (2011)
46. Pioro, M., Medhi, D.: Routing, Flow, and Capacity Design in Communication and Computer Networks. Morgan Kaufmann, San Francisco (2004)
47. Coley, D.: An introduction to genetic algorithms for scientists and engineers. World Scientific, Singapore (1999)
48. Corne, D., Oates, M., Smith, G. (eds.): Telecommunications Optimization: Heuristic and Adaptive Techniques. Wiley, New York (2000)
49. Donoso, Y., Fabregat, R.: Multi-Objective Optimization in Computer Networks Using Metaheuristics. Auerbach Publications, Boston (2007)
50. Goldberg, D.E.: Genetic Algorithms in Search, Optimization and Machine Learning, 1st edn. Addison-Wesley Longman Publishing Co., Inc., Boston (1989)
51. Yang, X.: Nature-Inspired Optimization Algorithms. Elsevier, London (2014)
52. Glover, F.: Tabu Search Fundamentals and Uses. Technical Report, University of Colorado (1995)
53. Kirkpatrick, Jr., S., Gelatt, C.D., Vecchi, M.P.: Optimization by simulated annealing. Science **220**, 671–680 (1983)
54. Resende, M.: Greedy randomized adaptive search procedures (grasp). Encycl. Optim. **2**, 373–382 (2001)
55. Talbi, E.G.: Metaheuristics—From Design to Implementation. Wiley, New York (2009)
56. Kmiecik, W., Walkowiak, K.: Metaheuristic algorithms for optimization of survivable multicast overlay in dual homing networks. Cybern. Syst. **44**(6–7), 606–626 (2013)
57. Walkowiak, K., Przewozniczek, M., Pajak, K.: Heuristic algorithms for survivable P2P multicasting. Appl. Artif. Intell. **27**(4), 278–303 (2013)
58. IBM. ILOG CPLEX optimizer. http://www.ibm.com
59. Vasseur, J., Pickavet, M., Demeester, P.: Network Recovery: Protection and Restoration of Optical, SONET-SDH, IP, and MPLS. Morgan Kaufmann Publishers Inc., San Francisco (2004)

60. Allen, J.D., Kubat, P.: Reliable video broadcasts via protected Steiner trees. IEEE Commun. Mag. **48**(2), 70–76 (2010)
61. Padmanabhan, V.N., Wang, H.J., Chou, P.A.: Resilient peer-to-peer streaming. In: Proceeding of the 11th IEEE International Conference on Network Protocols (ICNP2003), pp. 16–27 (2003)
62. Huang, F., Ravindran, B., Kumar, V.S.A.: An approximation algorithm for minimum-delay peer-to-peer streaming. In Proceedings of the IEEE Ninth International Conference on Peer-to-Peer Computing (P2P 2009), pp. 71–80 (2009)
63. Jiang, J.W., Zhang, S., Chen, M., Chiang, M.: Minimizing streaming delay in homogeneous peer-to-peer networks. In: Proceedings of the IEEE International Symposium on Information Theory Proceedings (ISIT 2010), pp. 1783–1787 (2010)
64. Wu, C., Li, B.: Optimal peer selection for minimum-delay peer-to-peer streaming with rate-less codes. In: Proceedings of the ACM Workshop on Advances in Peer-to-peer Multimedia Streaming (P2PMMS 2005), P2PMMS'05, pp. 69–78. ACM, New York (2005)
65. Cardwell, R.H., Monma, C., Tsong-Ho, Wu: Computer-aided design procedures for survivable fiber optic networks. IEEE J. Sel. Areas Commun. **7**(8), 1188–1197 (1989)
66. Kolar, D.J., Wu, T.-H.: A study of survivability versus cost for several fiber network architectures. In: Proceedings of the IEEE International Conference on Communications (ICC 1988), vol. 1, pp. 61–66 (1988)
67. Orda, A., Rom, R.: Multihoming in computer networks: a topology-design approach. In: Proceedings of the Seventh Annual Joint Conference of the IEEE Computer and Communications Societies (INFOCOM 1988), pp. 941–945 (1988)
68. Wu, T.-H., Kolar, D.J., Cardwell, R.H.: Survivable network architectures for broad-band fiber optic networks: model and performance comparison. J. Lightwave Technol. **6**(11), 1698–1709 (1988)
69. Wang, J., Vokkarane, V.M., Jothi, R., Qi, X., Raghavachari, B., Jue, J.P.: Dual-homing protection in IP-over-WDM networks. J. Lightwave Technol. **23**(10), 3111–3124 (2005)
70. Wang, J., Yang, M., Qi, X., Cook, R.P.: Dual-homing multicast protection. In: Proceedings of the IEEE Global Telecommunications Conference (GLOBECOM 2004), vol. 2, pp. 1123–1127 (2004)
71. Wang, J., Yang, M., Yang, B., Zheng, S.Q.: Dual-homing based scalable partial multicast protection. IEEE Trans. Comput. **55**(9), 1130–1141 (2006)
72. Yang, M., Wang, J., Qi, X., Jiang, Y.: On finding the best partial multicast protection tree under dual-homing architecture. In: Proceedings of the Workshop on High Performance Switching and Routing (HPSR 2005), pp. 128–132 (2005)
73. Kmiecik, W., Walkowiak, K.: Modeling of dual homing survivability for P2P multicasting. In: Proceedings of the Modelling and Simulation of Systems Conference (MOSIS 2011), pp. 71–78 (2011)
74. Kmiecik, W., Walkowiak, K.: Survivable P2P multicasting flow assignment in dual homing networks. In: Proceedings of the 3rd International Congress on Ultra Modern Telecommunications and Control Systems and Workshops (ICUMT 2011), pp. 1–7 (2011)
75. Kmiecik, W., Walkowiak, K.: Survivability aspects of overlay multicasting in dual homing networks. In: Proceedings of the 4th International Congress on Ultra Modern Telecommunications and Control Systems and Workshops (ICUMT 2012), pp. 786–792 (2012)
76. Kmiecik, W., Walkowiak, K.: Capacity and flow assignment in survivable overlay networks with dual homing architecture. In: Proceedings of the 5th International Congress on Ultra Modern Telecommunications and Control Systems and Workshops (ICUMT 2013), pp. 125–131 (2013)
77. Kmiecik, W., Walkowiak, K.: Flow assignment (FA) and capacity and flow assignment (CFA) problems for survivable overlay multicasting in dual homing networks. Telecommun. Syst. **1**, 1–13 (2015)
78. Alpaydin, E.: Introduction to Machine Learning, 2nd edn. The MIT Press, Cambridge (2010)
79. Przewozniczek, M., Walkowiak, K., Wozniak, M.: Optimizing distributed computing systems for k-nearest neighbours classifiers—evolutionary approach. Log. J. IGPL **19**(2), 357–372 (2011)

80. Walkowiak, K., Sztajer, S., Wozniak, M.: Decentralized distributed computing system for privacy-preserving combined classifiers—modeling and optimization. Computational Science and Its Applications (ICCSA 2011). Lecture Notes in Computer Science, vol. 6782, pp. 512–525. Springer, Berlin (2011)
81. Wozniak, M.: Hybrid Classifiers: Methods of Data, Knowledge, and Classifier Combination. Springer Publishing Company, Incorporated (2013)
82. Devroye, L.: On the inequality of cover and hart in nearest neighbor discrimination. IEEE Trans. Pattern Anal. Mach. Intell. **PAMI-3**(1), 75–78 (1981)
83. Kuncheva, L.: Combining Pattern Classifiers: Methods and Algorithms. Wiley-Interscience, New York (2004)
84. Walkowiak, K., Wozniak, M.: Decision tree induction methods for distributed environment. In Cyran, K., Kozielski, S., Peters, J.F., Stańczyk, U., Wakulicz-Deja, A. (eds.) Man-Machine Interactions. Advances in Intelligent and Soft Computing, vol. 59, pp. 201–208. Springer, Berlin (2009)
85. Walkowiak, K., Wozniak, M.: Modeling of network computing systems for decision tree induction tasks. In: Corchado, Emilio, Yin, Hujun (eds.) Intelligent Data Engineering and Automated Learning (IDEAL 2009). Lecture Notes in Computer Science, vol. 5788, pp. 759–766. Springer, Berlin (2009)
86. Kacprzak, T., Walkowiak, K., Wozniak, M.: Grasp algorithm for optimization of grids for multiple classifier system. In: Corchado, Emilio, Novais, Paulo, Analide, Cesar, Sedano, Javier (eds.) 5th International Workshop Soft Computing Models in Industrial and Environmental Applications (SOCO 2010). Advances in Intelligent and Soft Computing, vol. 73, pp. 137–144. Springer, Berlin (2010)
87. Kacprzak, T., Walkowiak, K., Wozniak, M.: Optimization of overlay distributed computing systems for multiple classifier system—heuristic approach. Log. J. IGPL **20**(4), 677–688 (2012)
88. Walkowiak, K., Kasprzak, A., Kosowski, M., Miziolek, M.: Scheduling and capacity design in overlay computing systems. In: Murgante, B., Gervasi, O., Misra, S., Nedjah, N., Rocha, A.M.A.C., Taniar, D., Apduhan, B.O. (eds.) Computational Science and Its Applications (ICCSA 2012). Lecture Notes in Computer Science, vol. 7336, pp. 514–529. Springer, Berlin (2012)
89. Walkowiak, K., Charewicz, W., Donajski, M., Rak, J.: A tabu search algorithm for optimization of survivable overlay computing systems. In: Herrero, Á., Snážel, V., Abraham, A., Zelinka, I., Baruque, B., Quintián, H., Calvo, J.L., Sedano, J., Corchado, E., (eds.) International Joint Conference CISISŠ12-ICEUTEt'12-SOCOt'12 Special Sessions. Advances in Intelligent Systems and Computing, vol. 189, pp. 225–234. Springer, Berlin (2013)
90. Walkowiak, K., Charewicz, W., Donajski, M., Rak, J.: Metaheuristic algorithms for optimization of resilient overlay computing systems. Log. J. IGPL **23**(1), 31–44 (2015)
91. Walkowiak, K., Rak, J.: 1+1 protection of overlay distributed computing systems: modeling and optimization. In: Murgante, B., Gervasi, O., Misra, S., Nedjah, N., Rocha, A.M.A.C., Taniar, D., Apduhan, B.O. (eds.) Computational Science and Its Applications (ICCSA 2012). Lecture Notes in Computer Science, vol. 7336, pp. 498–513. Springer, Berlin (2012)
92. Walkowiak, K., Rak, J.: Optimization issues in distributed computing systems design. In: Bock, H.G., Hoang, X.P., Rannacher, R., Schlöder, J.P. (eds.) Modeling, Simulation and Optimization of Complex Processes—HPSC 2012, pp. 261–272. Springer International Publishing, New York (2014)
93. Puchinger, J., Raidl, G., Pferschy, U.: The multidimensional knapsack problem: structure and algorithms. INFORMS J. Comput. **22**(2), 250–265 (2010)
94. Gunluk, O.: A branch-and-cut algorithm for capacitated network design problems. Math. Program. **86**, 17–39 (1998)
95. Marchand, H., Wolsey, L.: Aggregation and mixed integer rounding to solve MIPs. Op. Res. **49**(3), 363–371 (2001)
96. Barnhart, C., Hane, Ch., Vance, P.: Using branch-and-price-and-cut to solve origin-destination integer multicommodity flow problems. Op. Res. **48**(2), 318–326 (2000)

97. Kucharzak, M., Walkowiak, K.: File sharing-based heuristics for flow assignment in P2P systems. In: Proceedings of the 2nd International Logistics and Industrial Informatics (LINDI 2009), pp. 1–6 (2009)
98. Skowron, J., Walkowiak, K.: Heuristic algorithms for optimization of data distribution in P2P networks. In: Proceedings of the 2nd International Logistics and Industrial Informatics (LINDI 2009), pp. 1–6 (2009)
99. Walkowiak, K.:. On transfer costs in peer-to-peer systems: modeling and optimization. In: Proceedings of the 5th Polish-German Teletraffic Symposium (PGTS 2008), pp. 217–226 (2008)
100. Walkowiak, K.: Offline approach to modeling and optimization of flows in peer-to-peer systems. In: Proceedings of the 2nd Conference on New Technologies, Mobility and Security, (NTMS 2008), pp. 1–5 (2008)
101. Chmaj, G., Walkowiak, K.: Preliminary study on optimization of data distribution in resource sharing systems. In: Proceedings of the 19th International Conference on Systems Engineering (ICSENG 2008), pp. 276–281 (2008)
102. Arthur, D., Panigrahy, R.: Analyzing bittorrent and related peer-to-peer networks. In: Proceedings of the Seventeenth annual ACM-SIAM Symposium on Discrete Algorithm, pp. 961–969 (2006)
103. Ganesan, P., Seshadri, M.: On cooperative content distribution and the price of barter. In: Proceedings of the 25th IEEE International Conference on Distributed Computing Systems (ICDCS 2005), pp. 81–90. Washington (2005)
104. Killian, C., Vrable, M., Snoeren, A., Vahdat, A., Pasquale, J.: The overlay network content distribution problem. In: Technical Report CS2005-0824, UCSD (2005)
105. Qiu, D., Srikant, R.: Modeling and performance analysis of bittorrent-like peer-to-peer networks. In: Proceedings of the 2004 Conference on Applications, Technologies, Architectures, and Protocols for Computer Communications (SIGCOMM 2004), SIGCOMM '04, pp. 367–378. ACM, New York (2004)
106. Wu, C.-J., Li, C.-Y., Ho, J.-M.: Improving the download time of bittorrent-like systems. In: Proceedings of the IEEE International Conference on Communications (ICC 2007), pp. 1125–1129 (2007)
107. Yang, X., de Veciana, G.: Service capacity of peer to peer networks. In: Proceedings of Annual Joint Conference of the IEEE Computer and Communications (INFOCOM 2004), pp. 2242–2252 (2004)
108. Jansen, K., Muller, H.: The minimum broadcast time problem for several processor networks. Theor. Comput. Sci. **147**(1–2), 69–85 (1995)
109. Bharambe, A.R., Herley, C., Padmanabhan, V.N.: Analyzing and improving a bittorrent networks performance mechanisms. In: Proceedings of the 25th IEEE International Conference on Computer Communications (INFOCOM 2006), pp. 1–12 (2006)
110. Mundinger, J., Weber, R., Weiss, G.: Analysis of peer-to-peer file dissemination. ACM SIG-METRICS Perform. Eval. Rev. **34**(3), 12–14 (2006)
111. Locher, T., Moor, P., Schmid, S., Wattenhofer, R.: In: Proceedings of 5th Workshop on Hot Topics in Networks (HotNets 2006), pp. 85–90 (2006)
112. Bindal, R., Cao, P., Chan, W., Medved, J., Suwala, G., Bates, T., Zhang, A.: Improving traffic locality in bittorrent via biased neighbor selection. In: Proceedings of the 26th IEEE International Conference on Distributed Computing Systems (ICDCS 2006), pp. 66–75. IEEE Computer Society, Washington (2006)

# Chapter 5
# Conclusions

Due to the rapid evolution of cloud computing and content-oriented services, network operators and service providers are obliged to periodically rethink the way their networks are designed and controlled. This trend is enhanced by the fact that cloud computing and content-oriented services are diverse in terms of bandwidth and usage patterns. Therefore, there is a strong need to develop new effective approaches enabling modeling and optimization of cloud-ready and content-oriented networks.

This book covers these state-of-the-art topics and the latest results of research conducted by the author and his team. In particular, we present several optimization problems related to three various network layers which can be used to implement and deliver cloud computing and content-oriented services, namely, application layer, network layer and optical layer. The optimization problems concern up-to-date technologies and protocols currently used or planned to be applied in near future, including MPLS, Connection-Oriented Ethernet, Elastic Optical Networks, overlay and P2P systems. As well as presenting the problems as mixed-integer programming or integer programming models, numerous optimization methods are shown to solve the problems, including branch-and-cut, Lagrangian relaxation, heuristics and metaheuristics. In addition, results of extensive numerical experiments are reported and analyzed to show the performance of the optimization methods proposed and to report the impact of various parameters of the scenarios. Moreover, the experimental part of this book illustrates how to plan and run numerical experiments in the context of cloud-ready and content-oriented networks. It should be stressed that the problem formulations and algorithms are generic and can be modified and enhanced in order to address new constraints and challenges that can occur in the future as a consequence of the ongoing evolution of computer and communication networks.

The issues addressed in this book are still subject to research, although we can list several recent inventions and trends which we believe will receive much attention in the near future and serve to drive further developments in the field of computer and communication network optimization:

© Springer International Publishing Switzerland 2016
K. Walkowiak, *Modeling and Optimization of Cloud-Ready
and Content-Oriented Networks*, Studies in Systems, Decision
and Control 56, DOI 10.1007/978-3-319-30309-3_5

- **Software Defined Networks (SDN)**. The main idea behind SDN is to decouple the control and data planes, centralize network intelligence and abstract underlying network infrastructure from the applications. The centralization supported in SDN uses efficient algorithms to control the network, which can drive the development new optimization methods [1–4].
- **Network Function Virtualization (NFV)**. The concept of NFV separates software from hardware platforms using various virtualization technologies. The network is transformed from hardware appliances with customized application-specific integrated circuits (e.g., switches, routers, controllers) into software virtual machines run on common off-the-shelf (COTS) hardware. The key benefit of this approach is to make network services easy to use with increased flexibility and lower cost. The shift in network architecture caused by the introduction of the NFV concept generates novel optimization challenges including service chaining [1, 4–6].
- **Space Division Multiplexing (SDM)**. The SDM technology is perceived as the key approach for overcoming a potential capacity crunch in existing optical networks based on single-mode fibers and WDM technology. The main idea behind SDM is the ability to flexibly assign various spatial resources to different traffic demands. Specifically, the spatial resources may refer to strands of fiber in a fiber bundle, fiber cores in multi-core fibers (MCFs) or modes in multi-mode fibers (MMFs). Since additional degrees of freedom are introduced, the network complexity increases and in consequence efficient optimization approaches are required to design SDM networks [7–10].
- **5G networks**. The fifth generation (5G) represents the next major phase of mobile standards beyond the current 4G standards. 5G architecture has been developed in recent years as the key emerging technology addressing growing demand and improving mobile network performance in terms of capacity, bit-rate, end-to-end latency, usage cost and Quality of Experience. As well as new advancements in the radio domain, 5G networks will require many new solutions in the backhaul and backbone networks, including new optimization approaches [11–14].
- **Cross stratum optimization**. To improve performance of various network systems, a tighter coordination between network layers is needed. For instance, a cooperation between the application layer and the network layer allows for more efficient allocation of resources. The cross stratum cooperation also provides a better user experience and delivers more resilient service. However, cross stratum optimization leads to exceptionally complex problems, since multi-layer modeling and optimization are required [15, 16].

Forecasting the near future in the ICT domain is extremely challenging, since the majority of emerging network services and trends are initiated not by long-term efforts of the research community or business parties, but by individual users in an a prompt, independent and uncontrolled way. This book focuses on the modeling and optimization of cloud-ready and content-oriented networks, and we hope it provides a good starting point in the development of new efficient optimization approaches for various networking problems in the future.

# References

1. Jain, R., Paul, S.: Network virtualization and software defined networking for cloud computing: a survey. IEEE Commun. Mag. **51**(11), 24–31 (2013)
2. Jarschel, M., Zinner, T., Hossfeld, T., Tran-Gia, P., Kellerer, W.: Interfaces, attributes, and use cases: a compass for SDN. IEEE Commun. Mag. **52**(6), 210–217 (2014)
3. Sezer, S., Scott-Hayward, S., Chouhan, P.K., Fraser, B., Lake, D., Finnegan, J., Viljoen, N., Miller, M., Rao, N.: Are we ready for SDN? implementation challenges for software-defined networks. IEEE Commun. Mag. **51**(7), 36–43 (2013)
4. Wood, T., Ramakrishnan, K.K., Hwang, J., Liu, G.: Toward a software-based network: integrating software defined networking and network function virtualization. IEEE Netw. **29**(3), 36–41 (2015)
5. Han, B., Ji, L., Lee, S.: Network function virtualization: challenges and opportunities for innovations. IEEE Commun. Mag. **53**(2), 90–97 (2015)
6. Xia, M.: Optical service chaining for network function virtualization. IEEE Commun. Mag. **53**(4), 152–158 (2015)
7. Klonidis, D., Cugini, F., Gerstel, O., Jinno, M., Lopez, V., Palkopoulou, E., Sekiya, M., Siracusa, D., Thouenon, G., Betoule, C.: Spectrally and spatially flexible optical network planning and operations. IEEE Commun. Mag. **53**(2), 69–78 (2015)
8. Nakajima, K., Sillard, P., Richardson, D., Li, M.-J., Essiambre, R.-J., Matsuo, S.: Transmission media for an sdm-based optical communication system. IEEE Commun. Mag. **53**(2), 44–51 (2015)
9. Winzer, P.J.: Spatial multiplexing in fiber optics: the 10x scaling of metro/core capacities. Bell Labs Tech. J. **19**, 22–30 (2014)
10. Xia, T.J., Fevrier, H., Wang, T., Morioka, T.: Introduction of spectrally and spatially flexible optical networks. IEEE Commun. Mag. **53**(2), 24–33 (2015)
11. Bangerter, B., Talwar, S., Arefi, R., Stewart, K.: Networks and devices for the 5G era. IEEE Commun. Mag. **52**(2), 90–96 (2014)
12. Demestichas, P., Georgakopoulos, A., Tsagkaris, K., Kotrotsos, S.: Intelligent 5G networks: managing 5G wireless/mobile broadband. IEEE Veh. Technol. Mag. **10**(3), 41–50 (2015)
13. Ge, X., Cheng, H., Guizani, M., Han, T.: 5G wireless backhaul networks: challenges and research advances. IEEE Netw. **28**(6), 6–11 (2014)
14. Gupta, A., Jha, R.K.: A survey of 5G network: architecture and emerging technologies. IEEE Access **3**, 1206–1232 (2015)
15. Contreras, L.M., Lopez, V., De Dios, O.G., Tovar, A., Munoz, F., Azanon, A., Fernandez-Palacios, J.P., Folgueira, J.: Toward cloud-ready transport networks. IEEE Commun. Mag. **50**(9), 48–55 (2012)
16. Yang, H., Zhang, J., Zhao, Y., Ji, Y., Han, J., Lin, Y., Lee, Y.: CSO: cross stratum optimization for optical as a service. IEEE Commun. Mag. **53**(8), 130–139 (2015)

# Appendix A

## A.1 Network Topologies

### A.1.1 INT9

The INT9 network includes nine nodes and 46 directed links. The average link length is 1062 km (Fig. A.1 and Table A.1).

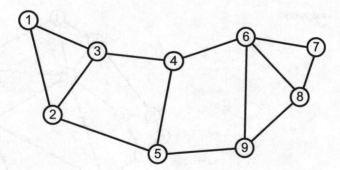

**Fig. A.1** INT9 network topology

© Springer International Publishing Switzerland 2016
K. Walkowiak, *Modeling and Optimization of Cloud-Ready and Content-Oriented Networks*, Studies in Systems, Decision and Control 56, DOI 10.1007/978-3-319-30309-3

**Table A.1** INT9 network topology, link lengths in km

|   | 1 | 2 | 3 | 4 | 5 | 6 | 7 | 8 | 9 |
|---|---|---|---|---|---|---|---|---|---|
| 1 | 0 | 1342 | 913 | 0 | 0 | 0 | 0 | 0 | 0 |
| 2 | 1342 | 0 | 1303 | 0 | 1705 | 0 | 0 | 0 | 0 |
| 3 | 913 | 1303 | 0 | 1330 | 0 | 0 | 0 | 0 | 0 |
| 4 | 0 | 0 | 1330 | 0 | 818 | 690 | 0 | 0 | 0 |
| 5 | 0 | 1705 | 0 | 818 | 0 | 0 | 0 | 0 | 1385 |
| 6 | 0 | 0 | 0 | 690 | 0 | 0 | 1400 | 905 | 1045 |
| 7 | 0 | 0 | 0 | 0 | 0 | 1400 | 0 | 278 | 0 |
| 8 | 0 | 0 | 0 | 0 | 0 | 905 | 278 | 0 | 700 |
| 9 | 0 | 0 | 0 | 0 | 1385 | 1045 | 0 | 700 | 0 |

## A.1.2 DT14

DT14 German national network that includes 14 nodes and 46 directed links. The average link length is 182 km. Network DT14 is taken from [1] (Figs. A.2 and A.3 and Table A.2).

**Fig. A.2** DT14 network topology

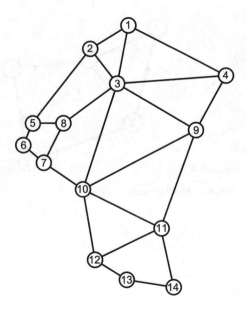

**Fig. A.3** DT14 network topology in geographical view

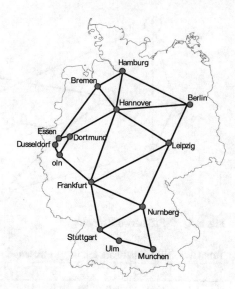

**Table A.2** DT14 network topology, link lengths in km

|    | 1   | 2   | 3   | 4   | 5   | 6  | 7   | 8   | 9   | 10  | 11  | 12  | 13  | 14  |
|----|-----|-----|-----|-----|-----|----|-----|-----|-----|-----|-----|-----|-----|-----|
| 1  | 0   | 115 | 161 | 306 | 0   | 0  | 0   | 0   | 0   | 0   | 0   | 0   | 0   | 0   |
| 2  | 115 | 0   | 121 | 0   | 279 | 0  | 0   | 0   | 0   | 0   | 0   | 0   | 0   | 0   |
| 3  | 161 | 121 | 0   | 295 | 0   | 0  | 0   | 220 | 257 | 314 | 0   | 0   | 0   | 0   |
| 4  | 306 | 0   | 295 | 0   | 0   | 0  | 0   | 0   | 173 | 0   | 0   | 0   | 0   | 0   |
| 5  | 0   | 279 | 0   | 0   | 0   | 37 | 0   | 37  | 0   | 0   | 0   | 0   | 0   | 0   |
| 6  | 0   | 0   | 0   | 0   | 37  | 0  | 41  | 0   | 0   | 0   | 0   | 0   | 0   | 0   |
| 7  | 0   | 0   | 0   | 0   | 0   | 41 | 0   | 84  | 0   | 182 | 0   | 0   | 0   | 0   |
| 8  | 0   | 0   | 220 | 0   | 37  | 0  | 84  | 0   | 0   | 0   | 0   | 0   | 0   | 0   |
| 9  | 0   | 0   | 257 | 173 | 0   | 0  | 0   | 0   | 0   | 353 | 275 | 0   | 0   | 0   |
| 10 | 0   | 0   | 314 | 0   | 0   | 0  | 182 | 0   | 353 | 0   | 224 | 207 | 0   | 0   |
| 11 | 0   | 0   | 0   | 0   | 0   | 0  | 0   | 0   | 275 | 224 | 0   | 189 | 0   | 181 |
| 12 | 0   | 0   | 0   | 0   | 0   | 0  | 0   | 0   | 0   | 207 | 189 | 0   | 87  | 0   |
| 13 | 0   | 0   | 0   | 0   | 0   | 0  | 0   | 0   | 0   | 0   | 0   | 87  | 0   | 146 |
| 14 | 0   | 0   | 0   | 0   | 0   | 0  | 0   | 0   | 0   | 0   | 181 | 0   | 146 | 0   |

## A.1.3 NSF15

The NSF15 network is a US backbone network that includes 15 nodes and 46 directed links. The average link length is 1022 km. Network NSF15 is taken from [2] (Fig. A.4 and Table A.3).

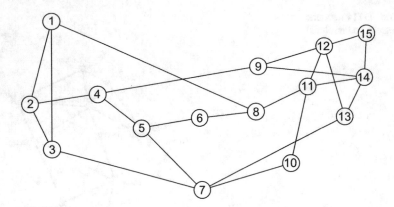

**Fig. A.4**  NSF15 network topology

**Table A.3**  NSF15 network topology, link lengths in km

|    | 1    | 2    | 3    | 4    | 5    | 6   | 7    | 8    | 9    | 10   | 11   | 12   | 13   | 14   | 15   |
|----|------|------|------|------|------|-----|------|------|------|------|------|------|------|------|------|
| 1  | 0    | 1131 | 1709 | 0    | 0    | 0   | 0    | 2831 | 0    | 0    | 0    | 0    | 0    | 0    | 0    |
| 2  | 1131 | 0    | 692  | 958  | 0    | 0   | 0    | 0    | 0    | 0    | 0    | 0    | 0    | 0    | 0    |
| 3  | 1709 | 692  | 0    | 0    | 0    | 0   | 2092 | 0    | 0    | 0    | 0    | 0    | 0    | 0    | 0    |
| 4  | 0    | 958  | 0    | 0    | 566  | 0   | 0    | 0    | 2340 | 0    | 0    | 0    | 0    | 0    | 0    |
| 5  | 0    | 0    | 0    | 566  | 0    | 732 | 1451 | 0    | 0    | 0    | 0    | 0    | 0    | 0    | 0    |
| 6  | 0    | 0    | 0    | 0    | 732  | 0   | 0    | 716  | 0    | 0    | 0    | 0    | 0    | 0    | 0    |
| 7  | 0    | 0    | 2092 | 0    | 1451 | 0   | 0    | 0    | 0    | 1135 | 0    | 0    | 1976 | 0    | 0    |
| 8  | 2831 | 0    | 0    | 0    | 0    | 716 | 0    | 0    | 0    | 0    | 704  | 0    | 0    | 0    | 0    |
| 9  | 0    | 0    | 0    | 2340 | 0    | 0   | 0    | 0    | 0    | 0    | 0    | 596  | 0    | 788  | 0    |
| 10 | 0    | 0    | 0    | 0    | 0    | 0   | 1135 | 0    | 0    | 0    | 834  | 0    | 0    | 0    | 0    |
| 11 | 0    | 0    | 0    | 0    | 0    | 0   | 0    | 704  | 0    | 834  | 0    | 365  | 0    | 451  | 0    |
| 12 | 0    | 0    | 0    | 0    | 0    | 0   | 0    | 0    | 596  | 0    | 365  | 0    | 385  | 0    | 446  |
| 13 | 0    | 0    | 0    | 0    | 0    | 0   | 1976 | 0    | 0    | 0    | 0    | 385  | 0    | 247  | 0    |
| 14 | 0    | 0    | 0    | 0    | 0    | 0   | 0    | 0    | 788  | 0    | 451  | 0    | 247  | 0    | 373  |
| 15 | 0    | 0    | 0    | 0    | 0    | 0   | 0    | 0    | 0    | 0    | 0    | 446  | 0    | 373  | 0    |

## A.1.4 Euro16

The Euro16 network is an European backbone network that embraces 16 nodes and 48 directed links. The average link length is 486 km. The Euro16 network is included in [3] as Core Topology and in [4] as COST266-CT network (Fig. A.5 and Table A.4).

**Fig. A.5** Euro16 network topology

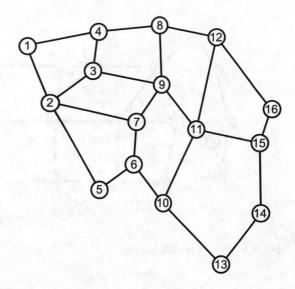

**Table A.4** Euro16 network topology, link lengths in km

|    | 1   | 2   | 3   | 4   | 5   | 6   | 7   | 8   | 9   | 10  | 11  | 12  | 13  | 14  | 15  | 16  |
|----|-----|-----|-----|-----|-----|-----|-----|-----|-----|-----|-----|-----|-----|-----|-----|-----|
| 1  | 0   | 514 | 0   | 540 | 0   | 0   | 0   | 0   | 0   | 0   | 0   | 0   | 0   | 0   | 0   | 0   |
| 2  | 514 | 0   | 393 | 0   | 594 | 0   | 600 | 0   | 0   | 0   | 0   | 0   | 0   | 0   | 0   | 0   |
| 3  | 0   | 393 | 0   | 259 | 0   | 0   | 0   | 0   | 474 | 0   | 0   | 0   | 0   | 0   | 0   | 0   |
| 4  | 540 | 0   | 259 | 0   | 0   | 0   | 0   | 552 | 0   | 0   | 0   | 0   | 0   | 0   | 0   | 0   |
| 5  | 0   | 594 | 0   | 0   | 0   | 507 | 0   | 0   | 0   | 0   | 0   | 0   | 0   | 0   | 0   | 0   |
| 6  | 0   | 0   | 0   | 0   | 507 | 0   | 218 | 0   | 0   | 327 | 0   | 0   | 0   | 0   | 0   | 0   |
| 7  | 0   | 600 | 0   | 0   | 0   | 218 | 0   | 0   | 271 | 0   | 0   | 0   | 0   | 0   | 0   | 0   |
| 8  | 0   | 0   | 0   | 552 | 0   | 0   | 0   | 0   | 592 | 0   | 0   | 381 | 0   | 0   | 0   | 0   |
| 9  | 0   | 0   | 474 | 0   | 0   | 0   | 271 | 592 | 0   | 0   | 456 | 0   | 0   | 0   | 0   | 0   |
| 10 | 0   | 0   | 0   | 0   | 0   | 327 | 0   | 0   | 0   | 0   | 522 | 0   | 720 | 0   | 0   | 0   |
| 11 | 0   | 0   | 0   | 0   | 0   | 0   | 0   | 0   | 456 | 522 | 0   | 757 | 0   | 0   | 534 | 0   |
| 12 | 0   | 0   | 0   | 0   | 0   | 0   | 0   | 381 | 0   | 0   | 757 | 0   | 0   | 0   | 0   | 420 |
| 13 | 0   | 0   | 0   | 0   | 0   | 0   | 0   | 0   | 0   | 720 | 0   | 0   | 0   | 783 | 0   | 0   |
| 14 | 0   | 0   | 0   | 0   | 0   | 0   | 0   | 0   | 0   | 0   | 0   | 0   | 783 | 0   | 400 | 0   |
| 15 | 0   | 0   | 0   | 0   | 0   | 0   | 0   | 0   | 0   | 0   | 534 | 0   | 0   | 400 | 0   | 376 |
| 16 | 0   | 0   | 0   | 0   | 0   | 0   | 0   | 0   | 0   | 0   | 0   | 420 | 0   | 0   | 376 | 0   |

## *A.1.5 UBN24*

The UBN24 network is a US backbone network containing 24 nodes and 86 directed links. The average link length is 998 km. Network UBN24 is taken from [5] (Fig. A.6 and Table A.5).

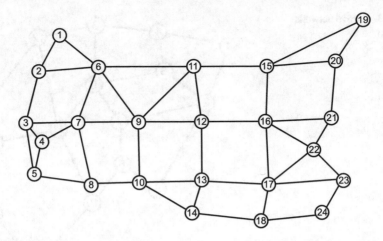

**Fig. A.6** UBN24 network topology

## A.1.6 Euro28

The Euro28 network is a European backbone network containing 28 nodes and 82 directed links. The average link length is 625 km. The Euro28 network is included in the SNDLib as nobel_eu [6] (Figs. A.7 and A.8 and Table A.6).

## A.1.7 US26

The US26 network is a US backbone network containing 26 nodes and 84 directed links. The average link length is 754 km. The US26 network is included in the SNDLib as janos-us [6] (Figs. A.9 and A.10 and Table A.7).

**Table A.5** UBN24 network topology, link lengths in km

| | 1 | 2 | 3 | 4 | 5 | 6 | 7 | 8 | 9 | 10 | 11 | 12 | 13 | 14 | 15 | 16 | 17 | 18 | 19 | 20 | 21 | 22 | 23 | 24 |
|---|---|---|---|---|---|---|---|---|---|---|---|---|---|---|---|---|---|---|---|---|---|---|---|---|
| 1 | 0 | 800 | 0 | 0 | 0 | 1000 | 0 | 0 | 0 | 0 | 0 | 0 | 0 | 0 | 0 | 0 | 0 | 0 | 0 | 0 | 0 | 0 | 0 | 0 |
| 2 | 800 | 0 | 1100 | 0 | 0 | 950 | 0 | 0 | 0 | 0 | 0 | 0 | 0 | 0 | 0 | 0 | 0 | 0 | 0 | 0 | 0 | 0 | 0 | 0 |
| 3 | 0 | 1100 | 0 | 250 | 0 | 0 | 1000 | 0 | 0 | 0 | 0 | 0 | 0 | 0 | 0 | 0 | 0 | 0 | 0 | 0 | 0 | 0 | 0 | 0 |
| 4 | 0 | 0 | 250 | 0 | 800 | 0 | 850 | 0 | 0 | 0 | 0 | 0 | 0 | 0 | 0 | 0 | 0 | 0 | 0 | 0 | 0 | 0 | 0 | 0 |
| 5 | 0 | 0 | 0 | 800 | 0 | 1000 | 0 | 1200 | 0 | 0 | 0 | 0 | 0 | 0 | 0 | 0 | 0 | 0 | 0 | 0 | 0 | 0 | 0 | 0 |
| 6 | 1000 | 950 | 0 | 0 | 1000 | 0 | 1000 | 0 | 1200 | 0 | 1900 | 0 | 0 | 0 | 0 | 0 | 0 | 0 | 0 | 0 | 0 | 0 | 0 | 0 |
| 7 | 0 | 0 | 1000 | 850 | 0 | 1000 | 0 | 1150 | 1000 | 0 | 0 | 0 | 0 | 0 | 0 | 0 | 0 | 0 | 0 | 0 | 0 | 0 | 0 | 0 |
| 8 | 0 | 0 | 0 | 0 | 1200 | 0 | 1150 | 0 | 0 | 900 | 0 | 0 | 0 | 0 | 0 | 0 | 0 | 0 | 0 | 0 | 0 | 0 | 0 | 0 |
| 9 | 0 | 0 | 0 | 0 | 0 | 1200 | 1000 | 0 | 0 | 900 | 1400 | 1000 | 0 | 0 | 0 | 0 | 0 | 0 | 0 | 0 | 0 | 0 | 0 | 0 |
| 10 | 0 | 0 | 0 | 0 | 0 | 0 | 0 | 900 | 900 | 0 | 0 | 0 | 950 | 850 | 0 | 0 | 1100 | 0 | 0 | 0 | 0 | 0 | 0 | 0 |
| 11 | 0 | 0 | 0 | 0 | 0 | 1900 | 0 | 0 | 1400 | 0 | 0 | 900 | 0 | 0 | 1300 | 0 | 0 | 0 | 0 | 0 | 0 | 0 | 0 | 0 |
| 12 | 0 | 0 | 0 | 0 | 0 | 0 | 0 | 0 | 1000 | 0 | 900 | 0 | 900 | 0 | 0 | 1000 | 0 | 0 | 0 | 0 | 0 | 0 | 0 | 0 |
| 13 | 0 | 0 | 0 | 0 | 0 | 0 | 0 | 0 | 0 | 950 | 0 | 900 | 0 | 650 | 0 | 600 | 0 | 0 | 0 | 0 | 0 | 0 | 0 | 0 |
| 14 | 0 | 0 | 0 | 0 | 0 | 0 | 0 | 0 | 0 | 850 | 0 | 0 | 650 | 0 | 0 | 0 | 0 | 1200 | 0 | 0 | 0 | 0 | 0 | 0 |
| 15 | 0 | 0 | 0 | 0 | 0 | 0 | 0 | 0 | 0 | 0 | 1300 | 0 | 0 | 0 | 0 | 600 | 1000 | 0 | 2600 | 1300 | 0 | 0 | 0 | 0 |
| 16 | 0 | 0 | 0 | 0 | 0 | 0 | 0 | 0 | 0 | 0 | 0 | 1000 | 600 | 0 | 600 | 0 | 0 | 0 | 0 | 0 | 1000 | 800 | 1000 | 0 |
| 17 | 0 | 0 | 0 | 0 | 0 | 0 | 0 | 0 | 0 | 1100 | 0 | 0 | 0 | 0 | 1000 | 0 | 0 | 800 | 0 | 0 | 0 | 850 | 0 | 0 |
| 18 | 0 | 0 | 0 | 0 | 0 | 0 | 0 | 0 | 0 | 0 | 0 | 0 | 0 | 1200 | 0 | 0 | 800 | 0 | 0 | 0 | 0 | 0 | 0 | 900 |
| 19 | 0 | 0 | 0 | 0 | 0 | 0 | 0 | 0 | 0 | 0 | 0 | 0 | 0 | 0 | 2600 | 0 | 0 | 0 | 0 | 1200 | 0 | 0 | 0 | 0 |
| 20 | 0 | 0 | 0 | 0 | 0 | 0 | 0 | 0 | 0 | 0 | 0 | 0 | 0 | 0 | 1300 | 0 | 0 | 0 | 1200 | 0 | 700 | 0 | 0 | 0 |
| 21 | 0 | 0 | 0 | 0 | 0 | 0 | 0 | 0 | 0 | 0 | 0 | 0 | 0 | 0 | 0 | 1000 | 0 | 0 | 0 | 700 | 0 | 300 | 0 | 0 |
| 22 | 0 | 0 | 0 | 0 | 0 | 0 | 0 | 0 | 0 | 0 | 0 | 0 | 0 | 0 | 0 | 800 | 850 | 0 | 0 | 0 | 300 | 0 | 600 | 0 |
| 23 | 0 | 0 | 0 | 0 | 0 | 0 | 0 | 0 | 0 | 0 | 0 | 0 | 0 | 0 | 0 | 1000 | 0 | 0 | 0 | 0 | 0 | 600 | 0 | 900 |
| 24 | 0 | 0 | 0 | 0 | 0 | 0 | 0 | 0 | 0 | 0 | 0 | 0 | 0 | 0 | 0 | 0 | 0 | 900 | 0 | 0 | 0 | 0 | 900 | 0 |

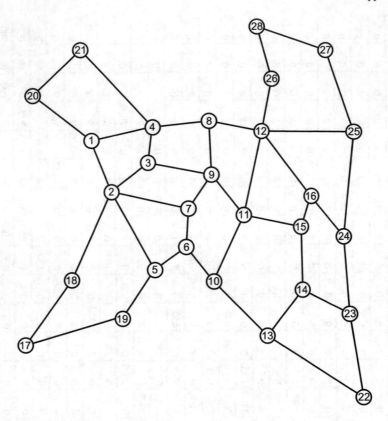

**Fig. A.7** Euro28 network topology

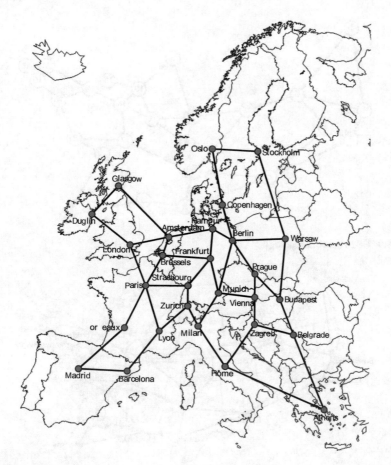

**Fig. A.8** Euro28 network topology in geographical view

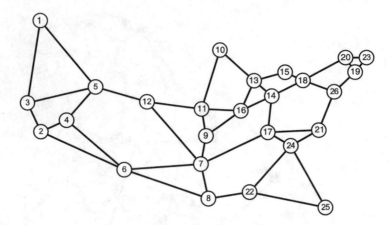

**Fig. A.9**  US26 network topology

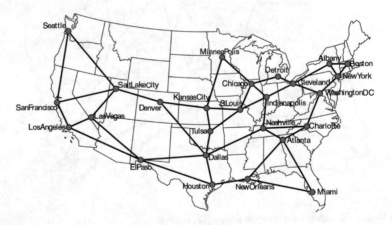

**Fig. A.10**  US26 network topology in geographical view

**Table A.6** Euro28 network topology, link lengths in km

| | 1 | 2 | 3 | 4 | 5 | 6 | 7 | 8 | 9 | 10 | 11 | 12 | 13 | 14 | 15 | 16 | 17 | 18 | 19 | 20 | 21 | 22 | 23 | 24 | 25 | 26 | 27 | 28 |
|---|---|---|---|---|---|---|---|---|---|---|---|---|---|---|---|---|---|---|---|---|---|---|---|---|---|---|---|---|
| 1 | 0 | 514 | 0 | 540 | 0 | 0 | 0 | 0 | 0 | 0 | 0 | 0 | 0 | 0 | 0 | 0 | 0 | 0 | 0 | 690 | 0 | 0 | 0 | 0 | 0 | 0 | 0 | 0 |
| 2 | 514 | 0 | 393 | 0 | 594 | 0 | 600 | 0 | 0 | 0 | 0 | 0 | 0 | 0 | 0 | 0 | 0 | 747 | 0 | 0 | 0 | 0 | 0 | 0 | 0 | 0 | 0 | 0 |
| 3 | 0 | 393 | 0 | 259 | 0 | 0 | 0 | 0 | 474 | 0 | 0 | 0 | 0 | 0 | 0 | 0 | 0 | 0 | 0 | 0 | 0 | 0 | 0 | 0 | 0 | 0 | 0 | 0 |
| 4 | 540 | 0 | 259 | 0 | 0 | 0 | 0 | 552 | 0 | 0 | 0 | 0 | 0 | 0 | 0 | 0 | 0 | 0 | 0 | 0 | 1067 | 0 | 0 | 0 | 0 | 0 | 0 | 0 |
| 5 | 0 | 594 | 0 | 0 | 0 | 507 | 0 | 0 | 0 | 0 | 0 | 0 | 0 | 0 | 0 | 0 | 0 | 0 | 796 | 0 | 0 | 0 | 0 | 0 | 0 | 0 | 0 | 0 |
| 6 | 0 | 0 | 0 | 0 | 507 | 0 | 218 | 0 | 0 | 327 | 0 | 0 | 0 | 0 | 0 | 0 | 0 | 0 | 0 | 0 | 0 | 0 | 0 | 0 | 0 | 0 | 0 | 0 |
| 7 | 0 | 600 | 0 | 0 | 0 | 218 | 0 | 0 | 271 | 0 | 0 | 0 | 0 | 0 | 0 | 0 | 0 | 0 | 0 | 0 | 0 | 0 | 0 | 0 | 0 | 0 | 0 | 0 |
| 8 | 0 | 0 | 0 | 552 | 0 | 0 | 0 | 0 | 592 | 0 | 0 | 381 | 0 | 0 | 0 | 0 | 0 | 0 | 0 | 0 | 0 | 0 | 0 | 0 | 0 | 0 | 0 | 0 |
| 9 | 0 | 0 | 474 | 0 | 0 | 0 | 271 | 592 | 0 | 0 | 0 | 0 | 0 | 0 | 0 | 0 | 0 | 0 | 0 | 0 | 0 | 0 | 0 | 0 | 0 | 0 | 0 | 0 |
| 10 | 0 | 0 | 0 | 0 | 0 | 327 | 0 | 0 | 0 | 0 | 456 | 0 | 720 | 0 | 0 | 0 | 0 | 0 | 0 | 0 | 0 | 0 | 0 | 0 | 0 | 0 | 0 | 0 |
| 11 | 0 | 0 | 0 | 0 | 0 | 0 | 0 | 0 | 0 | 456 | 0 | 757 | 522 | 0 | 534 | 0 | 0 | 0 | 0 | 0 | 0 | 0 | 0 | 0 | 0 | 0 | 0 | 0 |
| 12 | 0 | 0 | 0 | 0 | 0 | 0 | 0 | 381 | 0 | 0 | 757 | 0 | 0 | 0 | 0 | 420 | 0 | 0 | 0 | 0 | 0 | 0 | 0 | 0 | 775 | 540 | 0 | 0 |
| 13 | 0 | 0 | 0 | 0 | 0 | 0 | 0 | 0 | 0 | 720 | 522 | 0 | 0 | 783 | 0 | 0 | 0 | 0 | 0 | 0 | 0 | 1500 | 0 | 0 | 0 | 0 | 0 | 0 |
| 14 | 0 | 0 | 0 | 0 | 0 | 0 | 0 | 0 | 0 | 0 | 0 | 0 | 783 | 0 | 400 | 0 | 0 | 0 | 0 | 0 | 0 | 0 | 551 | 0 | 0 | 0 | 0 | 0 |
| 15 | 0 | 0 | 0 | 0 | 0 | 0 | 0 | 0 | 0 | 0 | 534 | 0 | 0 | 400 | 0 | 376 | 0 | 0 | 0 | 0 | 0 | 0 | 0 | 0 | 0 | 0 | 0 | 0 |
| 16 | 0 | 0 | 0 | 0 | 0 | 0 | 0 | 0 | 0 | 0 | 0 | 420 | 0 | 0 | 376 | 0 | 0 | 0 | 0 | 0 | 0 | 0 | 0 | 668 | 0 | 0 | 0 | 0 |
| 17 | 0 | 0 | 0 | 0 | 0 | 0 | 0 | 0 | 0 | 0 | 0 | 0 | 0 | 0 | 0 | 0 | 0 | 834 | 760 | 0 | 0 | 0 | 0 | 0 | 0 | 0 | 0 | 0 |
| 18 | 0 | 747 | 0 | 0 | 0 | 0 | 0 | 0 | 0 | 0 | 0 | 0 | 0 | 0 | 0 | 0 | 834 | 0 | 0 | 0 | 0 | 0 | 0 | 0 | 0 | 0 | 0 | 0 |
| 19 | 0 | 0 | 0 | 0 | 796 | 0 | 0 | 0 | 0 | 0 | 0 | 0 | 0 | 0 | 0 | 0 | 760 | 0 | 0 | 0 | 0 | 0 | 0 | 0 | 0 | 0 | 0 | 0 |
| 20 | 690 | 0 | 0 | 0 | 0 | 0 | 0 | 0 | 0 | 0 | 0 | 0 | 0 | 0 | 0 | 0 | 0 | 0 | 0 | 0 | 462 | 0 | 0 | 0 | 0 | 0 | 0 | 0 |
| 21 | 0 | 0 | 0 | 1067 | 0 | 0 | 0 | 0 | 0 | 0 | 0 | 0 | 0 | 0 | 0 | 0 | 0 | 0 | 0 | 462 | 0 | 0 | 0 | 0 | 0 | 0 | 0 | 0 |
| 22 | 0 | 0 | 0 | 0 | 0 | 0 | 0 | 0 | 0 | 0 | 0 | 0 | 1500 | 0 | 0 | 0 | 0 | 0 | 0 | 0 | 0 | 0 | 1209 | 0 | 0 | 0 | 0 | 0 |
| 23 | 0 | 0 | 0 | 0 | 0 | 0 | 0 | 0 | 0 | 0 | 0 | 0 | 0 | 551 | 0 | 0 | 0 | 0 | 0 | 0 | 0 | 1209 | 0 | 474 | 0 | 0 | 0 | 0 |
| 24 | 0 | 0 | 0 | 0 | 0 | 0 | 0 | 0 | 0 | 0 | 0 | 0 | 0 | 0 | 0 | 668 | 0 | 0 | 0 | 0 | 0 | 0 | 474 | 0 | 819 | 0 | 0 | 0 |
| 25 | 0 | 0 | 0 | 0 | 0 | 0 | 0 | 0 | 0 | 0 | 0 | 775 | 0 | 0 | 0 | 0 | 0 | 0 | 0 | 0 | 0 | 0 | 0 | 819 | 0 | 0 | 1213 | 0 |
| 26 | 0 | 0 | 0 | 0 | 0 | 0 | 0 | 0 | 0 | 0 | 0 | 540 | 0 | 0 | 0 | 0 | 0 | 0 | 0 | 0 | 0 | 0 | 0 | 0 | 0 | 0 | 0 | 722 |
| 27 | 0 | 0 | 0 | 0 | 0 | 0 | 0 | 0 | 0 | 0 | 0 | 0 | 0 | 0 | 0 | 0 | 0 | 0 | 0 | 0 | 0 | 0 | 0 | 0 | 1213 | 0 | 0 | 623 |
| 28 | 0 | 0 | 0 | 0 | 0 | 0 | 0 | 0 | 0 | 0 | 0 | 0 | 0 | 0 | 0 | 0 | 0 | 0 | 0 | 0 | 0 | 0 | 0 | 0 | 0 | 722 | 623 | 0 |

**Table A.7** US26 network topology, link lengths in km

| | 1 | 2 | 3 | 4 | 5 | 6 | 7 | 8 | 9 | 10 | 11 | 12 | 13 | 14 | 15 | 16 | 17 | 18 | 19 | 20 | 21 | 22 | 23 | 24 | 25 | 26 |
|---|---|---|---|---|---|---|---|---|---|---|---|---|---|---|---|---|---|---|---|---|---|---|---|---|---|---|
| 1 | 0 | 0 | 1375 | 0 | 1391 | 0 | 0 | 0 | 0 | 0 | 0 | 0 | 0 | 0 | 0 | 0 | 0 | 0 | 0 | 0 | 0 | 0 | 0 | 0 | 0 | 0 |
| 2 | 0 | 0 | 684 | 477 | 0 | 1415 | 0 | 0 | 0 | 0 | 0 | 0 | 0 | 0 | 0 | 0 | 0 | 0 | 0 | 0 | 0 | 0 | 0 | 0 | 0 | 0 |
| 3 | 1375 | 684 | 0 | 0 | 1209 | 1177 | 0 | 0 | 0 | 0 | 0 | 0 | 0 | 0 | 0 | 0 | 0 | 0 | 0 | 0 | 0 | 0 | 0 | 0 | 0 | 0 |
| 4 | 0 | 477 | 0 | 0 | 763 | 0 | 0 | 0 | 0 | 0 | 0 | 0 | 0 | 0 | 0 | 0 | 0 | 0 | 0 | 0 | 0 | 0 | 0 | 0 | 0 | 0 |
| 5 | 1391 | 0 | 1209 | 763 | 0 | 0 | 0 | 0 | 0 | 0 | 0 | 770 | 0 | 0 | 0 | 0 | 0 | 0 | 0 | 0 | 0 | 0 | 0 | 0 | 0 | 0 |
| 6 | 0 | 1415 | 1177 | 0 | 0 | 0 | 1135 | 1349 | 0 | 0 | 0 | 0 | 0 | 0 | 0 | 0 | 0 | 0 | 0 | 0 | 0 | 0 | 0 | 0 | 0 | 0 |
| 7 | 0 | 0 | 0 | 0 | 0 | 1135 | 0 | 440 | 481 | 0 | 0 | 1320 | 0 | 0 | 0 | 0 | 1257 | 0 | 0 | 0 | 0 | 0 | 0 | 0 | 0 | 0 |
| 8 | 0 | 0 | 0 | 0 | 0 | 1349 | 440 | 0 | 0 | 0 | 0 | 0 | 0 | 0 | 0 | 0 | 0 | 0 | 0 | 0 | 0 | 646 | 0 | 0 | 0 | 0 |
| 9 | 0 | 0 | 0 | 0 | 0 | 0 | 481 | 0 | 0 | 0 | 454 | 0 | 0 | 0 | 0 | 723 | 0 | 0 | 0 | 0 | 0 | 0 | 0 | 0 | 0 | 0 |
| 10 | 0 | 0 | 0 | 0 | 0 | 0 | 0 | 0 | 0 | 0 | 816 | 0 | 702 | 0 | 0 | 0 | 0 | 0 | 0 | 0 | 0 | 0 | 0 | 0 | 0 | 0 |
| 11 | 0 | 0 | 0 | 0 | 0 | 0 | 0 | 0 | 454 | 816 | 0 | 1093 | 0 | 0 | 0 | 478 | 0 | 0 | 0 | 0 | 0 | 0 | 0 | 0 | 0 | 0 |
| 12 | 0 | 0 | 0 | 0 | 770 | 0 | 1320 | 0 | 0 | 0 | 1093 | 0 | 0 | 0 | 0 | 0 | 0 | 0 | 0 | 0 | 0 | 0 | 0 | 0 | 0 | 0 |
| 13 | 0 | 0 | 0 | 0 | 0 | 0 | 0 | 0 | 0 | 702 | 0 | 0 | 0 | 367 | 507 | 523 | 0 | 0 | 0 | 0 | 0 | 0 | 0 | 0 | 0 | 0 |
| 14 | 0 | 0 | 0 | 0 | 0 | 0 | 0 | 0 | 0 | 0 | 0 | 0 | 367 | 0 | 0 | 462 | 496 | 552 | 0 | 0 | 0 | 0 | 0 | 0 | 0 | 0 |
| 15 | 0 | 0 | 0 | 0 | 0 | 0 | 0 | 0 | 0 | 0 | 0 | 0 | 507 | 0 | 0 | 0 | 0 | 188 | 0 | 0 | 0 | 0 | 0 | 0 | 0 | 0 |
| 16 | 0 | 0 | 0 | 0 | 0 | 0 | 0 | 0 | 723 | 0 | 478 | 0 | 523 | 462 | 0 | 0 | 0 | 0 | 0 | 0 | 0 | 0 | 0 | 0 | 0 | 0 |
| 17 | 0 | 0 | 0 | 0 | 0 | 0 | 1257 | 0 | 0 | 0 | 0 | 0 | 0 | 496 | 0 | 0 | 0 | 0 | 0 | 0 | 665 | 0 | 0 | 430 | 0 | 0 |
| 18 | 0 | 0 | 0 | 0 | 0 | 0 | 0 | 0 | 0 | 0 | 0 | 0 | 0 | 552 | 188 | 0 | 0 | 0 | 0 | 834 | 0 | 0 | 0 | 0 | 0 | 618 |
| 19 | 0 | 0 | 0 | 0 | 0 | 0 | 0 | 0 | 0 | 0 | 0 | 0 | 0 | 0 | 0 | 0 | 0 | 0 | 0 | 293 | 0 | 0 | 374 | 0 | 0 | 430 |
| 20 | 0 | 0 | 0 | 0 | 0 | 0 | 0 | 0 | 0 | 0 | 0 | 0 | 0 | 0 | 0 | 0 | 0 | 834 | 293 | 0 | 0 | 0 | 290 | 0 | 0 | 0 |
| 21 | 0 | 0 | 0 | 0 | 0 | 0 | 0 | 0 | 0 | 0 | 0 | 0 | 0 | 0 | 0 | 0 | 665 | 0 | 0 | 0 | 0 | 0 | 0 | 457 | 0 | 667 |
| 22 | 0 | 0 | 0 | 0 | 0 | 0 | 0 | 646 | 0 | 0 | 0 | 0 | 0 | 0 | 0 | 0 | 0 | 0 | 0 | 0 | 0 | 0 | 0 | 853 | 1327 | 0 |
| 23 | 0 | 0 | 0 | 0 | 0 | 0 | 0 | 0 | 0 | 0 | 0 | 0 | 0 | 0 | 0 | 0 | 0 | 0 | 374 | 290 | 0 | 0 | 0 | 0 | 0 | 0 |
| 24 | 0 | 0 | 0 | 0 | 0 | 0 | 0 | 0 | 0 | 0 | 0 | 0 | 0 | 0 | 0 | 0 | 430 | 0 | 0 | 0 | 457 | 853 | 0 | 0 | 1203 | 0 |
| 25 | 0 | 0 | 0 | 0 | 0 | 0 | 0 | 0 | 0 | 0 | 0 | 0 | 0 | 0 | 0 | 0 | 0 | 0 | 0 | 0 | 0 | 1327 | 0 | 1203 | 0 | 0 |
| 26 | 0 | 0 | 0 | 0 | 0 | 0 | 0 | 0 | 0 | 0 | 0 | 0 | 0 | 0 | 0 | 0 | 0 | 618 | 430 | 0 | 667 | 0 | 0 | 0 | 0 | 0 |

# References

1. DICONET Project Deliverable. Definition of dynamic optical network architectures. Technical report DICONET deliverable D2.1 Dec 2008, DICONET FP7 project (2008)
2. NLANR. Nsfnet-the national science foundation network. Technical report, National Laboratory for Applied Network Research (NLANR) Project. http://moat.nlanr.net/ (2007)
3. Inkret, R., Kuchar, A., Mikac, B.: Advanced infrastructure for photonic networks. Extended final report of cost action 266. Technical report, Faculty of Electrical Engineering and Computing, University of Zagreb (2003)
4. The COST 266 reference networks. Technical report, NoE e-Photon ONe FP6 project. http://opti.tmit.bme.hu/e1net/downloads_cost266.html (2006)
5. Xia, M., Tornatore, M., Zhang, Y., Chowdhury, P., Martel, C., Mukherjee, B.: Greening the optical backbone network: a traffic engineering approach. In: 2010 IEEE International Conference on Communications (ICC), pp. 1–5 (2010)
6. Orlowski, S., Pióro, M., Tomaszewski, A., Wessäly, R.: SNDlib 1.0—Survivable network design library. In: Proceedings of the 3rd International Network Optimization Conference (INOC 2007), Spa, Belgium. http://sndlib.zib.de (2007)

# Index

© Springer International Publishing Switzerland 2016
K. Walkowiak, *Modeling and Optimization of Cloud-Ready
and Content-Oriented Networks*, Studies in Systems, Decision
and Control 56, DOI 10.1007/978-3-319-30309-3

Printed in the United States
By Bookmasters